INSECTS

and Other Arthropods of

TROPICAL AMERICA

Paul E. Hanson
Kenji Nishida

A ZONA TROPICAL PUBLICATION

from

COMSTOCK PUBLISHING ASSOCIATES

a division of

Cornell University Press

Ithaca and London

First published 2016 by Cornell University Press

First printing, Cornell Paperbacks, 2016
Printed in China

Library of Congress Cataloging-in-Publication Data

Names: Hanson, Paul E., author. | Nishida, Kenji, 1972– photographer.
Title: Insects and other arthropods of tropical America / Paul E. Hanson, Kenji Nishida.
Description: Ithaca : Comstock Publishing Associates, a division of Cornell University Press, 2016. | "A Zona Tropical publication." | Includes bibliographical references and index.
Identifiers: LCCN 2015042449 | ISBN 9780801456947 (pbk. : alk. paper)
Subjects: LCSH: Insects—Latin America. | Insects—Tropics. | Arthropoda—Latin America. | Arthropoda—Tropics.
Classification: LCC QL476.5 .H36 2016 | DDC 595.7098—dc23
LC record available at http://lccn.loc.gov/2015042449

Cornell University Press strives to use environmentally responsible suppliers and materials to the fullest extent possible in the publishing of its books. Such materials include vegetable-based, low-VOC inks and acid-free papers that are recycled, totally chlorine-free, or partly composed of nonwood fibers. For further information, visit our website at www.cornellpress.cornell.edu.

Paperback printing 10 9 8 7 6 5 4 3 2 1

Book design: Gabriela Wattson

This book is for women and men throughout the world who dedicate their lives to conserving natural habitats and biodiversity.

CONTENTS

1. Introduction to Arthropods.....................1

2. Small Orders...13

3. True Bugs and Their Kin.......................59

4. Beetles...97

5. Wasps, Bees, Ants 153

6. Moths and Butterflies 196

7. Flics and Their Kin 269

8. Other Arthropods...............................302

Glossary 343
Bibliography 347
Acknowledgments.......................................357
Photo Credits 359
Index..361
About the Author and Photographer............375

1

INTRODUCTION
TO ARTHROPODS

Tropical forests are home to a large percentage of the earth's plants and animals. People who visit these forests generally come to see the birds, monkeys, orchids—in short, all the charismatic organisms that seem to symbolize the tropics. Aside from butterflies, however, insects usually do not make it on the list of things to see, which is a shame, since they are everywhere (and thus easy to see), often as beautiful as the gaudiest of birds, and have a fascinating natural history.

This book describes many of the readily observable interesting insects that one often encounters in a tropical forest in the Americas. In addition to descriptions of the principal insect groups, the reader will find a wealth of biological information that serves secondarily as an introduction to the natural history of insects (and other arthropods). Though written for those who have no prior knowledge of insects, it should also prove useful for those who study them. While most of the information is dedicated to insects, a final chapter provides a glimpse into the intriguing world of spiders, crabs, and other arthropods.

It is a lot more challenging to identify an insect than it is to identify a bird. For starters, most insects—at least 80% of species—are as yet unnamed. Another problem is that many species look virtually identical and can only be distinguished by a specialist using a microscope (differences in male genitalia, for example, are often a particularly useful means of distinguishing between closely related species). Finally, the sheer diversity—and number—of insects that occur in the tropics poses its own special challenge; for example, the weevils alone comprise more species than all vertebrates combined. Nonetheless, even the casual observer can learn to identify many insects to the family or subfamily level.

ARTHROPODS

The animal kingdom is divided into more than 30 major groups, or phyla. The subject of this book is the insects and other arthropods, all in the phylum Arthropoda (meaning *jointed legs*), which has more species than any other phylum. This group is generally divided into five main subgroups (subphyla), one of which—the trilobites—is long extinct. For the sake of clarity, it is important to note that our main focus is on insects, the subphylum Hexapoda. The final chapter of the book treats the other arthropods.

Currently, there are about 1.3 million named species of arthropod, distributed as follows: 1,100,000 species of Hexapoda; 115,000 species of Chelicerata; 73,000 species of Crustacea; and 12,000 species of Myriapoda.

Beetle (Scarabaeidae: *Chrysina optima*).

HEXAPODA

The subphylum Hexapoda (meaning *six legs*) contains more named species than any of the other three extant subphyla. It includes four classes: Collembola, Protura, Diplura, and Insecta. All members of the first three groups lack wings, whereas members of the class Insecta generally have two pairs of wings, although the two most ancient groups—Archaeognatha and Zygentoma—are wingless.

CRUSTACEA

Crab (Pseudothelphusidae).

In contrast to insects, which have just one pair of antennae, crustaceans have two pairs of antennae. The number of legs varies from group to group. Most species inhabit the oceans, but many occur in freshwater, and a few are terrestrial. In addition to the well-known groups that appear on the menus of seafood restaurants (lobster, shrimp, crab), crustaceans also include several microscopic groups, some of which (copepods, for example) are extremely important components of aquatic food webs. This book discusses only the terrestrial and freshwater crustaceans, and therefore excludes the vast majority.

Note that most experts now agree that the insects (Hexapoda) evolved from a group of crustaceans and that the insects and crustaceans should therefore be combined into a single group, the Pancrustacea.

MYRIAPODA

Millipede (Polydesmida).

This subphylum, characterized by one pair of antennae and numerous legs, includes the millipedes (class Diplopoda) and centipedes (class Chilopoda). The word *millipede* means one thousand feet, although no species has more than 750 (most have far fewer); they are distinguished not by the number of legs but by having two pairs of legs per segment, a result of each segment consisting of two fused segments. The word *centipede* means one hundred feet, although none has exactly 100 (the number varies from 30 to 382); they are distinguished from millipedes by having just one pair of legs per segment—and they are also much faster than millipedes.

CHELICERATA

Chelicerates do not have a head that is separate from the body, but rather a prosoma, which contains both the head and the four pairs of legs. In place of antennae they have chelicerae, which they use to eat in much the same way that other arthropods use their mandibles. Although the subphylum Chelicerata includes two marine groups (horseshoe crabs and sea spiders), more than 99% of its species belong to the class Arachnida, a predominately terrestrial group. Arachnids comprise 16 orders, the best known being the scorpions, spiders, and mites (which include ticks). Except for many mites, most species of arachnid are carnivores, and generally suck out the liquid contents of their prey. The group is named for a figure from Greek mythology, princess Arachne, whose boasting about her skills as weaver prompted Athena to transform her into a spider.

Jumping spider
(Salticidae: *Lurio solennis*).

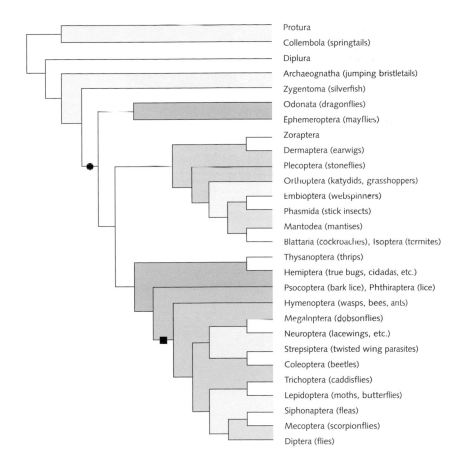

Figure 1-1. Evolutionary tree. This represents the most recent hypothesis of the evolutionary (phylogenetic) relationships between the insect orders. The figure excludes three orders not occurring in tropical America. The dark circle represents the origin of wings; the dark square represents the origin of complete metamorphosis.

ANATOMY AND NATURAL HISTORY

The nervous system of insects operates on the same biochemical basis as ours (although we obviously have a much larger brain). And, contrary to popular misconception, insects do have a heart, albeit one very different from ours. In many ways, of course, insects and humans are as different as could be. Insect blood (which is generally not red) flows through the cavities of the body instead of through capillaries. And rather than relying on blood to distribute oxygen, insects have a network of tiny tubes (tracheae) that carry oxygen directly to all parts of the body. Insects lack lungs; they breathe through tiny holes in the sides of their bodies (the openings to the tracheae, known as spiracles). This respiratory system is more efficient than ours, but it imposes physical limits on body size. Indeed, it is thought that the reason that some extinct insects were much larger (dragonflies with a wingspan of nearly a meter) than present day insects is that the atmosphere once contained more oxygen than it does today.

Unlike humans, who are visual creatures primarily, insects rely predominantly on their sense of smell (their small size imposes limits on visual

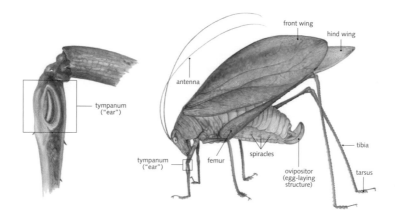

Figure 1-2. External anatomy of a katydid. The body of an insect is divided into three parts: head (with the eyes and antennae), thorax (with the legs and wings), and the abdomen (the rest of the body).

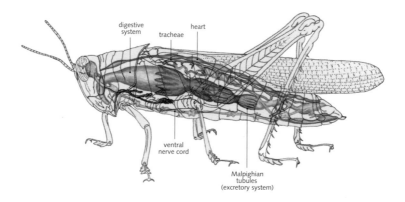

Figure 1-3. Internal anatomy of a grasshopper. Shown are the digestive system (orange), circulatory system (including the heart, red), respiratory system (including the tracheae, blue), nervous system (yellow), and excretory system (olive green).

and auditory reception). In place of a nose, they smell with their antennae, receptors so sensitive that humans have yet to develop artificial detectors to match them. Most insects communicate with other members of their species via odors (pheromones). Female moths, for example, waft sex pheromones to attract males; and ants, to cite another instance, lay down a chemical trail to guide nest mates to food. Like us, insects can taste with their mouthparts, though many of them can also taste with their feet!

Insects have auditory receptors located on various parts of their body, sometimes in bizarre places—crickets and katydids have "ears" on their front legs. Several larger insects can detect the high-frequency sounds of bats, thereby allowing them to evade those predators. Katydids, crickets, and cicadas can communicate via long-range, air-borne sounds (p. 77), but this is not true of most insect species, the vast majority of which are limited to short-range auditory signals (such as those transmitted by vibrations that carry over the surface of a plant).

Insect eyes are unlike human eyes, and their view of the world is surreally different from ours. Insects have compound eyes—each eye is made up of numerous independent photoreception units that offer up a very pixilated image. While compound eyes generally provide very poor image resolution, they are superb motion detectors. Humans do not detect the individual frames of a movie when shown at 25 per second; for insects, however, a movie would have to be shown at 200 frames per second to prevent them from deciphering individual frames. Most insects cannot distinguish the color red (which they perceive as black). And unlike us, they can see colors in the ultraviolet spectrum and can also detect polarized light.

The mating behavior of insects is incredibly diverse. Though insects generally reproduce sexually, a few species are asexual, either periodically or permanently. In asexual reproduction, the female exclusively produces female offspring, with no need for the participation of males (in some wasps and ants, asexual reproduction results from a bacterial infection that causes all eggs to produce females, thereby benefiting the bacteria, which are transmitted from mother to daughter). In the majority of species, where both sexes are present, males utilize various strategies for finding females;

Compound eyes of a deer fly (Tabanidae).

they sniff them out (via pheromones emitted by the females), cruise areas likely to contain females, and/or attempt to entice females into a territory that the male defends, to give just a few examples. Courtship generally relies on visual, acoustic, olfactory, and tactile stimuli, and is often quite elaborate, as males compete vigorously with one another to become the father of the female's offspring—and females are usually quite finicky. Mating lasts from a few seconds (e.g., in house flies) to many hours (e.g., some beetles), with different species employing a variety of positions, including male on top, female on top, and butt to butt. After mating, the female often has the option of discarding the male's sperm should she find a better mate. Most insects lay eggs, but aphids, some cockroaches, and a few flies forego an external egg stage and give birth to young juveniles. The majority of insects lay several hundred eggs during their lifetime, but in some species the number is smaller or much larger (in the thousands).

Insects and other arthropods (spiders, crabs, etc.) have the skeleton on the outside of the body. This exoskeleton is made of chitin, a very resilient material that helps protect the animal against predators and pathogens. As the insect grows, however, it must periodically molt (or shed) its exoskeleton, and replace it with a larger one in order to allow further growth. Note that in most insects molting only occurs during the juvenile stages, since in the vast majority of species the adults do not molt, and therefore do not grow (except for the female's abdomen, which often increases in girth after feeding or as it fills up with eggs). The number of molts is usually constant (more or less) within a species

Recently molted slug caterpillar (Limacodidae: *Acharia horrida*) with shed exoskeleton on right.

In grasshoppers, cockroaches, and many other insects, the changes are relatively minor; adults are very similar to juveniles except that they have fully developed wings (juveniles only have wing buds). This gradual change from juvenile to adult is known as incomplete metamorphosis. In contrast, insects with complete metamorphosis undergo a change so drastic that there is a quiescent pupal stage, during which the juvenile body is disassembled and the adult body is formed. Insects with this type of metamorphosis include beetles, wasps, moths, flies, and a few other smaller orders.

but varies considerably between insect groups; the majority of insects molt 3 to 6 times, but most mayflies molt 15 to 25 times; in primitive wingless insects, which continue to molt as adults, the total number of molts occasionally exceeds 50.

Like most animals, insects change in form as they grow; in other words, they undergo metamorphosis. However, there is considerable variation in the degree of change occurring between the juvenile stages and the adult.

Unfortunately, the terms used for the juvenile stages of insects have not been consistent. Traditionally (and in this book), juveniles of insects with incomplete metamorphosis are called nymphs, whereas those with complete metamorphosis are called larvae (or caterpillars and maggots in the case moths and flies, respectively). Some authors, however, use the term *larva* for the juvenile stages of all insects.

Larvae generally eat a much greater quantity of food than do adults; indeed, one can think of larvae as essentially eating machines and adults as primarily reproductive

Figure 1-4. Complete metamorphosis of a morpho butterfly (*Morpho helenor*). Upper row, left to right: egg, young larva (first instar), older larva (penultimate instar). Lower row, left to right: pupa, adult emerging from pupa, recently emerged adult drying its wings.

machines. The time required to develop from egg to newly emerged adult generally ranges from a couple of weeks to several months, but some species take several years. Adults of most insects probably live for a month or two, but longevity ranges from just a couple days to the extreme case of queen leafcutter ants, which are capable of living more than 20 years.

Insects have an extremely diverse diet. Among the species that feed on plants (what biologists call phytophagous species) are the leaf chewers, sap suckers, stem borers, seed borers, leaf miners (p. 202), gall formers (p. 280), and pollen and nectar feeders. Most carnivorous insects feed on other invertebrates, and they usually do so in one of two ways: as a predator that consumes multiple prey or as a parasitoid that feeds on a single individual (host) and eventually kills it. Predators include a very wide range of insects, whereas parasitoids occur primarily among the wasps and flies (and it is always just the larval stage that is the parasitoid). A third group of carnivorous insects includes those that feed on vertebrates and they do so as parasites, which generally do not kill their host (unlike parasitoids); examples include lice, fleas, and many flies.

Many insects feed on fungi, and some species, including leafcutter ants and some bark beetles, even cultivate them in order to assure a ready supply. Numerous beetle and fly larvae are detritivores, feeding on dead plant parts, animal waste, and animal carcasses. Various aquatic insects feed by filtering small particles of nutriment from the water. Insects that feed throughout their lives on wood, plant sap, or blood (lice and kissing bugs) find it difficult to obtain a balanced diet and usually obtain missing nutrients from symbiotic bacteria that they harbor inside their bodies.

Insects, in turn, are a major food source for many other animals. A majority of birds eat insects, as do anteaters, many bats, lizards, frogs, and freshwater fish. To protect themselves from predation, insects rely on a several kinds of defensive strategies. Many have what biologists call cryptic coloration, which is to say that the color or pattern of the insect matches that of its surroundings. Some species show disruptive coloration, an irregular pattern that makes it hard for predators to trace the exact outline of the insect. Many insects contain nasty chemicals, and those that do also have a bright or otherwise distinctive coloration to warn off predators. Bees, wasps, and many other insects can sting. Some insects rely on an erratic flight pattern to help them escape from predators. In addition to being the victims of predators, insects are also parasitized by nematodes, fungi, protozoans, bacteria, and viruses, some of which are used in the biological control of insect pests (p. 166).

Insects also participate in numerous mutualistic interactions. For example, insects pollinate flowers and, in return, receive nectar, pollen, and sometimes oil or resin. However, there are numerous other fascinating instances of mutualistic interactions, many of them involving ants (p. 183).

Insects interact with humans in ways both detrimental and beneficial. A few bite or sting, while others devastate our crops and stored grains. Yet we could not live without insects. They pollinate our crops, help control populations of pest insects, recycle nutrients, and offer us numerous other "services"; a small handful of domesticated insects provide us with dyes (p. 95), honey (p. 182), and silk (p. 200).

INSECTS IN THE TROPICS

Insects that inhabit the tropics are generally not all that different from insects that live in temperate regions, although the number of species in the tropics is much higher than elsewhere, a fact that is equally true, for the most part, for other animals and for plants. On a single rainforest tree, for example, you will find more species of ant than in all of England. Most insect species found in the tropics are restricted to a certain kind of habitat. While some species have a very limited geographical distribution, others have a much broader distribution; a few tropical American species, for example, also occur in North America.

Human alteration of the environment in tropical America continues to have a negative impact on many insect species. At the same time, a few insects (mostly pests) have spread around the world (often by human agency) and thrive in landscapes altered by human activities.

It is a general belief that insects in the tropics are larger and more colorful than insects in temperate regions, though it's also true that there's more of everything in the tropics, including small, drab insects. To our eyes, brightly colored insects are beautiful (p. 243), but to predators such as birds, bright colors are a warning sign that indicate the insect is likely to contain nasty chemical compounds (p. 223). Some perfectly innocuous insects mimic the colors of toxic species, thereby scaring off potential predators through subterfuge (a phenomenon known as Batesian mimicry). At the same time, numerous nasty insects all resemble one another, making it easier for predators to remember—and avoid—them (a phenomenon known as Mullerian mimicry). With these two kinds of mimicry at play, there are a lot of distinct species that resemble one another, which makes it difficult to correctly identify, say, a tropical butterfly. It remains to be seen whether mimicry is more common in the tropics—perhaps due to greater predation pressure—or whether the percentage of mimetic species is the same as in temperate regions.

Unlike temperate regions, lowland tropical rainforests are dominated by eusocial insects (p. 175), which are insects that live in colonies consisting of one or more queens and numerous workers (for reasons not entirely clear, eusocial insects disappear at higher altitudes). These include all termite and ant species, some wasp species (those that build paper nests and sting), and some bee species (mainly stingless bees and honeybees). Although they represent fewer than 1% of insect species, eusocial insects occur in large numbers and account for more than half of the insect biomass (cumulative weight) in lowland tropical forests. Indeed, the first Europeans to settle in tropical America, often referred to the region as the kingdom of the ants.

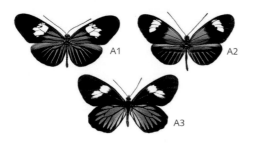

A1: *Heliconius erato venustus* (Heliconiinae; Bolivia)
A2: *H. melpomene aglaope* (Ecuador)
A3: *Archonias brassolis negrina* (Pieridae; Peru).

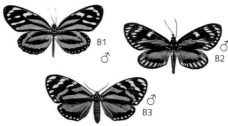

B1: *Lycorea halia* (Danaini; Costa Rica)
B2: *Zegara columbina* (Castniidae; Costa Rica)
B3: *Chetone angulosa* (Arctiinae; Costa Rica)

C1: *Heraclides torquatus* (Papilionidae; Costa Rica)
C2: *Archonias brassolis approximata* (Pieridae; Panama)
C3: *Dysschema jansonis* (Arctiinae; Costa Rica)

D1: *Oleria paula* (Ithomiini; Costa Rica)
D2: *Dismorphia theucharila fortunata*
(Pieridae; Costa Rica)

Figure 1-5. Examples of mimicry in butterflies and moths.

DANGEROUS ARTHROPODS OF TROPICAL AMERICA

Many travelers to the tropics fear encounters with a snake, insect, or spider, though in reality the number of potentially dangerous arthropods (and snakes) is very small. Many of them are merely bothersome rather than dangerous. Foremost among these are chiggers (Trombiculidae), which are mites (p. 326) that produce intense itching around the ankles, beltline, and wherever else clothing presses tightly against the body. Once you have chigger bites the only thing you can do is apply an anti-itch cream. These barely visible mites inject digestive enzymes into the skin, suck up the dissolved cells and lymph, and then drop from the body—contrary to popular thinking, they do not burrow into the skin. The chigoe flea, also known as the sand flea or jigger, does burrow into the skin, but is rarely encountered. Other very annoying but generally innocuous insects are biting midges, black flies, and deer flies. Then there is the human bot fly (p. 301), which, though not dangerous, may inflict psychological trauma—think *Alien* without the physical damage.

Some arthropods sting, others bite (though the vast majority does neither). A sting invariably involves the injection of venom, whereas some species of biting arthropods do not inject anything, relying solely on the pain caused by the bite for defense. Biting arthropods such as some spiders do inject venom, and when mosquitos "bite" they inject an irritating saliva (the stuff that causes an itch). Arthropods that inject saliva are out for blood and their saliva occasionally harbors disease-causing microbes that enter the bloodstream (fortunately, not all individuals of a species *capable* of carrying a microbe actually do).

The most dangerous arthropods are those that transmit diseases. The mosquitos *Aedes aegypti* and *A. albopictus* are vectors of three viruses, yellow fever, dengue fever, and chikungunya (p. 272); there is a vaccination for the first but not for the second and the third. The best way to avoid getting dengue or chikungunya is to use a repellent to ward off the mosquito vectors, which are active primarily during the day. Some species of mosquito in the genus *Anopheles* are vectors of the malaria protozoan. In general malaria is less prevalent than dengue and a prophylactic treatment (chloroquine) is available for those travelling to high-risk areas. Moreover, the fatality rate from malaria is much lower in tropical America than in Africa. Within the sand fly family (Psychodidae), some species in the genus *Lutzomyia* are vectors of the protozoan that transmits leishmaniasis (p. 271), a leprosy-like malady. And finally, some kissing bugs (family Reduviidae, subfamily Triatominae) are vectors of the protozoan that causes the potentially fatal Chagas disease (p. 65). Although Chagas is a serious disease in tropical America, most travelers to the region need not be overly concerned since kissing bugs are mostly encountered in houses with dirt floors or thatched roofs.

The remaining biting and stinging arthropods that occur in tropical America pose less of a threat, as they do not seek out humans and do not transmit diseases. The danger posed by scorpions (p. 303) and spiders (p. 307) is greatly exaggerated; fewer than 1% of the spider species have a venom that is potentially dangerous to humans—mostly to babies, the elderly, or those suffering from a pre-existing medical condition. Some species of wasp, ant, and bee also sting, of course, though it is generally not the venom *per se* that is dangerous, but rather the allergic reaction it provokes in some people. A few species of caterpillar and, sometimes, even their cocoons, have either venom-filled spines or urticating hairs. Such species are confined to relatively few families, principally flannel moths, slug caterpillars (p. 204), and giant silk moths (p. 209).

TROPICAL AMERICA

The tropics are most simply defined as that part of our planet that is situated between the Tropic of Cancer, the most northerly position at which the Sun appears directly overhead during the June solstice, and the Tropic of Capricorn, the most southerly position during the December solstice. The locations of these two latitudes, one on each side of the Equator, are of course determined by the tilt of the Earth's axis of rotation, although this tilt fluctuates slightly over time; for example, the

Tropic of Cancer currently lies at 23° 26′ 16″, but in 1917 it was at 23° 27′ north latitude.

One can also define the tropics as those areas harboring a substantial number of tropical plants and animals, though this is a rather circular definition and can lead to some puzzling claims. By this definition, for example, the Florida Keys, which lie north of the Tropic of Cancer, are more tropical than are the mountains surrounding Mexico City, which lie to the south. In the Southern Hemisphere, most of the Atacama Desert, the driest desert on earth, lies to the north of the Tropic of Capricorn. Regardless of how it is defined, tropical America encompasses a wide diversity of climates and habitats. In this book the exact definition is relatively unimportant since we use a broad brush in describing the insect inhabitants of the region.

Because the tropics lie within twenty some degrees of the equator, the length of a day does not vary appreciably from month to month. Moreover, at any given location the temperature is relatively constant throughout the year, though the average temperature decreases with altitude. Seasonality in the tropics, then, is more a function of variation in rainfall. While many parts of tropical America experience heavy rainfall more or less throughout

the year, certain areas have a very pronounced dry season; for example, the Pacific coast of Mexico and northern Central America, and also a wide belt running from easternmost Brazil (Caatinga vegetation) to eastern Bolivia (the Gran Chaco). While the populations of many insect species show little apparent seasonality, some species have a life cycle that is cued to the coming of the rains. In the case of insects that feed on plants, the insects' life cycles are often synchronized with the seasonal cycle of their host plants.

The insect diversity of tropical America results not only from the present-day mosaic of topographies and climates but from a complex geological history. About 250 million years ago, today's continents were joined together in a supercontinent known as Pangea; later, during the heyday of the dinosaurs (in the Jurassic period, beginning 200 million years ago), Pangea began breaking apart, giving rise to Laurasia (North America, Europe, and Asia) and Gondwana (South America, Antarctica, Australia, and Africa). It should be noted that North America was separated by water from South America.

South America began its final separation from Africa about 100 million years ago, but retained a connection with Antarctica and

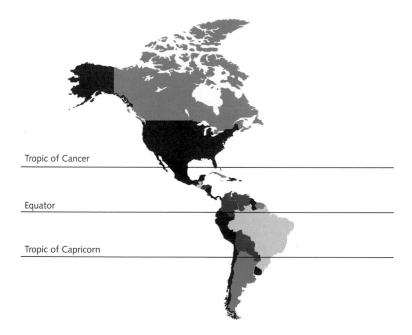

Figure 1-6. Map indicating tropical regions of the Americas.

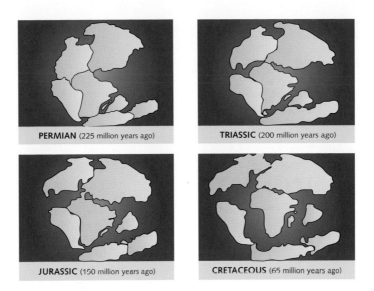

Figure 1-7. The breakup of Gondwana. The separation of the continents resulted from plate tectonics (the movement of the earth's plates).

Australia for a much longer period of time (the connection with Antarctica was severed between 40 and 30 million years ago). For this reason, South America shares various groups of plants and animals with Australia. After the breakup of Gondwana, South America remained isolated from other continents, thereby encouraging the evolution of indigenous fauna and flora, including sloths, anteaters, armadillos, hummingbirds, toucans, tyrant flycatchers, leafcutter ants, orchid bees, and bromeliads.

Even though South America was isolated from other continents by water, plants from North America did manage to colonize this southern landmass; and two other groups of colonizers arrived from Africa: monkeys and the ancestors of caviomorph rodents (porcupines, agoutis, and capybaras). While Africa was at that time much closer to South America than it is today, these colonizing animals wouldn't have been able to swim the distance, and so how they arrived is still a bit of a mystery. Some insects also traveled from Africa to South America at about the same time, including butterflies in the tribe Acraeini (related to passion-vine butterflies, p. 258).

Presently, the natural history of tropical South America is largely defined by the Andes (the world's longest mountain range) and the Amazon River (with greater water flow than any other river in the world). The Andes have resulted from the downward movement of oceanic crust below the western edge of the South American plate. Although this has been going on for the last 100 million years, the Andes only began to take on their modern form about 30 million years ago, and the most intense period of mountain building taking place only in the last 12 million years.

Just as the Andes have changed over time, so too has the Amazon River. During most of South America's existence as a separate continent, the Amazon as we know it did not exist; instead, water from what is now the western Amazonian region flowed northward into the Caribbean Sea, circumscribing a region that for millions of years was dominated by large lakes, with occasional incursions of seawater from the Caribbean. The Amazon River began flowing eastward across the continent about 11 million years ago, but for the first few million years it was not the mighty river it is today. With the gradual rise of the Andes, and the consequent increase in surface runoff, the river grew progressively larger, reaching its current size around 2.4 million years ago.

The final grand episode in the formation of tropical America was the emergence of a land bridge between North and South America. Between 25 and 20 million years ago, northwestern South America began colliding with the easternmost tip of the Central American volcanic arc, probably allowing for some limited exchange of plants and animals between the two continents. However, it was not until about 3 million years ago that the exchange became dramatic—in an event labeled the Great American Biotic Interchange—whereby North American animals began flooding into South America, and vice versa. It is possible that a continuous (or nearly continuous) land bridge existed long before, and it was a change in climate on the Panamanian isthmus, from wet and hot to savanna-like conditions (due to ice sheet formation in Canada), that finally allowed a notable increase in biotic interchange, especially of land mammals. Immigrants from North America included tapirs, llamas (both now extinct in their homeland), peccaries, squirrels, coatis, and members of the cat and dog families. Immigrants from South America into North America included opossums, giant ground sloths (now extinct), armadillos, and porcupines. Many South American immigrants only made it as far as north as Central America and the Caribbean islands. While we can most easily trace the past movements of mammals (they leave the clearest fossil record), many insects and other arthropods undoubtedly showed the same dispersal patterns.

A NOTE ON NAMES AND CLASSIFICATION

Relying on common names to identify animals is fraught with difficulties, especially in the case of arthropods, the majority of which have no common name! In many cases, the same common name may be used to refer to a variety of species. And, the common names used in one language will of course differ from the common names used another. Nonetheless, the common names of the principal groups of arthropods are more familiar to most people than are the scientific names; therefore, in this book, common names are used together with the scientific names. Many Spanish names are also included, but only when these are widely used in the region. Portuguese names are not included; for those interested in knowing Portuguese names, we refer readers to one of the best entomology textbooks to be produced in Latin America in recent years, *Insetos do Brasil* (Rafael et al. 2012), which contains Portuguese names.

In contrast to the common name, the scientific name of a given animal is its standardized unique name, recognizable in any language. The species taxon is the most fundamental unit of life; its scientific name, which by convention is italicized, consists of two parts: genus and species. The scientific name for the house fly, for example, is *Musca domestica*, in which *Musca* is the name of the genus and *domestica* is the tag name that indentifies a unique species within that genus. This binomial system of naming organisms was established by Carl Linnaeus, an 18th century Swedish biologist whose image still appears on 100 kronor bills.

Linnaeus also devised a manner whereby all organisms can be classified in a hierarchical system. Each kingdom (animals, plants, fungi, etc.) is divided into phyla, each phylum into classes, each class into orders, each order into families, each family into tribes, each tribe into genera, and each genus into species. When needed, each of these ranks can be preceded by the prefix super- or sub- in order to add more levels to the hierarchy. For example, each order can be divided into suborders, each suborder into superfamilies, each superfamily into families, and each family into subfamilies. Some names have standardized suffixes: in animals, the name for a superfamily always ends in -oidea, families in -idae, subfamilies in -inae, and tribes in -ini. Finally, it should be noted that taxonomic classifications frequently change; although this can be confusing, these changes reflect our growing understanding of the evolutionary relationships among groups.

In this book we provide the number of named species for each of the major groups. While nearly all of these numbers will continue to increase in the coming years, the current numbers are given in order to provide an idea of the group's relative diversity. For groups in which the number of named species for tropical America was not readily available, the number of world species is given. Finally, it bears repeating that the majority of arthropod species are still unnamed.

2

SMALL ORDERS

Insects are classified into more than 30 orders. Five of these, which are very large groups, are given detailed coverage in chapters 3 through 7. This chapter covers the smaller orders, no less fascinating even if they do contain fewer species. Not included are ice crawlers (Grylloblattodea) and African rock crawlers (Mantophasmatodea), two orders that are entirely absent from tropical America, and the snakeflies (Raphidioptera), which, in tropical America, occur only at high altitudes in Mexico and northern Guatemala.

WINGLESS INSECTS

Insects crawled over the surface of the earth long before the arrival of the dinosaurs. The first insects lacked wings and molted their exoskeleton not only as juveniles but also as adults (most modern insects only molt when juveniles). Instead of copulating, males generally deposited a packet of sperm on the ground or on a silken thread, and the female then retrieved it. Descendants of these primitive creatures are still with us today, in five groups of wingless insects (not to be confused with insects such as lice and fleas, which are wingless, but descend from winged ancestors).

The first three groups (Collembola, Protura, Diplura) are so different from all other insects that they are now classified as three separate classes. The other two groups of wingless insects (Archaeognatha and Zygentoma), together with all winged insects, are placed in the Class Insecta.

Springtail (Neanuridae) on the surface of water.

CLASS COLLEMBOLA
(Springtails)

The common name *springtail* derives from the fact that most species have a tail-like appendage (below the abdomen) that is bent forward under the body, where it is held in place, under tension, by a clamp; when the tail is released from this clamp, the insect springs upward. Many springtails are white while others, especially those that live in more exposed habitats, are quite colorful; in some instances these

Entomobryidae (springtail).

colors are due to tiny scales that cover the body, in other instances they are due to pigmentation of the cuticle. Unlike most other insects, springtails generally breathe through their cuticle, which has a non-wetting surface that allows them to continue breathing when their surroundings become inundated with water.

Springtails are the most widespread and abundant insects in the world, but they are seldom noticed due to their small size. They occur virtually everywhere—in soil, under bark and rocks, on pond surfaces, in the intertidal zone, and up in tree canopies—feeding on decaying vegetation and fungi, unicellular algae, arthropod feces, or small invertebrates (depending on the species). A cubic meter (a little more than a cubic yard) of soil can harbor up to 200,000 individuals, a density rivaled only by mites. Some species live in unusual habitats: members of the family Coenaletidae, for example, are confined to the shells of terrestrial hermit crabs (p. 335); a few springtails occur in volcanic vents with temperatures as high as 48°C (118° F); and other species inhabit glaciers.

Despite their abundance and importance in soil ecosystems, springtails are poorly known. Worldwide, there are more than 8,000 named species, which are classified in four orders.

Many Entomobryidae springtails have colorful scales. This species twirls slowly around on the upper surface of leaves.

CLASS PROTURA

Proturans are minute (smaller than 2.5 mm/0.1 in), white insects that lack antennae and eyes. Because they use the front pair of legs as sensory appendages (as a substitute for antennae), they walk on just four legs.

Proturans live in soil, leaf litter, and rotting wood, where they probably feed on fungi; some species are known to feed on mycorrhizal fungi, which are symbiotic on plant roots. It is still unknown whether proturans molt as adults, as do the other wingless groups discussed in this section. It is possible that they utilize direct sperm transfer (though their biology is still very poorly understood). There are about 800 named species of proturans worldwide.

CLASS DIPLURA

Like proturans, diplurans are whitish, eyeless inhabitants of soil and rotting wood, but they differ by having antennae and two appendages (cerci) at the tip of the abdomen (Diplura means *two tails*). Members of the superfamily Campodeoidea are very small (usually less than 6 mm/0.2 in), have long, filamentous cerci and generally feed on decaying plant material; in contrast, members of the superfamily Japygoidea can reach lengths up to 50 mm (2 in), have short, pincer-like cerci (as in earwigs), and many are predators of other small arthropods. There are about 1,000 named species of diplurans worldwide.

CLASS INSECTA

The vast majority of insects belong to this class, which is characterized by a number of morphological characteristics; the tarsi on the legs are subdivided into segments (technically, not true segments), and the females possess an egg-laying organ, the ovipositor (in several groups, however, this has been lost over evolutionary time). All of the remaining orders in this chapter belong to the class Insecta.

ORDER ARCHAEOGNATHA
(Jumping Bristletails)

Jumping bristletails superficially resemble silverfish, but are more cylindrical, have relatively huge eyes, and are capable of jumping. They are mostly nocturnal and feed on algae, lichens, and decaying vegetation. Many live in the leaf litter but some live up in the trees; if an arboreal species falls from the tree it can glide toward the trunk using its middle tail-filament to steer. There are more than 500 named species of jumping bristletails worldwide.

Meinertellidae, the only family of jumping bristletails in South America.

ORDER ZYGENTOMA
(Silverfish)

This is perhaps the best known group of wingless insects, since a few species of silverfish have taken up residence in our houses (usually *Acrotelsa collaris* in the tropics).

With long antennae, three long tail-filaments, and amazing speed, silverfish sometimes startle people, though most species are seldom noticed since they inhabit leaf litter and rotting wood. They feed on a wide variety of dead plant material, and at least some are capable of digesting cellulose, which allows them to feed on moldy paper and book bindings. A few species live in the nests of termites or ants, where they apparently feed on detritus. There are nearly 600 named species of silverfish worldwide.

Acrotelsa collaris (Lepismatidae), a common silverfish in houses.

INSECT FLIGHT

Female *Orthemis discolor* dragonfly (Libellulidae) flying over puddle.

Powered flight has evolved four times; first in the insects (nearly 400 million years ago) and later in pterosaurs (now extinct), birds, and bats. Insect wings probably arose from extensions of the upper thorax in silverfish-like insects that engaged in controlled gliding from tall plants. Today the oldest winged insects inhabiting the earth are the mayflies, dragonflies, and damselflies, whose fossil records date back to the Paleozoic (about 300 million years ago). These insects are known as Paleoptera ("old wings"), as opposed to all other winged insects, Neoptera ("new wings"), which have the ability to fold their wings back over the abdomen.

Unlike vertebrate wings (which are modified front legs that contain muscles along their length), insect wings only have muscles that attach to tiny plates at the base of the wing; these muscles function to control the tilt of the wing by pulling the leading edge up or down. Most of the power for flight comes from huge thoracic muscles that flex the top of the thorax up and down, thereby indirectly moving the wings (except on dragonflies and damselflies, with muscles attached directly to the base of the wing—which design allows each wing to

move independently). Insects do not simply flap their wings up and down, but instead move them in a figure eight pattern whose aerodynamic complexity boggles the mind.

Insects generally have two pairs of wings (i.e., a total of four). In many species, the front and hind wings are interlocked and move as a single unit; beetles fly with just their hind wings, dedicating the use of their front wings as protective shields; flies use only their front wings, having lost through evolution the hind pair.

The speed at which many wasp, bee, and fly species beat their wings far exceeds the speed at which a nerve impulse travels. Mosquitoes beat their wings up to 600 times per second (some biting midges beat their wings up to 1,000 times per second), compared with a mere 50 to 80 times per second for a hummingbird. Such high rates of muscle activity require prodigious amounts of energy. Bumble bees are among the most profligate guzzlers of calories. To put this in perspective, a human runner burns the equivalent of a chocolate candy bar in one hour, whereas a foraging bumble bee burns a proportionately similar amount of calories in 30 seconds.

WINGED AQUATIC ORDERS

Though not otherwise closely related to one another, the five insect orders in this section are composed entirely of species that spend a part of their lives in water (note that several primarily terrestrial orders of insect include among their ranks numerous freshwater species).

It is the juvenile stages that are aquatic, obtaining oxygen from water via gills; depending on the species, they require from a month to more than a year to reach the adult stage. Some species inhabit running water, other species live in bodies of standing water, ranging from lakes to the small puddles of water that accumulate in tree holes, bromeliads, etc. Adults are terrestrial and usually live for relatively short periods of time (a few days to a month or more). Adult females eventually return to the water in order to lay eggs; some species are deceived by reflections from dark colored cars, perceiving them as a body of water, and sometimes even lay eggs on them.

Among the aquatic orders, caddisflies have the most species. Shown here is *Macrostemum* sp. (Hydropsychidae).

ORDER EPHEMEROPTERA
(Mayflies)

The majority of mayfly species have three long filaments at the tip of the abdomen (several species have only two). They are the only insects that molt from one winged stage to another. The fully grown nymph emerges from the water (usually during the early hours of the night), molts into a subimago (with milky colored wings), and about ten hours later molts into an adult (with transparent wings). In the tropics, adults emerge asynchronously and essentially throughout the year; in temperate regions, adults emerge simultaneously and at specific times of the years.

Adult mayflies lack functional mouthparts, do not eat, and, depending on the species, live just minutes (less than an hour) to a few days (one of two etymological roots in the order name is *emphemera*).

Generally under the cover of darkness, males form swarms above streamside plants. When a female enters the swarm, one of the males grabs her with his elongate front legs, and, while in flight, copulates with her by means of an unusual double penis. In the most common family, Baetidae, males have divided eyes that probably help them navigate when there is little light (twilight) and pick out females in a swarm.

Male Baetidae, with a divided eye, the upper part very enlarged.

Female mayflies deposit eggs while fluttering just above the surface of the water, which makes them an easy target for fish. Juvenile mayflies, which, like adults, also have three terminal filaments (two in some species), usually feed on algae or detritus. Although the nymphs of a few species are capable of short bursts of swimming (via dorsal-ventral undulations of the body), most attach themselves to the undersides of rocks and other submerged substrates, or burrow into sediment.

There are more than 3,200 named species of mayfly worldwide, of which about 600 occur in Central and South America.

Baetodes sp. (Baetidae). Mayflies have long filaments at the tip of the abdomen.

Baetidae nymph with inflated wing buds, indicating that it is ready to emerge from the water.

Mayfly nymphs (such as this Baetidae) have gills on the sides of the abdomen.

ORDER ODONATA

(Dragonflies and Damselflies)

The order Odonata is made up of two similar but distinct suborders, the dragonflies and the damselflies (Table 2-1). Adults generally capture flying insects, either by cruising (many dragonflies) or by darting out from a perch (dragonflies in the families Libellulidae and Gomphidae, and most damselflies). Giant damselflies (Pseudostigmatidae), with wingspans of up to 18 cm (7.1 in), pluck insects and spiders from vegetation or from spider webs.

Dragonflies and damselflies mate unlike all other species of insect. The male transfers sperm from the tip of his abdomen to a specialized organ at the base of the abdomen, thus allowing him to use the tip of his abdomen to grab a female by the neck. Copulation requires female consent, however, since she must curve her abdomen upward to contact the base of the male's abdomen. The result is a "mating wheel." In some species the male continues to grasp the female by her neck as she ventures out to lay eggs in the water, and by doing so the male prevents her from taking a detour into another male's territory and mating with him. Although this strategy ensures the male of his paternity, he risks losing his own territory to another male while he is accompanying the female to the water.

Female damselflies and some female dragonflies have ovipositors that they use to insert their eggs into submerged vegetation, rotting wood, or mud; in most dragonfly species, the female lacks an ovipositor and, rather than concealing her eggs inside some substrate, she places them on the surface of an object submerged in water, or she simply allows them to sink to the bottom of the water. The nymphs of damselflies and dragonflies lurk in the bottom sediments or on submerged vegetation; they have an elbowed appendage (labium) beneath the head, which is normally folded up but can be extended forward in a split second in order to capture prey. Some damselfly nymphs live in the water that collects at the base of bromeliads. They prey on animals that eat detritus; when the nymphs defecate, they thus return nitrogen to the miniature ecosystem of the bromeliad.

For reasons that are poorly understood, a few dragonfly species undertake mass migrations, especially during the rainy season.

There are about 6,000 named species of dragonfly and damselfly worldwide, of which roughly 1,650 occur in Central and South America.

Table 2-1. Differences between the two suborders of Odonata

	Dragonflies	Damselflies
Head	Eyes spherical (and close together, except in Gomphidae, whose species have eyes that are well separated)	Eyes wide (and well separated)
Body	Stocky	Slender
Flight	Rapid	Relatively slow
Perching posture	Wings outstretched	Wings above body (except in Lestidae and Megapodagrionidae, where the wings are outstretched)
Ovipositor	Present only in Aeshnidae and Cordulegastridae	Present in all species
Nymph	Wide and flattened, with gills inside the rectum	Elongate, with three leaf-like gills at the tip of the abdomen

Male *Orthemis* dragonfly (Libellulidae). Females of this genus are more brownish colored.

Immature male of *Erythrodiplax fervida* (Libellulidae).

Male *Erythrodiplax funerea* (Libellulidae).

Dragonfly nymph (Libellulidae).

Female *Neocordulia batesii* dragonfly (Corduliidae).

Male *Argia* damselfly (Coenagrionidae).

Male *Leptobasis vacillans* (Coenagrionidae).

Male *Argia* (Coenagrionidae).

Megaloprepus caerulatus (Coenagrionidae).

Close-up of *Argia* head.

Hetaerina damselfly nymph (Calopterygidae).

Female damselflies (Coenagrionidae) laying eggs. Note three males grasping females.

ORDER PLECOPTERA
(Stoneflies)

Both nymph and adult stoneflies have two filaments (cerci) at the tip of the abdomen. Adults are about 3 cm (1.2 in) in length. They are weak fliers; when resting they hold their wings flat over the abdomen. Adults live for only a week or two and feed primarily on nectar and honeydew.

In several species, males attempt to attract females by drumming the substrate with a special structure (known as a hammer) that is located on the underside of the abdomen. Under the cover of darkness, the female deposits a gelatinous mass of eggs in or near water; during her short life, she can lay from 2 to 5 egg masses, with a total of up to 1,000; the eggs require nearly a month to hatch.

Nymphs have gills at the base of their legs (mayflies, p. 18, have gills on the abdomen). Initially nymphs eat algae and detritus, but as they age they become carnivorous; they require from three to six months to reach the adult stage. They occur in well-oxygenated rivers and streams, either in dead leaves in the stream bed or under rocks, depending on the species.

This is one of the few groups of insect that is less diverse in the tropics than in temperate zones. There are about 3,800 named species of stonefly worldwide, of which only about 500 are known from Central and South America; 300 of these belong to the genus *Anacroneuria*, in the family Perlidae.

Anacroneuria (Perlidae), the most diverse genus of stonefly in tropical America. The red spot is a parasitic mite.

Río Savegre, San Gerardo de Dota, Costa Rica.

In most countries, environmental agencies monitor water in rivers and streams to ensure compliance with legally mandated water quality standards. The problem with using chemical tests to check for pollutants, however, is that pollutants wash down streams very quickly—and it is thus difficult to detect the source of the pollutant.

But pollutants often have lasting effects on the organisms that live in water, so that studies of aquatic populations can provide reliable information about water quality. For these studies, biologists often prefer to analyze aquatic insects, since many species are fairly easy to sample, and also because these are the most diverse group of macroscopic organisms living in freshwater; because some species can only live in pristine waters while other species are capable of enduring very high levels of pollution—and because distinct species sometimes respond differently to the same pollutant—biologists can learn a lot simply by knowing what species inhabit a given body of water.

Historically, coffee processing plants were major polluters of rivers in many parts of tropical America. While the pulp from coffee berries is not toxic, any drastic increase in the organic matter entering freshwater can lead to an explosion in the populations of certain bacteria, which in turn can deplete the dissolved oxygen present in the water (a process known as eutrophication). Fortunately, coffee pulp is now generally either put to good use in composting piles or segregated in containment ponds. Today, most water pollution in the region is due to both agriculture—from the use of pesticides and fertilizers as well as river sedimentation caused by runoff—and to expanding urban areas that add to the amount of sewage and other wastewater.

ORDER MEGALOPTERA
(Dobsonflies)

This order of insect is closely related to the order Neuroptera (p. 53). It is composed of just two families, both of which occur in tropical America: the dobsonflies (Corydalidae) and the infrequently seen alderflies (Sialidae).

Dobsonflies are 8 to 12 cm (3.1 to 4.7 in) in length. Males in the genus *Corydalus* have extremely long, tusk-like mandibles that they use to joust with one another. Adult dobsonflies eat very little; males probably subsist on nothing but water, and females are thought to consume only nectar or fruit juice, depending on availability. Females lay their eggs in a large cluster (sometimes more than a 1,000 in a single cluster) that is encased in a white protective material and is attached to plants and other objects that overhang water. Each female deposits two or three egg masses during her short lifetime. The larvae (known as hellgrammites) are voracious predators of other aquatic insects; when mature, they leave the water to pupate either in soil under rocks or in rotting logs. In some parts of Latin America, people fear dobsonfly larvae (probably due to their large size), but some people in the Amazonian region of Peru actually eat the larvae.

There are about 350 named species of Megaloptera worldwide, of which 70 are known from tropical America.

Head of male *Corydalus* (Corydalidae). Despite the formidable mandibles, it poses no danger to humans.

Chloronia species (Corydalidae) are generally smaller than those of *Corydalus* and are pale yellowish green.

Corydalus (Corydalidae) female laying eggs on exposed rock in river (white spots are egg masses).

ORDER TRICHOPTERA
(Caddisflies)

Caddisflies are close relatives of moths and butterflies (one way to distinguish between the two groups is to compare the wings; on caddisflies, the wings are covered with hairs, on moths and butterflies, with scales). Although most species are nocturnal and have drab coloration, some are diurnal, with relatively bright coloration. This is the second largest group of aquatic insect after the non-biting midges (p. 276).

In nocturnal species, females often emit pheromones to attract males. In some diurnal species, the situation is quite different; in these, males fly in swarms to attract females.

Females lay an egg mass. Depending on the species, they place the egg mass either beneath the water surface or on a rock or vegetation that overhangs the water (on hatching, the larvae drop into the water below). The diet of the larvae is quite diverse, depending on the species. Like moth caterpillars, caddisfly larvae secrete silk; all species use silk as anchoring lines to attach themselves to the substrate and for constructing a case in which to pupate. Some species also use silk to make submerged nets for filtering food particles from the water. In other species, the larvae gather sand grains or small bits of plant material and bind them together with silk in order to make a portable case around their body for protection; the resulting pattern of the constructed case often typifies a genus. (It is thought that the name *caddisfly* derives from the Old English *cadice men*, people who pinned bits of cloth to their coats in order to advertise their fabric wares.) However, several caddisfly larvae build neither nets nor cases.

Adults probably feed on nectar and other liquids. Depending on the species, larvae feed on algae, detritus, or other aquatic insects.

There are about 15,000 named species of caddisfly worldwide, of which more than 2,000 are known from tropical America.

Leptonema (Hydropsychidae) is a common caddisfly genus of relatively large-sized individuals with a foul odor.

Neptopsyche (Leptoceridae). Many species in this genus have scales (in addition to hairs) on the wings.

The larva of *Phylloicus* (Calamoceratidae) makes its protective case by cutting pieces of dead leaves and sewing them together with silk.

Larva of *Leptonema* (Hydropsychidae) with a silken net for filtering food particles from streams.

WINGED TERRESTRIAL ORDERS

With just a few exceptions, the following orders consist entirely of terrestrial insects. These orders are not all closely related to one another; barklice, lice, and thrips are closely related to true bugs, whereas antlions, lacewings, mantisflies, twisted-winged parasites, scorpionflies, and fleas belong to the group of insects that undergo complete metamorphosis, a group that also includes beetles, wasps, moths, and flies.

Among the well-known terrestrial orders are grasshoppers. Lubber grasshopper (*Taeniopoda*).

ORDER ORTHOPTERA
(Katydids, Crickets, and Grasshoppers)

The order Orthoptera (meaning *straight wing*) is characterized by insects with hind legs adapted for jumping. It includes some of the most vocal insects (p. 33), and, as these insects are ancient—they appeared before the dinosaurs—it is likely that they produced some of the first songs on our planet. The group comprises about 24,000 species worldwide and is divided into two suborders, the Ensifera (katydids and crickets) and the Caelifera (grasshoppers).

Microcentrum species (Phaneropterinae) are called angle-wing katydids.

Table 2-2. Differences between the two suborders of Orthoptera

	Katydids and crickets	**Grasshoppers**
Antennae	Usually longer than the length of the body (> 30 segments)	Less than half the length of the body (< 30 segments)
Daily activity cycle	Most species are nocturnal	Nearly always diurnal
Dominant method of sound production	Rubbing forewings together. Sound-producing structures are located at the base of each wing	Rubbing hind legs against a vein on the forewing. (Many grasshoppers do not sing.)
Location of auditory organs ("ears")	Front legs and first pair of respiratory spiracles	Base of abdomen
Egg deposition	Prominent ovipositor; eggs not laid in a pod	Ovipositor reduced; eggs laid in a pod

KATYDIDS

Katydids (family Tettigoniidae) are known as bushcrickets in Europe. The name *katydid* derives from the North American species *Pterophylla camelifolia*, whose call, *ka-ty-did*, purportedly makes reference to a folktale about a woman named Katy who killed her lover. The Spanish common name is *esperanza* (hope).

Most katydids are herbivores, but some species are opportunistic predators and a few are strictly predatory. For example, members of the subfamily Listroscelidinae use their spiny legs to spear small insects. The rhinoceros katydid (*Copiphora rhinoceros*) is virtually omnivorous, feeding on flowers, fruits, seeds, caterpillars, other katydids, snails, frog eggs, and even small lizards.

The majority of species lack chemical defenses (though species of *Moncheca* are some of the few katydids known to possess them). But because most species do lack chemical defenses and because katydids are generally large (and thus make a hearty meal), they are the favorite food of birds, monkeys, and bats. To survive, many katydid species are incredibly well camouflaged, resembling leaves or lichens.

Males sing to attract females, but, by doing so, they also attract predatory bats. To avoid being eaten, males have developed a variety of strategies; some sing when bats are least active; several species sing from hidden sites (e.g., inside a spiny bromeliad); yet others keep their song brief, and, on detecting a female, continue to woo her by (silently) vibrating the plant she is on. These adaptations are most common in the American tropics, where foliage-gleaning bats that cue in on male katydid songs are relatively abundant. Despite the measures taken by katydids to avoid being detected, the bats use counter measures such as turning off their echolocation so that the katydids can't hear them coming.

During copulation the male deposits his sperm in a large edible packet (the spermatophylax), and leaves it protruding from the female's rear end; after mating the female eats the packet. The female's ovipositor (egg-laying organ) varies from species to species, and its shape often offers a clue to where the female deposits her eggs; the ovipositors are usually flattened from side to side, whether long and sword-like or short and up-curved.

Tettigoniidae is the largest family in the order, with more than 6,500 named species worldwide; katydids are especially well represented in tropical America, where there are about 2,000 species (and perhaps an equal number of species that do not yet have names).

Cocconotini (Pseudophyllinae).

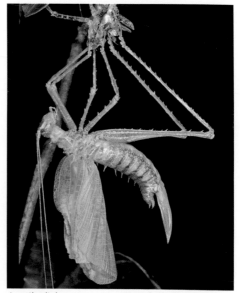

Acanthodiphrus conspersus (Pseudophyllinae) recently emerged from exoskeleton of last nymphal stage.

Platyphyllini (Pseudophyllinae) female with spermatophore (sperm packet) beneath her body.

Mimetica crennulata (Pseudophyllinae) resembling a dead leaf.

Diyllus maximus (Pseudophyllinae) female cleaning her ovipositor.

Aganacris insectivora (Phaneropterinae), a mimic of *Pepsis* spider wasps (p. 169)

Empty eggs of *Philophyllia* (Phaneropterinae).

Markia hystrix (Phaneropterinae) resembles lichens.

Camouflaged Lichenomorphus (Phaneropterinae) feeding on lichens.

Orophus species (Phaneropterinae) come in several colors.

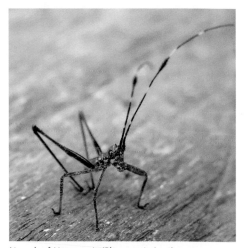

Nymph of *Harroweria* (Phaneropterinae).

Male *Copiphora rhinoceros* (Conocephalinae) produce one of the dominant night sounds in lowland forests of southern Central America.

Moncheca elegans (Conocephalinae). Not visible in this photo are the turquoise colored front wings.

CRICKETS

Crickets (Gryllidae) can usually be distinguished from katydids by having wings that wrap around the body (an adaptation for hiding in burrows) or are held horizontally rather than vertically (as in most katydids), by having much longer cerci (filaments at the rear end), and a thin, needle-like ovipositor. Some species have reduced wings, thus making it difficult to distinguish adults from nymphs.

Some crickets live within vegetation; other species occur, variously, in leaf litter or in tree holes or other kinds of cavity. Many species are omnivorous, although tree crickets (Oecanthinae) are primarily predaceous.

Like male katydids, male crickets are prolific singers. During copulation, female tree crickets feed on the males' thoracic glands; in some species in the subfamily Nemobiinae, the females feed on enlarged spines on the males' hind tibia.

There are nearly 5,000 named species of Gryllidae worldwide. (Despite their name, mole crickets belong to a different family, Gryllotalpidae, which comprises 100 species worldwide. They dig burrows in the ground with their shovel-like front legs and feed on plant roots or small invertebrates.)

Female *Anurogryllus* (Gryllinae) lay their eggs in a burrow; young nymphs are fed by their mother for a few weeks.

Amphiacusta (Phalangopsinae) comprises more than 70 species of flightless crickets; some (including this species) live in caves.

Anaxipha agaea (Trigonidiinae). Like many other crickets, species in this genus can be identified by the male's song.

Mole cricket (Gryllotalpidae) digging in soil.

INSECT SONGS

Singing male *Amphiacusta saba* cricket (Phalangopsinae).

While many insects sing, relatively few species produce sounds that are audible to humans. Yet nearly all species of cricket and cicada (p. 77) produce songs that are within the human range of hearing. Only a handful of katydid species, on the other hand, make songs we can hear, as the majority produce high frequency sounds far above the human range of hearing; in fact, some species produce songs of frequencies above 100,000 Hz (the upper range of human hearing is 22,000 Hz).

Most insects produce sounds by rubbing one body part against another, a phenomenon named stridulation. Although katydids, crickets, and grasshoppers belong to the same insect order, singing evolved independently in each of these groups, as evidenced by the distinct method that each uses to produce sounds. In katydids, the left front wing is moved over the right wing, causing a file on the underside of the left wing to be rubbed across a scraper on the upper surface of the right wing. In crickets, the situation is reversed:

the right wing is moved over the left wing. In both groups, the type of sound produced is a function of the size and density of the ridges on the file, the rate at which the file and scraper are moved against each other, and the position of a specialized membrane on the wing that is used to amplify the sounds. The sounds range from the raspy buzzes of katydids to the melodic trills and chirps of crickets. In grasshoppers, each hind leg is rubbed against the corresponding forewing, and the sound produced by each leg is almost always out of phase with that of the other.

Insects have other means of producing sounds: they drum or scrape the substrate; rapidly expel air from the mouth or spiracles; or can vibrate a specialized plate called a tymbal (cicadas, the loudest of the insect songsters, rely on a tymbal to produce sounds).

Insects produce sounds for a variety of reasons. Many species use them to defend against predators. Males of some Orthoptera species produce aggressive calls to ward off rival males attempting to enter their territory. But the most common instance is when males sing to attract females. The female typically moves towards the male with the most enticing song. In most species, females are usually silent, though female cicadas, grasshoppers, and some katydids (e.g., Phaneropterinae) often respond to males with a brief tick that initiates a duet that lasts until the male locates the female.

Sexual selection generally involves females choosing a male on the basis of his song, but some of the information that the male's song conveys—about his size and vigor, for example—is also of interest to rival males. After all, a large vigorous male is apt to be a good fighter. The problem confronting males is that sexual selection favors long and conspicuous songs, but natural selection (predators can more easily locate males with conspicuous songs) operates in the opposite direction, favoring males with less noticeable songs.

Many male katydids and crickets sing in synchrony. Some biologists have speculated that they do so cooperatively, either to make it difficult for predators to hone in on a given insect's song or to better attract females. However, thus far the evidence suggests that synchronous singing results from competition between males, since the females of at least some species prefer a male whose song begins a few milliseconds before that of other males. A male whose song is delayed therefore resets his calling rhythm when he hears a neighboring male and a synchronous chorus results when all males sing at a similar rate. In some cases, males may simply lurk silently near a singing male and attempt to intercept females attracted to his song. Of course, the mating strategy adopted by a male also depends on population density; if lots of females are in the area, a male might better dispense with singing entirely and simply search for a female.

PYGMY MOLE CRICKETS AND PYGMY GRASSHOPPERS

Despite the name, pygmy mole crickets are actually grasshoppers (families Ripipterygidae and Tridactylidae). The similarly small pygmy grasshoppers are in the family Tetrigidae. Members of these two groups of grasshoppers do not sing and they lack auditory organs. Species in both groups are generally quite small (less than 15 mm/0.6 in). Many species live on moist soil, where they feed on diatoms and other algae. Some pygmy mole crickets inhabit foliage and feed on the debris that rains down from the canopy. These foliage inhabitants sometimes move about in a jerky, wasp-like manner. Among the most beautiful pygmy mole crickets are several species that are dark blue with white antennal tips.

Worldwide, there are about 200 named species of pygmy mole cricket and 1,250 species of pygmy grasshopper.

Pygmy grasshopper (Tetrigidae).

Mating pygmy mole crickets (*Ripipteryx limbata*).

Ripipteryx pygmy mole cricket (Scrofulosa group). This species is often found on rocks along streams.

LUBBER GRASSHOPPERS

Lubber grasshoppers (Romaleidae) derive their name from those species that are flightless, the "landlubbers." The nymphs of some species congregate in groups. Both nymphs and adults have various defensive mechanisms for protecting themselves against predators. Some species sport striking colors in order to advertise their distastefulness.

Several species secrete a frothy repellent from the hind pair of thoracic spiracles (respiratory openings), and make a hissing sound as they do so.

Species in the genus *Tropidacris* are the world's largest grasshoppers; some species reach 13 cm (5.1 in) in length, with a wingspan of 25 cm (9.8 in). Lubber grasshoppers occur only in the Americas and comprise nearly 500 named species.

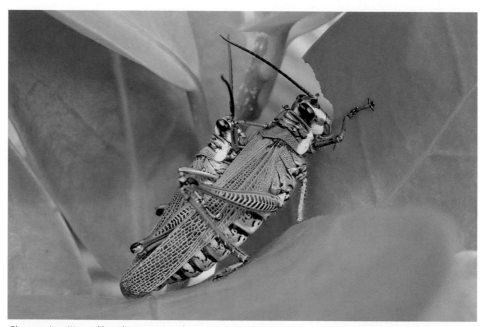

Chromacris psittacus, like other species in the genus, feeds almost entirely on *Solanum* (Solanaceae).

Taeniopoda reticulata feeds on a wide range of plants; the nymphs live in groups.

Taeniopoda reticulata exposes its red hind wings when threatened.

GRASSHOPPERS

The "true" grasshoppers, in the family Acrididae, can be divided into two main groups (and a third minor group).

The first is composed of subfamilies that are restricted to tropical America. These grasshoppers usually live in forest gaps, do not sing, and are often brightly colored (the bright colors advertise to potential predators that the grasshoppers possess nasty chemicals, which they have sequestered from their host plant). Species essentially live only on a specific host plant, where they feed and also lay their eggs. Interestingly, three species feed on floating aquatic plants: *Cornops aquaticum* (which occurs in Central America and South America), *Marellia remipes* (South America), and *Paulinia acuminata* (South America).

The second group is made up of subfamilies that occur primarily in northern temperate regions, though some species are found in the tropics. These grasshoppers live in grassy or disturbed areas, some of them (e.g., members of the subfamily Gomphocerinae) sing, and most are drably colored. Species in this group tend to have a more diverse diet and the majority lay their eggs in the soil. Some, such as the band-winged grasshoppers (Oedipodinae), make a loud crackling sound when they fly; this sound is created by rapidly flexing taut areas in the hind wings, a behavior used in courtship and territorial displays. When they take to the air, the crackling sound can be startling; a potential predator probably focuses on the flashy hind wing, only to have the grasshopper disappear from sight when it lands and blends into the background.

Of the 6,000 grasshopper species in the world, fewer than 20 are considered locusts, a special type of grasshopper. What makes locusts unique is their Jekyll and Hyde behavior; they pass through many generations as normal, solitary grasshoppers, but when

Metaleptea brevicornis (Acridinae) lives in long grass in damp areas.

favorable environmental conditions allow them to increase their numbers, the crowded conditions that result yield several generations of grasshoppers that band together in spectacular migrating swarms of millions of individuals. After a few generations, they revert back again to their solitary behavior. Some locusts belong to the second group mentioned above, but others belong to a third group of grasshopper, the subfamily Cyrtacanthacridinae, whose members occur in warm regions around the world. Locusts of tropical America belong to this subfamily and are generally restricted to seasonally dry forests: *Schistocerca cancellata* (northwestern Argentina), *S. interrita* (Peru), and *S. piceifrons* (Central America). The genus *Schistocerca* comprises about 50 mostly New World species (the majority of which are not locusts); only *S. gregaria* (the desert locust of Biblical fame) occurs in the Old World.

Drymophilacris (Proctolabinae) species are specialists on plants in the nightshade family (Solanaceae).

Oedipodinae, a subfamily that is more common in North America and the Old World.

Grasshoppers in the family Episactidae (*Episactus tristani*) hold their hind legs perpendicular to the body.

ORDER PHASMIDA
(Stick Insects)

Stick insects, also known as walkingsticks, are close relatives of grasshoppers (p. 37) and katydids (p. 29). They are, like many species of katydid, remarkable mimics of twigs and leaves (species that mimic leaves are called leaf insects). Indeed the scientific name of the order, Phasmida, means *apparition*. This order also includes some very large insects; *Heteronemia grande*, a South American species that measures up to 26 cm (10.2 in), is the longest insect in the Americas (the world record goes to a Malaysian species, at over a half a meter, or nearly 2 feet).

Stick insects have defensive glands in the front of the thorax that spray nasty chemicals at potential enemies. All species feed on plant foliage and some species can change color to match their background.

Many species are wingless. In those species that have wings, the front wings are often short and leathery while the hind wings are fanlike (and folded tightly against the body); in some species, both front wings and hind wings have evolved to resemble leaves. Yet others have brightly colored hind wings that are normally concealed but can be rapidly unfolded to startle a predator.

Although most species reproduce through sex between male and female, in many species the female is able to reproduce asexually if males are absent; indeed, in some species only females are known. Depending on the species, the female inserts her eggs into a crevice, glues them to the host plant, or flicks them to the ground. Eggs that fall to the ground often bear a cap that is attractive to ants. The ants carry the eggs back to the nest, cut off the caps, and feed them to their larvae. The egg itself is then thrown into a garbage dump (inside or outside the nest), where it is presumably safer from predators and parasitoids than if it had remained under the tree where it fell. Upon hatching the nymphs climb up a nearby plant.

There are more than 3,000 species of stick insect worldwide.

Female *Phanocles costaricensis* (Diapheromeridae) on *Inga* (Fabaceae).

ORDER EMBIOPTERA
(Webspinners)

Caterpillars and other insects that produce silk usually do so with their salivary glands. Webspinners are unique in that they secrete silk from glands located in their swollen front legs. With this silk both juveniles and adults construct tent-like tunnels on tree trunks, earthen banks, or in leaf litter.

Females lack wings, whereas males usually have wings. Males do not eat (females and juveniles feed on dead plant material), and thus live short lives. After mating inside a silken tent, the female lays her eggs, guards them, expands the tent, and stays with the juveniles until they are about half grown, at which point the female leaves the tent and then usually dies. Sometimes a colony (a mother with her offspring) unites with another colony to form a larger aggregation.

There are nearly 500 named species of webspinner worldwide.

Webspinner adult.

Webspinner nymph showing swollen front tarsi, which contain silk glands.

ORDER ZORAPTERA

Due to their small size (3 mm/0.1 in), non-descript appearance, and habit of living concealed in dead wood, zorapterans are generally quite difficult to observe in the field. They live in small colonies and feed on fungi, nematodes, or minute arthropods. In each species there are two types of adult: those without wings and eyes (the most common form), or those with wings and eyes (this second form, however, sheds its wings after dispersing to a new nesting site).

Zoraptera includes a mere 40 named species worldwide; moreover, it is the smallest order of insects in tropical America, where only 16 species are found.

Silk tent produced by a colony of webspinners.

ORDER DERMAPTERA
(Earwigs)

The Spanish name for these insects is *tijerillas*, meaning "little scissors." Perhaps because of their menacing pincers, earwigs are much maligned, unjustly so. In Latin America, they are commonly thought to be venomous—they are not. And in many parts of the world, people believe that earwigs crawl into human ears; while they do like to hide in nooks and crannies, earwigs are no more likely to enter one's ear than any other insect.

The pincers have several functions. They are used to fold and unfold the hind wings; in self-defense, to inflict a moderately painful pinch; to capture prey; and in mating.

Earwigs tend to be nocturnal. They usually inhabit leaf litter, river banks, rotting logs, and other habitats that afford hiding places. Most species are likely omnivores, though a few are exclusively predators.

In some species (e.g., Anisolabididae), males have two penises, the supplementary penis apparently serving as a spare should the other penis become damaged. The female builds a nest and cares for her eggs, licking them to protect them from fungal infection. In some species, the female periodically leaves the nest in order to eat; in other species, she remains in her nest the entire time that she cares for her eggs. She also protects the newly hatched nymphs and makes periodic forays to bring food back to them. The amount of time the nymphs remain in the nest depends on the species.

There are about 2,000 species of earwig worldwide.

Anisolabidae. Many members of this family are black, sometimes with orange front wings.

Spongiphoridae (previously known as Labiidae), the largest family of earwigs.

This *Metresura* species (Forficulidae) often feeds on nectar.

MANTISES, COCKROACHES, AND TERMITES

Evidence from internal anatomy and DNA shows that these three apparently disparate orders are in reality closely related to one another. In fact, there is now solid evidence that the termites evolved from certain species of cockroach, and some entomologists therefore classify the termites with the cockroaches (leaving only two orders).

ORDER MANTODEA
(Mantises)

Mantises are exclusively predatory. The majority of species eat insects but some of the larger species sometimes eat small vertebrates.

The name *praying mantis* comes from the seemingly devout manner in which members of this order hold their raptorial front legs. The Greek word *mantis* means "soothsayer" or "diviner." Various cultures attribute occult powers to these insects, perhaps because of the human-like way in which they rotate their heads.

Mantises lay their eggs in packets (oothecae), which vary in shape from a globular mass (that is wrapped around a plant stem) to a narrow-necked flask. The number of nymphs that hatch from each egg packet varies from 20 to more than 200 depending on the species.

Mantises are often well camouflaged. Each species mimics the dominant vegetation in its habitat, whether leaves, twigs, flowers, or tendrils. As do actual leaves, mantis species in the genus *Choeradodis* harbor on their bodies living lichens and liverworts; given that lichens and liverworts take two years to develop, biologists reason that the mantis species in this genus (and perhaps in others) must live longer than two years.

Tithrone roseipennis (Acanthopidae) has a single green front wing and one red front wing; when the wings are closed, the green wing folds over the red wing, thus concealing it. Such asymmetry is very rare in nature and probably serves to confuse predators.

On the underside of the thorax, between the hind legs, mantises have an ear that is used to detect the ultrasounds emitted by bats. This ability is especially important for males flying at night in search of a female's scent. During copulation, female mantises sometimes bite off the head of their mate, although, even in a headless state, the male can continue to copulate for more than an hour.

There are about 500 named species of mantis in tropical America.

Female *Acontista multicolor* (Acanthopidae).

Liturgusa (Liturgusidae). Species in this genus live on the bark of trees, especially those with smooth bark.

Female *Mantoida* (Mantoididae).

Male *Phyllovates* (Mantidae).

A well camouflaged *Callimantis antillarum* (Mantidae).

Nymph of *Stagmomantis* (Mantidae).

Choeradodis (Mantidae).

Empty egg case (ootheca) of *Stagmomantis* (Mantidae).

Molted exoskeleton of *Choeradodis* nymph (Mantidae).

ORDER BLATTARIA
(Cockroaches)

Repulsive to most people, domestic cockroaches represent only a small group—about a half a dozen African and Asian species—in the entire order. In point of fact, the vast majority of species occur only in the forest and some of these are actually beautiful. For example, a few species that apparently mimic distasteful beetles have red and yellow markings, and several species in the genus *Panchlora* (Blaberidae) are a lovely pale green. Nonetheless, a few wild species may perhaps cause panic in humans because of their large size. For example, some species in the genus *Blaberus* (Blaberidae) reach a length of 8 cm (3.1 in); and one Colombian species in the genus *Megaloblatta* (Ectobiidae) reaches a length of 10 cm (3.9 in).

Cockroaches (in Spanish, *cucarachas*) inhabit leaf litter, rotting wood, forest canopy, and even the nests of leafcutter ants. Species that live on the forest floor often become active at particular hours during the night and climb to a specific height on the understory vegetation; it is not entirely clear why they do this, but such behavior makes it easier for females and males to find one another. Often considered to be omnivorous scavengers that feed on a diversity of decomposing materials, they are probably in fact somewhat selective in what they eat.

When they urinate, most terrestrial animals expel nitrogenous wastes either as urea or uric acid, but cockroaches are able to retain uric acid in their bodies; when necessary, their endosymbiotic bacteria (*Blattabacterium*) convert this uric acid into useable nitrogen. In fact, the males of some species even pass along uric acid to the female after mating, as a nuptial gift.

Female cockroaches are remarkable for the diversity of ways in which they give birth. Many deposit their eggs in a beanlike packet (ootheca), while some carry the egg case partially protruding from the tip of the abdomen until the youngsters hatch; others (some Blatellinae and all Blaberidae) retain the egg case entirely inside the abdomen.

There are more than 4,600 species of cockroach worldwide.

Paratropes bilunata (Ectobiidae), an inhabitant of the canopy. Some species mimic net-winged beetles (p. 120).

Pseudomops (Ectobiidae, previously known as Blatellidae).

Researcher Rossy Morera holds up a *Megaloblatta blaberoides* (Ectobiidae).

Egg case of *Paratropes bilunata* (Ectobiidae).

Archimandrita tesselata (Blaberidae) can live up to two years.

Blaberus giganteus (Blaberidae) lives in groups inside hollow trees.

Panchlora (Blaberidae). Species in this genus are strong fliers.

Adult *Capucina* (Blaberidae) recently emerged from final nymphal stage.

Panchlora (Blaberidae) cleaning its antenna.

Egg case (ootheca) of *Eurycotis biolleyi* (Blattidae).

ORDER ISOPTERA
(Termites)

All termite species live in colonies in which the queen's task is it to lay eggs while workers, which make up a majority of individuals in the colony, build the nest and gather food. This division of labor between a reproductive and non-reproductive caste is known as eusocial behavior (p. 175), a phenomenon that also occurs in ants, paper wasps, and certain bees. Unlike these other eusocial insects, however, in which workers are non-reproductive adults, termite workers are nymphs; termite adults are dedicated exclusively to reproduction. Another difference is that termite societies are bisexual, with workers of both sexes, and have a resident king, an adult male who's only job is to repeatedly copulate with the queen (in ant, wasp, and bee societies, workers are all female; adult males leave their natal colony, copulate with new queens emerging from other colonies, then die). Finally, in nearly all termite species the colony contains a second sterile caste, the soldiers (specialized nymphs of either sex), which defend the colony against predators (a few ants have specialized workers that act as soldiers, but a true soldier class *per se* is absent).

During certain times of the year many of the nymphs in the colony molt to the adult stage and hundreds to thousands of virgin females and virgin males fly away from their natal colony in search of a mate (usually from another colony). After mating, each couple (the young king and queen) initiates a new colony, shedding their wings in the process. Their colony slowly grows, eventually reaching a population of a few hundred to over a million, depending on the species. If at some point the king or queen dies, one or more of the juveniles can molt into a reproductive caste known as a neotenic reproductive, a sexually mature juvenile that lacks wings.

Four families of termite are found in tropical America: Kalotermitidae, Rhinotermitidae, Serritermitidae (just 3 species, in Brazil and Guyana), and Termitidae (the largest family, with nearly 2,000 species worldwide). While most termites feed on wood, some species of Termitidae eat leaf litter and soil—and in tropical rainforests they actually process more dirt than do earthworms. Kalotermitidae nest within wood, which is also their food source; most members of Rhinotermitidae and Serritermitidae build a nest separate from their food source and must therefore leave the nest to forage, often utilizing covered walkways.

Among the most commonly noticed nests in tropical America are those of the genus *Nasutitermes* (Termitidae)—dark brown, basketball-sized structures on trees or fence posts. Some other members of this family, including species in the genus *Cornitermes*, also build visually arresting nests; in the savannas (*cerrados*) and pastures of central and southern Brazil, for example, their tower-like mounds can reach densities as high as 300 per hectare (120 per acre).

In Kalotermitidae, one of the most primitive of termite families, the soldiers rely exclusively on their mandibles for defending the colony; soldiers in the other families, however, employ chemical artillery as a supplement to (or substitute for) physical weaponry. In the family Termitidae, for example, soldiers belonging to the genus *Nasutitermes* lack mandibles and instead use their nozzle-like heads to squirt enemies (anteaters and ants) with a noxious glue.

There are nearly 3,000 named species of termite worldwide, but only a very small minority of these are pests.

Termite with shed wings.

Adult Kalotermitidae on dry wood in an urban area.

Nasutitermes (Termitidae) soldiers lining a tear in their nest.

Nasutitermes (Termitidae) soldiers.

Nest of *Nasutitermes* (Termitidae).

Trail of *Nasutitermes* (Termitidae) protected by roof made of debris.

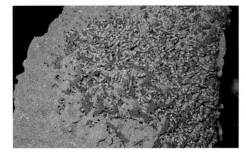

Nasutitermes workers (pale colored) expanding the nest at night.

EATING WOOD

Nasutitermes (Termitidae) trail and damage.

Termite pellets.

Wood consists primarily of cellulose, hemicellulose, and lignin. Cellulose, a polymer found in the cell walls of plants, is the most abundant biological compound on the planet, occurring not only in wood but also in leaves and leaf litter.

When you eat what is commonly called fiber to limber up your digestive tract, you are actually eating cellulose, which itself is indigestible. Although it is difficult to break down cellulose into digestible components, various insects manage to do so by producing enzymes (cellulases), by harboring symbiotic microbes that produce these enzymes (the strategy employed by cattle and other ruminants), or by doing both. The larvae of longhorn beetles (p. 130), for example, produce cellulase in their midgut.

The reason termites are so efficient at digesting wood is that they complement their own midgut cellulase with other types of cellulase produced by some of the protozoans and bacteria that live in their hindgut. A single termite sometimes harbors up to 20 species of protozoan and 200 species of bacteria. When a termite molts, it loses its microbes and must lick the anus of a nest mate in order to inoculate itself once again.

Another major ingredient of wood is lignin, a molecule that is much more complex than cellulose—and even more difficult to break down. The more lignin a wood contains, the greater its durability. Within dead vegetation, it is one of the components that most slowly decomposes. If you've ever wondered why old newspapers turn yellow, it's because they contain lignin (which is removed from finer grades of paper). Until recently, biologists thought that termites were incapable of digesting lignin, but it now appears that at least some species can break down this very refractive compound. The only organisms known for certain to have this capability are some fungi; biologists speculate that termites may harbor fungi in their intestines, although this is still under investigation.

When cellulose is digested, a key byproduct is hydrogen gas. The microbes living inside termites can produce up to 2 liters (2.1 quarts) of hydrogen gas from a single sheet of paper, making termites one of the planet's most efficient bioreactors. Since hydrogen can be used to run our automobiles, there is a great deal of interest in deciphering the bacterial genes responsible for these chemical reactions. Apart from potential commercial applications, termites play major roles in tropical ecosystems, where they constitute about 10% of all animal biomass and are an essential component of the carbon cycle. Some of the bacteria living in termite guts fix atmospheric nitrogen, thereby making more nitrogen available to the ecosystem.

ORDER PSOCOPTERA
(Barklice)

Barklice are small, soft-bodied insects with relatively large heads and a bulbous face. They inhabit the bark or foliage (especially older leaves) of woody plants; leaf litter; and the nests of birds and mammals. Barklice feed primarily on fungi and terrestrial algae, and share with lice a unique mechanism for absorbing water vapor inside the mouth. Most species produce silk from their salivary glands, which may be used to cover the eggs or as a protective sheet over the nymphs and adults. Many barklice live in groups, which range from nomadic herds, to small stationary family groups (with or without a silken covering), to the enormous silken cites of some Archipsocidae that cover an entire tree trunk. A few wingless species, known as booklice (*Liposcelis* spp.), are the barely visible critters you sometimes see scurrying across the pages of a musty book, and they are also found in moldy plant material.

There are nearly 6,000 named species of barklice worldwide.

Poecilopsocus (Psocidae).

Cerastipsocus (Psocidae) nymphs live in large nomadic herds.

Silk covering produced by web-spinning barklice (Archipsocidae: *Archipsocus*) on *Cecropia* tree trunk.

Barklouse (Psocoptera).

ORDER PHTHIRAPTERA
(Lice)

Regular lice (in Spanish, *piojos*) evolved from barklice. They are the only insects that spend their entire life cycle on the bodies of birds and mammals, as permanent parasites, and they are very host specific.

The eggs, glued to feathers or hairs, are known as nits, hence the expression "nit picking."

Lice are placed into two ecological groups. The larger group is the chewing lice (suborders Amblycera and Ischnocera), with most species feeding on birds but some on mammals. More specifically, chewing lice feed mostly on feathers, hair, and skin. (Birds remove lice and other ectoparasites by preening with their bill and scratching with their feet; a comparison of Peruvian birds showed that species with

a longer overhang on the upper mandible of their bill had fewer lice.)

Members of the second group, the sucking lice (order Anopleura), feed exclusively on the blood of mammals. Two species of sucking lice are specific to humans. The crab (or pubic) louse (*Pthirus pubis*), whose closest relative occurs on gorillas, and the human louse (*Pediculus humanus*), whose closest relative occurs on chimpanzees. Since we humans have very little body hair, the human louse was once restricted to the head, but, with the advent of clothing, a new race evolved that is capable of inhabiting our fabrics.

Lice not only carry a social stigma—one looks lousy or has the crabs—but sometimes the human louse transmits the bacterium *Rickettsia typhi*, which causes typhus, a disease that has historically killed large numbers of people (e.g., in Napoleon's army). There are more than 5,100 named species of lice worldwide.

Human head louse (*Pediculus humanus*).

ORDER THYSANOPTERA
(Thrips)

Thrips are very abundant (and diverse), though they are seldom seen because of their diminutive size. The order name Thysanoptera means "fringed wings," referring to the long hairs on the margins of their slender wings.

Thrips have several unusual characteristics. Unlike any other group of insect, they have a single mandible (on the left) rather than two; they feed by puncturing the food item with the mandible, injecting saliva, and then sucking up the liquid contents (by means of two maxillary stylets). Another unusual feature of thrips is the way in which the sex of the offspring is determined; in most animals sexual identity is determined by sex chromosomes (in humans the x and y chromosomes), but in thrips male offspring result from unfertilized eggs while female offspring result from fertilized eggs. This system, whereby males have half the number of chromosomes as do females, is very rare in animals, although it also occurs in all wasps, ants, and bees. Finally, their development is somewhat peculiar in that two nymphal stages are followed by two or three non-feeding stages, after which they finally molt into the adult stage.

About half the species in this order feed on fungi; most species feed on fungal filaments (hyphae), but some ingest whole spores (the reproductive units that are usually dispersed by wind). Among fungal feeders, in many species males are much larger than females and they often fight with rival males.

Of the species that do not feed on fungi, the majority feed on the sap of plant cells in leaves or flowers, and a few are predators. Two species (*Frankliniella diversa* and *F. insularis*) are the principal pollinators of the Panama rubber tree (*Castilla elastica*), a plant that during pre-Columbian times was the main source of rubber in Central America.

An easy way to find thrips is to vigorously shake a flower over a plastic plate and look for the tiny insects that adhere to it by means of minute bladder-like structures on the tips of their legs.

Two species from southeast Asia (*Gynaikothrips ficorum* and *G. uzeli*) that were introduced (unintentionally) to tropical America cause leaf galls (folded leaves) on ornamental figs (*Ficus microcarpa* and *F. benjamina*, respectively), which were also introduced (though by design) to the region. There are about 6,000 named species of thrips worldwide.

Echinothrips thrips (Thripidae) on *Bocconia* (Papaveraceae).

Heliothrips haemorrhoidalis (Thripidae) feeds on the leaves of various trees.

Phlaeothripidae (probably *Pseudophilothrips*) on *Croton draco* (Euphorbiaceae).

Phlaeothripidae (probably *Pseudophilothrips*) nymphs on *Croton draco.*

Mixothrips (Phlaeothripidae) adults and nymphs in cecidomyiid galls on *Topobea* (Melastomataceae).

Non-feeding stage of thrips ("pupa") inside its cocoon.

ORDER NEUROPTERA

With fossils of extinct families that date back to the Permian (before the age of the dinosaurs), the order Neuroptera is one of the oldest groups of insects that undergo complete metamorphosis. The larvae are carnivores that suck out the contents of their prey by means of elongate, sickle-shaped mandibles and maxillae (a pair of mouthparts immediately behind the mandibles). On completing their development, the larvae spin a cocoon, but unlike most other silk-producing insects, which produce silk in the salivary glands, neuropterans produce silk in the Malpighian tubules (in insects, the equivalent of kidneys) and secrete it from the anus.

There are nearly 6,000 species of Neuroptera worldwide. The following sections present three groups: antlions and owlflies, green lacewings, and mantisflies.

ANTLIONS AND OWLFLIES

Antlions (Myrmeleontidae) and owlflies (Ascalaphidae) resemble dragonflies but have longer antennae, especially so in the case of owlflies. In most species, adults are aerial predators, usually flying at dusk or at night and resting during the day. Antlions usually lay eggs singly in the soil, whereas many species of owlfly lay their eggs in a row along the length of a twig.

The larvae of both families are oval-shaped, flattened, and well camouflaged—and are seldom seen. Owlfly larvae have lateral projections, while antlion larvae lack these projections.

Antlion and owlfly larvae are ambush predators; most species wait in soil or rotten wood for passing prey, although some owlfly larvae inhabit the branches of trees. Antlion larvae (sometimes known as doodlebugs) often bury themselves in soil by walking backwards, submerging themselves little by little. The best known antlions are those belonging to the genus *Myrmeleon*, which build craters in loose, fine-grained soil in places that are relatively sheltered from rain. The larva buries itself at the bottom of the crater and when a small insect or spider stumbles in, the larva attempts to grab it with its mandibles; if the prey begins to escape, the antlion sometimes tosses soil particles up the sides of the crater in order to create a miniature landslide, thereby dragging the hapless prey back to the bottom of the pit.

Worldwide, there are about 2,000 named species of antlion and 450 species of owlfly.

Antlion larva (*Myrmeleon*) pushing backwards to bury itself in the soil.

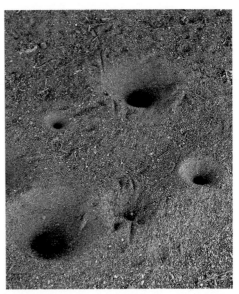

Craters in soil constructed by antlion (*Myrmeleon*) larvae.

Owlfly larva. Larvae in this family have projections on the side of the body, unlike antlions.

Owlfly adult. Adults of antlions are similar, but have much shorter antennae.

GREEN LACEWINGS

With delicate green wings and fluttering flight, green lacewings (Chrysopidae) are fairly distinctive insects. At night, they are often seen fluttering around lights.

Most species have an auditory membrane on each front wing that allows them to escape from bats by detecting the ultrasonic sounds they make. Another defensive capacity found in some species is the ability to secrete a foul-smelling substance (skatole) from the thorax, although this is not effective against one group of predators, the orb-weaver spiders. If entangled in a web, a lacewing can slowly work itself free, aided by hairs on the wings.

In some species, adults prey on small insects, while in other species adults feed exclusively on honeydew and pollen. The latter species harbor in their digestive tract a symbiotic yeast that aids in digestion.

Males of some species court females by drumming the plant surface with the abdomen, creating low-frequency vibrations that carry a distance of only a meter. Each species has a unique "song" and in some cases these songs have proven valuable in distinguishing between species that are very similar in appearance. In a few species, the male and female perform a duet, each sending an identical vibratory song to the other.

The female lays her eggs, which are stalk-shaped, on foliage, either singly or in groups. Some species coat the egg stalk with droplets that contain an ant repellent (the newly hatched larva laps up the droplets as it crawls down the stalk). The larvae prey on soft-bodied insects; most larvae species have the unusual habit of covering their body with detritus, inedible bits of prey, and the like, all of which serves as camouflage.

Brown lacewings (Hemerobiidae) are generally smaller and have drabber coloration than green lacewings. Their larvae never cover themselves with detritus.

There are about 1,500 named species of green lacewing worldwide.

Green lacewing (*Ceraeochrysa montollana*).

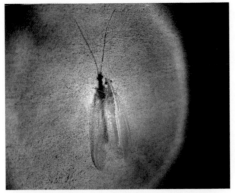

Green lacewing (cf. *Plesiochrysa*) at night light.

Green lacewing (*Leucochrysa*) larva covered with debris.

Green lacewing (Chrysopinae) larva covered with debris.

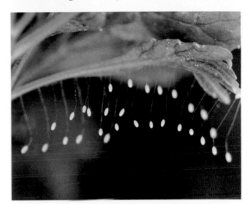

Green lacewing (Chrysopinae) eggs.

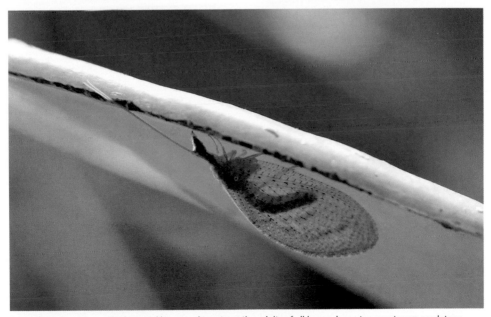

Brown lacewing (Hemerobiidae). Unlike green lacewings, the adults of all brown lacewing species are predatory.

Mantisfly (*Climaciella*) that mimics certain species of paper wasps (p. 173) in the genus *Polistes*.

MANTISFLIES

Generally, mantisflies (Mantispidae) resemble mantises (p. 42), though their front wings are membranous rather than leathery. Several species are mimics of stinging wasps (p. 173).

Depending on the species, adults are either diurnal or nocturnal. They are predators but also imbibe nectar from flowers. During courtship the male and female face one another while moving their raptorial front legs. Females lay discrete clutches that contain very large numbers of eggs, with each egg being attached by a small stalk to the substrate and the entire clutch often arranged in a curved row.

The larvae of most species prey on spider eggs, obtained in one of two ways. Some species first locate an egg sac and then burrow directly through the silk to get at the eggs. Other species attach themselves to a female spider, wait until she matures, and then enter her egg sac as it is being constructed; in the interim, the larvae feed on the female spider's blood.

Regardless of the strategy used to enter the egg sac, the larva eventually sucks out the contents of all the eggs contained within it. Upon completing its development, the larva spins a cocoon right inside the spider's egg sac. There are about 400 named species of mantisfly worldwide.

ORDER STREPSIPTERA
(Twisted-winged Parasites)

Though rarely seen, twisted winged-parasites are among the strangest of all insects. The larvae are endoparasites of other insects; although they generally do not kill the host they often prevent it from reproducing. Depending on the species, they parasitize crickets, katydids,

Strepsiptera male, with reduced front wings.

Strepsiptera female

Strepsiptera female protruding from planthopper (Delphacidae).

mantises, true bugs, planthoppers, leafhoppers, paper wasps or bees.

Adult females lack wings, legs, eyes and antennae —and and they never leave their host. The tiny (4 mm/0.2 in), winged males fly off in search of a host that contains a female (she produces pheromones to attract males). The female's head protrudes slightly between the host's abdominal segments and the male copulates with her via a brood canal located behind her head. She then produces thousands of tiny larvae that leave the maternal host to search for another host. There are more than 600 named species of Strepsiptera worldwide.

ORDER MECOPTERA
(Scorpionflies)

Like Neuroptera (p. 53), scorpionflies are one of the most ancient groups of insects that undergo complete metamorphosis. The adults resemble crane flies but have two pairs of wings instead of just a single pair. They spend much of their time hanging by their legs from vegetation, and feed on small insects that they capture with their hind legs.

The caterpillar-like larvae live in leaf litter, where they are thought to feed primarily on dead insects.

There are about 400 named species of scorpionfly worldwide. The order includes nine families, though just a single family (Bittacidae) occurs in tropical America (two small families occur in the temperate regions of southern Chile and Argentina). Worldwide, Bittacidae comprises about 160 species, of which 65 occur in tropical America.

Bittacus (Bittacidae). Members of this family are known as hanging scorpionflies because they hang from vegetation rather than perch.

ORDER SIPHONAPTERA
(Fleas)

Fleas (in Spanish, *pulgas*) evolved from scorpi-onflies. They are ectoparasites, but it is usually only the adult stage that sucks blood.

The maggot-like larvae live in animal burrows or nests, or in our carpets, where they feed on organic debris and sometimes the dried feces of adult fleas. Eventually they spin a cocoon and transform into a pupa.

The adults are wingless and laterally flattened. Though usually very mobile (and known for their ability to jump), they spend much of their time on the host. Eggs are laid in the host's burrow, or on the host, from which they then drop.

In the vast majority of species, adults feed on mammals (unlike most lice species, which feed on birds), and they are usually less host specific than lice. Three quarters of the species are associated with rodents, while the rest utilize a diversity of other hosts, ranging from armadillos to bats.

Throughout many parts of the Americas, the cat flea (*Ctenocephalides felis*) is one of the most common species. A native of North Africa and the Middle East, this flea sucks blood from various mammals, including humans. Another polyphagous species is the so-called human flea (*Pulex irritans*), which probably originated in the Andes and later spread to North America, then to Eurasia. The chigoe flea (*Tunga penetrans*), native to tropical America, is unusual in that the female burrows into the skin; in the case of humans, who generally acquire the flea by walking barefoot, it often burrows beneath a toenail.

Some fleas transmit diseases, the most notorious being bubonic plague, caused by *Yersinia pestis*, a bacterium that occurs naturally in certain burrowing rodents, which themselves do not become particularly sick but are capable of passing the disease (via their fleas) to a wide range of other mammals. This microbe was originally restricted to the Old World but is now permanently established in certain rodent populations in drier regions of North and South America.

There are more than 2,000 named species of flea worldwide.

Ctenocephalides canis dog flea (Pulicidae). This species is less common than the cat flea, even on dogs.

3

TRUE BUGS
AND THEIR KIN
(order Hemiptera)

The order Hemiptera, with over 100,000 named species worldwide, is divided into three suborders: the Heteroptera (true bugs), Auchenorrhyncha (cicadas, planthoppers, treehoppers, and their kin), and Sternorrhyncha (plant lice, scale insects, etc.). What they all have in common are mouthparts adapted for piercing and sucking. Whether needle-like or thread-like, the mouthparts allow these insects to feed on the fluid contents of other insects or plants, which they generally inject with saliva before ingesting.

Auchenorrhyncha and Sternorrhyncha were previously placed in a separate order, the Homoptera. Members of these two groups feed almost exclusively on plant sap, either the sap within the main internal tissue, which is composed of mesophyll, or from the sap flowing through the veins and other conducting tissues. One set of conducting tissues, composed of xylem, carries water through the plant; another set, composed of phloem, principally carries the product of photosynthesis, sugar.

Because xylem sap consists mostly of water, it must be imbibed in large quantities in order to extract the needed nutrients; in a single day, one of these insects may consume 300 times its own body weight. As xylem sap is generally maintained under strong negative pressure, it takes considerable effort to suck it out; cicadas, spittlebugs, and some leafhoppers do this using a large muscular pump in the front of the head, which lends their faces a bulging appearance.

Although phloem sap has relatively high concentrations of sugars—and is easier to suck out—insects still require specialized mouthparts to extract it (and a specialized digestive tract to process it). Phloem feeders are fairly easy to spot in the field because their sugary excrement (honeydew) attracts ants.

Plant-feeding hemipterans generally harbor in their bodies endosymbiotic bacteria that provide them with the nutrients they lack in their diet. In the case of true bugs, these bacteria reside freely in the intestine, whereas other hemipterans have specialized cells that contain the bacteria (these bacteria have become so dependent on their host insect that they have lost much of their genome). In most hemipterans, the bacteria are passed from generation to generation directly, inside the eggs; in stink bugs, however, the mother deposits bacteria on the outer surface of the eggs (after they have been laid), and upon hatching, the young nymphs then ingest them.

opening to stink gland

Most true bugs, including this stink bug, secrete defensive compounds.

SUBORDER HETEROPTERA
(True Bugs)

The word *bug* derives from the Middle English *bugge*, meaning spirit or ghost. Back then, waking up in the morning with itching red welts meant a nocturnal visit from a *bugge* or, factually, a bed bug. Nowadays people use the word *bug* for almost any insect, and so entomologists employ the term *true bug* to distinguish members of the suborder Heteroptera from all other insects.

Most true bugs have front wings that are relatively thick and leathery throughout most of their length, but that also have a thin membranous tip. This group includes both aquatic and terrestrial insects, and carnivorous and herbivorous insects (unlike the other two groups of hemipterans, whose members are nearly all terrestrial and herbivorous). Almost all true bugs secrete defensive compounds from specialized scent glands that are either located on the top of the abdomen (in nymphs and sometimes adults) or on each side of the thorax (in adults).

WATER STRIDERS AND RIFFLE BUGS

Water striders (Gerridae) and riffle bugs (Veliidae) skate on the surface of water using their middle legs as oars. The point of contact between the tip of the leg and the water surface creates a dimple, which functions as an oar blade. The sculling action creates two vortices

that propel the insect forward. These insects are extremely fast.

They feed mostly on other insects that fall into the water, using surface vibrations to detect the prey. They also use surface vibrations to communicate with one another, especially during territorial defense and courtship. Females of both groups usually glue their eggs to floating objects or to vegetation.

Most water striders inhabit still waters, including those of mangroves; a few species in the genus *Halobates* complete their entire life cycle on the open ocean, which is extremely unusual among insects. Many riffle bugs are adapted to fast flowing streams, others to living in water in bromeliad axils, tree holes, or even crab holes along the coast.

In both groups, many species show wing polymorphism; in other words, through a part of the year the majority of adults lack wings and during the rest of the year they do have wings. Adults of the genus *Tachygerris*, on the other hand, are always fully winged. To further complicate matters, some species use their legs to snap off half of each wing along preformed lines of breakage. This means that they will never fly again, but presumably this provides them with advantages for their life on the water surface.

Worldwide, there are about 750 named species of water strider and 970 species of riffle bug.

Water striders (Gerridae) showing dimples where the legs contact the water surface.

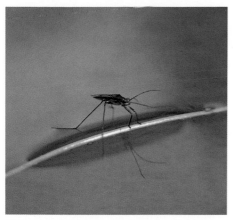

Water strider (Gerridae). Members of this family have very long mid and hind legs.

Rhagovelia riffle bug (Veliidae). The tuft of hairs at the tips of the mid legs allows species in this genus to skate over the surface of running water.

AQUATIC HETEROPTERA

Members of the four families of aquatic bugs (Belostomatidae, Naucoridae, Notonectidae, and Corixidae) have hind legs adapted for swimming underwater. Both nymphs and adults live in water. Adults can fly and sometimes leave one body of water in search of another; some species are attracted to lights and a few species occasionally colonize lighted swimming pools. Most aquatic bugs are predators and some are capable of inflicting a painful bite.

Giant water bugs (Belostomatidae) reach lengths of up to 11 cm (4.3 in) and are capable of feeding on tadpoles and even small fish.

They are among the few animals in which males care for the eggs. The latter are relatively large and cannot obtain sufficient oxygen in stagnant water. In all species in the genera *Abedus* and *Belostoma*, females deposit their egg mass on the back of the male, who ensures that the eggs receive sufficient oxygen by periodically rising to the water surface. In the genus *Lethocerus*, the female deposits her egg mass on emergent vegetation (where the eggs obtain more oxygen), and the male keeps them moist by using his body to drip water over them. By leaving the task of egg care to the male, the female can lay a greater number of eggs during her lifetime.

The other aquatic bugs are smaller than the giant water bugs and include the creeping water bugs (Naucoridae), backswimmers

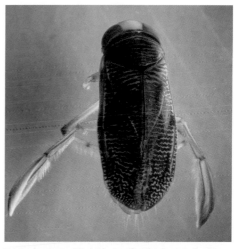

Water boatman (Corixidae), a family characterized by transverse zigzag markings on the back.

Toad bugs (Gelastocoridae) are mostly terrestrial, but are related to the aquatic Heteroptera.

(Notonectidae), and water boatmen (Corixidae). Females either attach their eggs to submerged objects or insert them within the interstices of submerged vegetation. Water boatmen are the only aquatic bugs that are not strictly predators, many of them feeding on algae and detritus.

Aquatic bugs display a fascinating array of strategies for breathing underwater. In many species, nymphs obtain oxygen directly through their cuticle, which serves as a kind of gill. Most adults, however, breathe by means of a layer of air that is trapped among body hairs and under the wings. This air layer acts as a gill that can absorb oxygen from the surrounding water, but over time it slowly collapses, and the bug must return to the surface of the water in order to renew the layer of air. Backswimmers (Notonectidae) carry this air layer on the ventral surface; as

this makes their belly much lighter than their back, they must therefore swim upside down. Some backswimmers of the genus *Buenoa* are among the very few insects to synthesize large amounts of hemoglobin, which they store in their tracheae and use to replenish the oxygen stored in their ventral air supply. This allows them to stay submerged longer. Instead of using a normal air layer as a gill, which requires periodic renovation, creeping water bugs in the genus *Cryphocricos* utilize a very thin, permanent film of air (called a plastron), which allows these bugs to remain submerged indefinitely, provided that the water is well oxygenated.

Worldwide there are 160 named species of giant water bug (Belostomatidae), 390 species of creeping water bug (Naucoridae), 400 species of backswimmer (Notonectidae), and 600 species of water boatmen (Corixidae).

Mating backswimmers (Notonectidae).

Limnocoris, a creeping water bug (Naucoridae).

Male giant water bug (Belostomatidae) carrying eggs on his back.

Lethocerus (Belostomatidae). This genus is characterized by the widened hind tibia and tarsus.

ASSASSIN BUGS

Assassin bugs (Reduviidae) are mostly predators, as one might guess from their name. Generally possessing an elongate head, they also usually have a stout beak that ensheathes the lance-like mouthparts that are used to stab prey, inject it with salivary enzymes, and suck up the liquefied contents. Although most are food generalists, some species of assassin bug show certain preferences. For example, species in the subfamily Ectrichodiinae feed exclusively on millipedes, while a species (*Salyavata variegata*) in another subfamily "fishes" for termites by dangling a dead termite from its beak; when another termite comes to inspect the cadaver, the bug grabs a fresh victim. Many assassin bugs prefer soft-bodied prey such as caterpillars (Harpactorini) or flies (Emesinae), but most species in the subfamily Peiratinae go after hard-bodied beetles and grasshoppers.

The means of capture can be as diverse as the prey. While many species ambush prey, others chase after the intended victim, pounce, and then use specialized pads at the tips of their

Zelurus formosus (Reduviinae).

Zelurus spinidorsis (Reduviinae) resembles a *Pepsis* wasp (p. 169).

Arilus gallus (Harpactorini). Members of this genus are known as "wheel bugs."

front legs to grab it. Members of the tropical American tribe Apiomerini apply sticky plant resins to their front legs to facilitate the capture of fast-moving prey such as stingless bees. The only members of this family that are not predators are the kissing bugs, which feed on the blood of vertebrates.

Depending on the species, assassin bugs either lay clusters of eggs in exposed areas or lay a single egg in crevices. Females of many Apiomerini apply a protective coating of resins to their egg masses. To avoid being eaten, some nymphs (Reduviinae) cover themselves with detritus, while others (Harpactorini) feign death.

Assassin bugs are unusual among true bugs in that both nymphs and adults produce sounds, which they make by rubbing the beak across tiny ridges in a furrow on the underside of the thorax. These barely audible sounds are emitted both during courtship and when an individual is threatened. Another defensive option is stink glands. Although the thoracic stink glands on most assassin bugs are generally quite small or even absent, nearly all assassin bugs possess a pair of glands (Brindley's glands) in the front of the abdomen that secrete defensive acids. Other defensive strategies include warning coloration, and, in a few species, mimicry of spider wasps or of stingless bees.

There are about 6,500 named species of assassin bug worldwide.

Heza (Harpactorini) sucking the bodily fluids from a fly.

Apiomerus pictipes (Apiomerini). Members of this tribe are known as "bee assassins" or "resin bugs."

Ghilianella (Emesinae). Members of this subfamily differ from other assassin bugs by their very slender body.

Egg mass of an assassin bug.

KISSING BUGS AND CHAGAS DISEASE

Triatoma dimidiata (Triatominae).

Nymph of *T. dimidiata*.

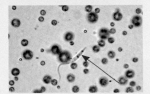

Protozoan (*Trypanosoma cruzi*) that causes Chagas disease.

Kissing bugs (Triatominae), which occur primarily in the New World tropics, are a group of assassin bugs that feed on vertebrate blood, including that of humans. Their name derives from the habit of biting on the face (though they can bite anywhere), but more salient is the fact that they are capable of transmitting the parasite *Trypanosoma cruzi*, which causes Chagas disease, a malady often fatal to humans.

The World Bank classifies Chagas disease, which occurs from the southern United States to Argentina, as the most serious parasitic disease of the Americas, in terms of both its social and economic impacts. It is estimated that around 10 million people have the disease, most of whom do not realize it, and that about 50,000 die every year from its long-term effects.

What makes Chagas so pernicious is that there are generally no obvious symptoms, and the disease is very difficult to treat except within the first year after contracting it. Its effects generally go unnoticed until many years later, when about a quarter of those infected develop damaged heart muscles, with a smaller percentage developing an enlarged esophagus or colon.

Unlike most other diseases that are spread by insects, the protozoan that causes Chagas disease is not injected with the saliva of the bug. Instead, it is excreted with the bug's feces and urine. The most common way of obtaining the disease is by inadvertently scratching the bite while sleeping, thereby contaminating one's fingers with fecal droplets, and then rubbing one's nose, mouth, or eyes, where the protozoan can readily cross the mucous membrane. Only a few kissing bugs are significant vectors of Chagas disease in humans—generally those that occur in or near human dwellings and those that begin defecating very soon after commencing their blood meal (slow defecators are less likely to leave feces on one's body). Among the 130 some species of triatomine bug that occur in the Americas only about a dozen species have both characteristics.

Not all kissing bugs carry the Chagas protozoan, only individuals that have fed on a mammal carrying the protozoan in its blood. In most mammals, including humans, the protozoans only circulate in the blood stream for relatively brief intervals of time, and it is only then that the kissing bug can acquire them. However, the protozoans are more or less permanently present in the blood of opossums, which are therefore key reservoirs of the parasite.

The triatomine species that are associated with human dwellings are most frequently found in houses that have dirt floors, piles of firewood, or other niches in which the bugs can hide during the day. Chagas disease thus tends to be most prevalent among people living in poor housing.

PLANT BUGS

Among the true bugs, the plant bugs compose the largest family (Miridae), with about 10,000 named species worldwide. As numerous as they are, however, plant bugs are usually overlooked because they are so tiny.

Most plant bugs suck sap from plant leaves and shoots, several species are scavengers or predators, and other species feed on both plant and animal tissue, which is rare among insects. Among those that feed on plants, some cause damage to crops; species in the genus *Monalonion*, for example, feed on cacao pods.

Plant bugs have various adaptations for clinging to the surface of plants. A few species, for example, have feet (tarsi) that allow them to walk on plants with sticky hairs (where they often feed on insects that become stuck to the plant). The nymphs of many plant bugs have a rectum that can be turned outward, so that, should the individual become dislodged from a plant, it can project its rectum and use it to stick to another leaf before hitting the ground. Even the predatory species, which in many groups of animals range far and wide in search of prey, are adapted to living on just a narrow range of plants.

Plant bugs of the genus *Ranzovius* (Phylinae) are unusual in that they feed on prey captured in spider webs.

Eccritotarsini (Bryocorinae) on *Heliconia*.

Mating plant bugs (Deraeocorinae).

Orthotylini on *Hamelia patens* (Rubiaceae) flowers.

LACE BUGS

Lace bugs (Tingidae) are small (no bigger than 1 cm/0.4 in), cryptically colored bugs. In many species, the wings have delicate reticulated patterns that resemble lace.

To feed, lace bugs suck sap from plant tissues. They are somewhat host specific, generally residing only on the plants that they prefer to eat. Lace bugs typically live in groups on the undersides of leaves, where they excrete fecal droplets that look like black, varnish-like dots. Most species insert their eggs into the plant, but species of the genus *Gargaphia* cement a cluster of eggs to the leaf surface, in which case the mother stands guard over the eggs or leaves them in the care of another female. Nymphs generally bear numerous setae, some of which secrete a clear, slightly viscous fluid that probably helps protect them from predators.

A few tropical American lace bugs are pests. The species *Pseudacysta perseae* feeds on avocado. Another species, *Dictyla monotropidia*, is a pest on *Cordia alliodora*, a valuable timber tree. Yet another species, *Teleonemia scrupulosa*, feeds on the lantana plant, and has been used as a biological control agent where this plant is an invasive weed, in Hawaii and in parts of the Old World.

There are about 2,000 named species of lace bug worldwide.

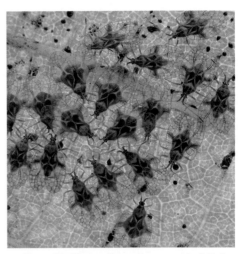

Lace bugs (Tingidae) on *Persea* (Lauraceae) leaf. Note the black fecal drops.

Phymacysta (Tingidae) on *Malpighia glabra* (Malpighiaceae) leaf.

We share an intimate history with bed bugs (*Cimex*).

No larger than an apple seed, these flattened bugs with vestigial wings (Cimicidae) can hide in a suitcase and readily move from country to country. Centuries ago, two species that commonly feed on human blood were introduced to the New World: *Cimex hemipterus*, which occurs primarily in tropical regions, and *Cimex lectularius*, which occurs throughout the world. Although bed bugs can live for a year or more without feeding, they typically seek blood every five to ten days. Generally most active after midnight, they return to their hiding places after feeding for about ten minutes. The bites are similar to those of mosquitoes but often appear in a row. Fortunately, bed bugs do not generally transmit diseases.

Bed bugs copulate in a bizarre manner called traumatic insemination. The female generally mates after feeding, often with a series of males. The male bedbug pierces the female's abdomen with his hypodermic penis, ejaculates into the body cavity (rather than into the genital tract), and then the sperm follow an alternate route to the eggs.

STINK BUGS

The large pentagonal-shaped stink bugs (Pentatomidae) range in color from bright green to dull brown. While most true bugs have odor glands that produce defensive compounds, stink bugs are especially foul smelling since their relatively large bodies (1 to 2 cm/0.4 to 0.8 in) come equipped with proportionately large odor glands.

The vast majority of species feed on plants, especially immature fruits but also plant sap. A few are crop pests, including the cosmopolitan and omnivorous southern green stink bug (*Nezara viridula*) and the small rice stink bug (*Oebalus poecilus*), which is one of the most important rice pests in South America. The Neotropical brown stink bug (*Euschistus heros*) and the small green stink bug (*Piezodorus guildinii*) are quite omnivorous, but are especially fond of Brazilian soybeans. On the other hand, the species in one subfamily (Asopinae) are predators, feeding on slow-moving insects such as caterpillars.

Courtship in stink bugs often involves a combination of chemical attractants (pheromones) and acoustic signals transmitted via the plant surface. The female glues her eggs to a leaf or stem. Some species lay small unguarded egg clusters, while other species lay a hexagonal-shaped egg mass that the female protects. In species where the female stands over her egg mass, she will kick away predators and parasitic wasps, though the latter are very persistent and often succeed in parasitizing eggs around the perimeter of the egg mass. In early stages, nymphs often feed in groups whereas older nymphs usually become more solitary.

There are 4,100 named species of stink bug worldwide.

Edessa arabs (Edessinae). The genus *Edessa* is a large, heterogenous group with hundreds of species yet to be named.

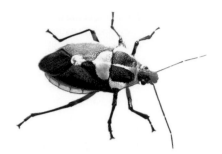

Euthyrhynchus floridanus stink bug (Asopinae). Members of this subfamily are predators.

Female *Chlorocoris isthmus* stink bug (Pentatminae) with her nymphs.

Stink bug nymphs with empty eggs from which they emerged.

Burrower bug (Cydnidae: *Cyrtomenus teter*). This family is related to stink bugs and its members feed on plant roots (the spiny legs are used for digging).

SHIELD-BACKED BUGS

Members of the family Scutelleridae are known as shield-backed bugs, a name that refers to the greatly prolonged portion of the thorax (the scutellum), which completely covers the abdomen and wings (giving them a beetle-like appearance). They are generally quite rounded in shape, and many species are very brightly colored. Like their close relatives, the stink bugs, the females of many species stand guard over their egg mass.

There are 450 named species of shield-backed bugs worldwide

Dystus puberulus shield-backed bug (Scutelleridae) is usually found on fig trees.

Pachycoris torridus (Scutelleridae) feeds on plants in the family Euphorbiaceae.

SEED BUGS

The twelve families of seed bug compose a superfamily (Lygaeoidea).

Not surprisingly, most seed bugs feed on seeds, either seeds still on the plant or seeds that have fallen to the ground. Species in the subfamily Lygaeinae sequester into their bodies plant compounds from the seeds they eat, and these compounds provide them with chemical protection from potential predators; the bugs advertise their toxicity through bright coloration. Species in the genus *Oncopeltus*, for example, sequester cardiac glycosides from milkweeds, and are a bright orange and black. However, the majority of seed bugs, in the family Rhyparochromidae, bear drab, cryptic coloration, including those that feed on fig seeds (*Ozophora*).

Seed bugs in the family Rhyparochromidae are usually drab in coloration. Some are ant mimics.

In the case of *Neacoryphus bicrucis*, which feeds on *Senecio* species (Asteraceae), the males are very aggressive—they attempt to copulate with almost anything that moves—and they often displace other bug species from the plants they inhabit.

Chinch bugs (Blissidae), which get their name from the Spanish word *chinche* (bug), do not feed on seeds, but rather on the sap of various grasses, and some species are notorious pests of cereal crops. They often migrate from their native host plant to a crop plant, causing much more damage to the latter as it lacks the tolerance of the native host plant. There are about 4,400 named species of seed bug worldwide.

LEAF-FOOTED BUGS

Leaf-footed bugs, all in the family Coreidae, are some of the largest terrestrial bugs, measuring up to 4 cm (1.6 in). Some species have irregular leaf-like expansions on the hind tibia that they raise above their bodies, like a warning flag, when disturbed. In other species, the males have enlarged hind femora which they use to defend territories from other males.

Leaf-footed bugs feed on plant sap; some species are generalists, others specialize on a specific kind of plant. Among the specialists are species that feed on woody legumes, heliconia (*Leptoscelis tricolor*), or passion fruit (*Diactor bilineatus*), to cite just a few examples. Interestingly, the species *Piezogaster*

Seed bugs in the family Lygaeidae often show more striking coloration.

Leptoscelis tricolor feeds on fruits and flowers of *Heliconia*.

chontalensis feeds on *Vachellia collinsii* trees, which are defended by ants (*Pseudomyrmex spinicola*); apparently it survives by mimicking the odor of the ants, but some other mechanism may also come into play.

In some species both nymphs and adults have bright warning coloration. Such nymphs often form feeding aggregations and if disturbed they pulsate aggressively, spray jets of anal fluid into the air, and exude noxious secretions from glands on the upper side of the abdomen

Worldwide there are about 1,800 named species of leaf-footed bug.

Hypselonotus concinnus.

Nymph of *Thasus*. Adults can reach 4 cm (1.6 in).

Anisoscelis species feed on passion fruit (*Passiflora*) and exemplify the family name *leaf-footed bugs.*

Aggregation of recently emerged leaf-footed bug nymphs. Empty egg shells on bottom right.

Leptoglossus zonatus feeds on various plant species.

Nymph of a broad-headed bug (Alydidae), which is related to the leaf-footed bugs. Species in several families of true bug mimic ants.

Nymph of *Acanthocephala*.

Cicada (Auchenorrhyncha) sucking plant sap.

SUBORDER AUCHENORRHYNCHA (Planthoppers, Cicadas, and Their Kin)

Many species in this group are excellent jumpers, and nearly all are strictly terrestrial and herbivorous. Both nymphs and adults usually feed on plant sap, be it watery xylem sap (in the case of cicadas, spittlebugs, and one subfamily of leafhoppers), sugary phloem sap, or mesophyll.

As do many insects that live on plants, species in this group use the plant surface to communicate with one another (e.g., during mating), moving the plant either by flicking the wings or vibrating the abdomen (in some species, the vibrating abdomen moves the plant without actually making contact). Most auchenorrhynchan species are unique in that they also use a pair of specialized acoustical organs (tymbals) at the base of the abdomen to vibrate the plant. These organs are usually best developed in adult males, but in some cases they are also present in females and sometimes even in nymphs. Only in cicadas are these organs capable of transmitting sound through air rather than relying on vibrations sent over the plant surface.

PLANTHOPPERS

Planthoppers (superfamily Fulgoroidea) comprise a dozen families. Most species feed on phloem sap from stems and leaves, although nymphs in the family Cixiidae feed below ground on roots, especially those of grasses, and nymphs in the families Achilidae and Derbidae feed on fungi that grow on rotten wood.

Most planthoppers are quite small, but several members of the family Fulgoridae are large and have bizarre extensions in front of their heads. The most spectacular examples are the three species of peanut-headed bug (in Spanish, *la machaca*), in the genus *Fulgora*; these are among the largest (9 cm/3.5 in) members of the order Hemiptera. They feed on the sap of certain trees, particularly species such as *Hymenaea courbaril* (Fabaceae) and *Protium* (Burseraceae), which have resinous or bitter sap. In much of Latin America, these insects are mistakenly believed to be venomous, and according to local lore, if bitten, a person will die unless he or she has sexual intercourse within 24 hours.

A few of the smaller planthoppers can be serious plant pests. Some species in the family Delphacidae feed on corn and rice. *Myndus crudus* (Cixiidae) is a vector of a bacterial disease called lethal yellowing that destroys palms.

Females of some planthopper species insert their eggs into plant tissue or soil, whereas other species lay a mass of eggs on the plant surface and then coat it with a sticky wax to protect the eggs against predators. Nymphs of most species also produce a protective waxy covering over their bodies. In a few species the adults have very long wax filaments trailing from their abdomen, but the function of the wax in these cases is less clear.

There are about 12,000 named species of planthopper worldwide.

This planthopper nymph (Fulgoridae) resembles liverwort or moss.

Peanut-headed bug *Fulgora laternaria* (Fulgoridae). Arrow indicates parasitic moth larva (Epipyropidae).

Peanut-headed bug, close-up of wing with white wax.

During the day groups of *Phrictus quinquepartitus* (Fulgoridae) feed and rest on tree trunks. They fly at night.

Scalaris neotropicalis (Fulgoridae).

Peanut-headed bug *Fulgora lampetis* (Fulgoridae).

Odontoptera carrenoi (Fulgoridae), with false eye spot on front wing.

Egg mass of *Diareusa* (Fulgoridae).

Colpoptera (Issidae).

Nymph of *Colpoptera*. When threatened it spreads the wax filaments at the rear end.

Nymph of Flatidae covered with wax secretion. When threatened it abandons the front mass and uses the two hind masses to glide away.

Adult Flatidae on fig tree (Moraceae: *Ficus*).

CICADAS

During the day, the dominant sound in a tropical rainforest is the unrelenting and sometimes deafening buzz of male cicadas (Cicadidae) singing to attract females. Each species has its unique song, and males of a particular species often synchronize their song. Such choruses are generally sung at dawn and dusk, though many species also sing during the day. The sound is produced by buckling a ribbed membrane (tymbal) located at the base of the abdomen, one on each side. Muscle contractions cause the tymbal to buckle inward and when the muscle is relaxed the tymbal pops back to its original position (these muscle contractions are so rapid that some cicadas risk overheating, which they avoid by evaporating water from the surface of their body). Large air sacs in the abdomen and thorax serve as resonators and the sound is radiated outward through a pair of thin diaphragms on the underside of the thorax.

If a female is interested in a male, she responds by producing a brief snapping sound with a flick of her wings. Once impregnated, the female inserts her eggs into either a branch (live or dead) or a grass stem (e.g., sugarcane), depending on the species. When the eggs hatch, the tiny nymphs fall to the ground and use their stout front legs to burrow into the soil, where they suck sap from rootlets of plants, especially those of legume trees. Little is known about how long tropical species stay underground, though they probably require at least a year to complete their nymphal development.

Upon reaching maturity, the nymphs emerge from the ground, crawl up the closest vertical object, and cast off their exoskeleton; the newly emerged adults then fly up into the canopy. The time of the year when nymphs emerge from the ground—and when males sing—varies from species to species.

Both adults and nymphs suck xylem sap, which is consumed and excreted in great quantities. Indeed, one can often feel a fine mist raining down from a canopy full of cicadas.

There are about 1,200 named species of cicada in the world.

The genus *Taphura* includes the smallest (1 cm/0.4 in) cicadas.

In order to attract females, male *Zammara* often congregate on certain trees to sing together.

Male *Carineta* cicada showing the sound-producing organ (tymbal) at base of abdomen. At dawn and dusk, they sing in flight.

Zammara cicada camouflaged on tree trunk.

Empty nymphal exoskeleton of cicada in forest understory. Note the large front legs used for digging.

Cicada adult, recently emerged from nymphal exoskeleton.

SPITTLEBUGS

Spittlebugs belong to the superfamily Cercopoidea, which consists of four families, three of which occur in the Americas.

In Latin America, a number of the common names for spittlebugs make reference to *spit*, as in *baba de culebra* ("snake drool"). Adults and nymphs feed on the watery xylem sap of plants. The nymphs put the excess water to good use, excreting the water and then creating a frothy mass within which they hide from predators. Usually there is just one nymph per spittle mass, although a few species (e.g., *Cephisus siccifolius* in the family Aphrophoridae) form large communal masses with dozens of nymphs. Adult spittlebugs (sometimes called froghoppers) do not produce spittle and instead simply flick away the excreted water droplets.

Adults protect themselves using different strategies. Some species (Aphrophoridae, most Clastopteridae) wear drab camouflage colors; many species in the family Cercopidae display warning coloration and are clad in yellows, reds, or oranges—or, in some cases, in a bold black-and-white pattern. If threatened, these brightly colored species secrete unpleasant smelling droplets from the tips of their feet.

Because xylem sap generally contains very little nitrogen, spittlebugs tend to seek out plant species whose xylem sap contains relatively high amounts of nitrogen. Many such plant species harbor nitrogen fixing bacteria that convert atmospheric nitrogen into compounds that can be used by the plant (and hence by the spittlebugs). These include not only legumes but also grasses and several other plants. In general, spittlebugs are not very host specific.

There are about 3,000 named species worldwide; common examples from tropical America include species in the genus *Mahanarva* that are found in the floral bracts of heliconias, and species in the genera *Aenolamia, Prosapia,* and *Zulia,* which are pests of pasture grasses and sugarcane.

This *Clastoptera* (Clastopteridae) has a false head (top); actual head is at bottom.

Iphirhina perfecta (Cercopidae). Species in this genus feed on *Piper*.

Mahanarva costaricensis (Cercopidae). Species in this genus feed on *Heliconia*.

Spittle masses are often found on or near the ground.

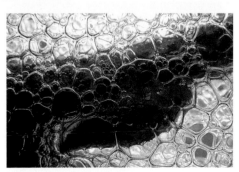

Spittlebug nymph inside spittle.

LEAFHOPPERS

With about 10,000 named species worldwide, Cicadellidae is one of the largest families in the order Hemiptera. In addition to their diversity, leafhoppers are often among the most abundant plant-feeding insects, though most are quite small and seldom noticed by the casual observer.

Leafhoppers come in almost every imaginable color, from brilliant purple to drab brown. Adults—and nymphs even more so—have a characteristic behavior of quickly shuffling from one side of a plant to the other when they detect danger. The larger species, in the subfamily Cicadellinae, feed on xylem sap, while some of the smallest species, in the subfamily Typhlocybinae, feed on mesophyll sap; many other species feed on phloem sap or on some combination of the three sap types.

Females insert their eggs into the host plant. Many species of leafhopper are unique in that they secrete after each molt a protein substance from the anus that is then spread over the body with their legs. When this secretion dries, it turns into a residue of minute spherical granules that serve to waterproof the body. In the subfamily Cicadellinae, the females of many species store large globs of this material on their wings (appearing as a white spot on each side of the body) and use it to cover their eggs to protect them against parasitic wasps.

A few species are vectors of crop diseases. For example, *Dalbulus maidis*, a species that is found only on corn, transmits a viral disease. Several of the larger species, in the subfamily Cicadellinae, transmit a bacterial disease (*Xylella*) to coffee, citrus species, avocado, and various other trees.

Coronigonalia spectanda standing on one pair of legs, using the first pair to clean its head.

Pseudophera (Proconiini) using hind legs to cover her eggs (row in lower left) with white proteinaceous granules (brochosomes) stored on front wings.

Translucent leafhopper nymph with adult in background.

Paromenia isabellina feeding on plant sap.

Baleja flavoguttata.

Baleja.

Dilobopterus instratus.

Dilobopterus lineosus.

Agrosoma bispinella.

Ladoffa leafhoppers often spread their wings.

Graphocephala coronella.

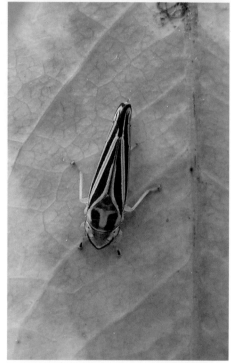

Graphocephala albomaculata sucking plant sap.

Graphocephala coccinea.

Beirneola coripana males (red stripes) and females (yellow stripes).

Barbinola costaricensis.

TREEHOPPERS

Species in this family (Membracidae) come in an amazing diversity and strangeness of shapes. Some treehoppers bear ornate dorsal protuberances, some resemble plant spines, and others resemble caterpillar excrement.

The female deposits eggs on the surface of branches or into the living plant tissue itself. In many species that deposit eggs on the branch surface, the female covers the eggs with a frothy substance that hardens upon drying and thus provides protection from predators and parasitoids. Although some species are solitary, most treehoppers in the lowlands of tropical America show varying degrees of maternal care. While in some species the mother merely guards the egg mass, in others she also remains with the nymphs throughout their development. In a few species, she uses her ovipositor to make incisions into the twig to make it easy for the nymphs to access the sap. If threatened by a predator, the nymphs produce alarm calls (using vibrations transmitted via the plant surface), and the mother responds by buzzing her wings or kicking with her hind legs in order to drive away the intruder.

Treehoppers feed on sugar-rich phloem sap, and therefore produce a sugary excrement, which is called honeydew. This honeydew often attracts ants, and many treehopper species, in a symbiotic relationship, are in turn protected by the ants. Such species tend to live in larger groups, since, with more individuals, a greater amount of honeydew is produced, which attracts a greater number of ants—and thus increases the safety of the treehopper group. The mother may either remain with the nymphs or leave them to be cared for by the ants.

Worldwide there are about 3,400 named species of treehopper. They are especially diverse in tropical America, which is where the group is thought to have originated.

Female *Umbonia ataliba* with her nymphs.

Aggregation of young adults of *Umbonia ataliba*.

Young females of *Umbonia crassicornis*. Species in this genus often feed on Fabaceae.

Female *Umbonia crassicornis* with her eggs.

Aconophora.

Membracis mexicana with a chalcidoid wasp (Mymaridae) parasitizing her eggs.

Membracis dorsata nymphs and very young adults; the latter will soon disperse.

Female *Ennya pacifica* with her eggs.

Mating *Ennya chrysura* (the male is dark).

Hyphinoe nymph on Cucurbitaceae.

Campylocentrus feeding on Cucurbitaceae.

Polyglypta species often feed on Asteraceae.

Polyglypta male (top) performing courtship vibrations.

Erechtia feeding on *Hamelia patens* (Rubiaceae).

Vestistilus feeding on plant sap.

Bocydium on Conostegia (Melastomataceae). The function of the bizarre projection is unknown.

Nymph of Darninae on Ficus colubrinae (Moraceae).

Smerdalea resembles moss.

Bolbonota treehoppers resemble caterpillar excrement.

Cladonota on Ipomoea (Convolvulaceae).

Cladonota frontal view.

Cyphonia on Asteraceae. This treehopper mimics an ant with open mandibles (left).

Female Guayaquila with her egg mass and an ant (Ectatomma) in search of honeydew.

HONEYDEW

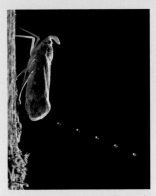

Enchophora sanguinea (Fulgoridae) with a snail feeding on its honeydew and an ant (*Camponotus*) licking honeydew from the body of the snail.

Enchophora ejecting a drop of honeydew (single droplet exposed multiple times with a stroboscopic flash).

Treehopper nymphs (*Antianthe expansa*) on Solanaceae being tended by ants (*Crematogaster*).

Plants produce three types of sugar-rich foods: nectar, fleshy fruits, and phloem sap. Phloem circulates through a plant's internal plumbing system, carrying the sugars produced by photosynthesis. Many animals consume nectar and fruits, but the only animals that regularly consume phloem sap are certain members of the order Hemiptera: some true bugs, planthoppers, treehoppers, leafhoppers, aphids, psyllids, whiteflies, and scale insects.

Rich in sugars but poor in essential amino acids, phloem sap is not an ideal food. But insects that specialize on the sap overcome this deficiency by harboring in specialized cells endosymbiotic bacteria that produce these missing ingredients. The insects excrete the excess sugar as droplets from their rear ends, and this sugary excretion is known as honeydew.

Some insect species that ingest phloem sap excrete prodigious quantities of honeydew, sometimes excreting in just one hour an amount equivalent to their own body weight. When a plant harbors a large population of such insects, the amount of sugar present in the honeydew sometimes far exceeds the amount present in the plant's fruits. This then attracts secondary phloem feeders, animals that either imbibe the honeydew as it emerges from the anus of the insect or that scavenge the honeydew accumulated on the leaves and surrounding plants. Ants are avid honeydew feeders and often protect the phloem sucking insects, tending them like dairy cattle and sometimes even using detritus to construct a roof over them. The quantity of honeydew harvested by ants in the canopies of tropical forests is

difficult to measure but is thought to be substantial. Other animals that imbibe honeydew—including various wasps, stingless bees, flies, and even hummingbirds—generally do not protect the phloem-sucking insects that produce it.

Plants harboring phloem-sucking insects can be recognized not only by an abundance of energized ants, but also by the presence of sooty molds that eventually colonize the excess honeydew that accumulates on leaves or at the base of a tree. Even before the sooty molds take over, however, the presence of too much sticky honeydew can cause problems for the insects producing it. To minimize these problems, many species have developed anal musculature that allows them to flick the droplets of honeydew away from their body. A few species, however, have become so dependent upon ants to remove the honeydew that their anal musculature has become atrophied.

The quality and quantity of phloem sap can vary between plants and, even on a given plant, the amount and quality of sap can vary depending on the time of day or season. Moreover, different species of phloem-feeding insects produce different amounts—and kinds—of honeydew. While most insects that harvest honeydew are probably generalists, it is likely that a few species target specific kinds of honeydew. In Costa Rica, for example, the honeydew from nocturnal planthoppers (Fulgoridae) seems to be the favorite kind of honeydew for a rather bizarre assemblage that includes cockroaches, moths, and even snails.

SUBORDER STERNORRHYNCHA
(Plant Lice, Scale Insects and Their Kin)

All members of this group are strictly terrestrial and herbivorous. Most species are quite small, measuring less than 1 cm (0.4 in), and are seldom noticed except when populations undergo explosive growth. Many excrete honeydew, thereby attracting ants, and thus the best indication of their presence is often a concentrated grouping of ants on a plant. They extract plant sap in a fascinating way. Their thread-like mouthparts penetrate the plant surface and then maneuver around individual plant cells and extract the sap.

As mentioned in the introduction to this chapter, plant-feeding hemipterans harbor endosymbiotic bacteria that provide the insects with nutrients that are lacking in their diet. In addition to these primary endosymbiotic bacteria, Sternorrhyncha also harbor another category of endosymbionts, known as secondary or facultative endosymbionts. All individuals contain primary endosymbionts, but an individual insect may or may not harbor a particular secondary endosymbiont. The role that these other bacteria play in the insect is largely unknown, although in aphids some of the bacteria appear to increase the possibility of surviving an attack by a parasitic wasp or a fungal pathogen.

JUMPING PLANT LICE

Jumping plant lice are in the superfamily Psylloidea. Adults resemble bark lice (p. 49) but lack the bulging face. Adults also resemble aphids (p. 91) but are much more active, often jumping when disturbed. The nymphs are often covered in waxy secretions.

Most species of jumping plant lice live exposed on the foliage of a plant, but some live enclosed (or partially enclosed) in galls (p. 280) that they induce on the plant; examples of plants harboring galls in tropical America include Anacardiaceae, Araliaceae, Burseraceae, certain Fabaceae (e.g., *Lonchocarpus*), Lauraceae, Moraceae, Myrtaceae, and Rutaceae. Unlike other sternorrhynchans, jumping plant lice do not feed exclusively on phloem, but rather a combination of saps. Each species feeds on either just a single plant species or a very small number of closely related species. Relatively few species feed on crop plants, one of the most notable being *Diaphorina citri*, a species recently introduced to tropical America that spreads citrus greening disease (a bacterium), one of the most serious diseases that afflict citrus trees.

Worldwide there are approximately 3,000 named species of jumping plant lice. There are likely more species in tropical America (and temperate South America) than in any other region.

Aphids are among the best known Sternorrhyncha. *Aphis nerii* on milkweed.

Diclidophlebia lucens jumping plant louse (Psyllidae) on *Miconia calvescens* (Melastomataceae).

Trichochermes (Triozidae) on *Pseudolmedia mollis* (Moraceae).

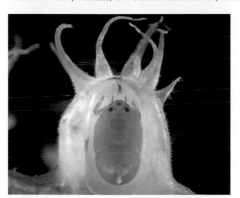

Trichochermes nymph inside gall on *Pseudolmedia*.

Ciriacreminae (Psyllidae) nymph on Annonaceae, with wax secretions resembling cork-screws.

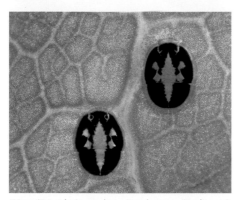

Trioza (Triozidae) nymphs on *Dendropanax* (Araliaceae).

Gall on *Lonchocarpus* (Fabaceae) induced by *Euphalerus* (Psyllidae).

WHITEFLIES

Whiteflies are placed in the family Aleyrodidae, which comprises about 1,600 named species.

Adult whiteflies secrete a white powdery wax-like substance from the underside of the abdomen and then use their legs to spread this substance over the rest of the body, which presumably helps waterproof them. Depending on the species, the female lays her wax covered eggs in a circle, semicircle, or spiral, usually on the undersides of leaves. Interestingly, fertilized eggs produce female offspring while unfertilized eggs yield male offspring.

The nymphs resemble the nymphs of scale insects in that only the first nymphal stage is capable of movement; in the succeeding stages, the nymphs remain immobile, attached to the plant, from which they extract sap. (In most other insect species, all nymphal stages are mobile.) Mature nymphs stop feeding, become quiescent and undergo dramatic morphological changes, and then finally transform into adults. The difference between the final

Whitefly nymphs on *Citrus*, with long protruding wax filaments.

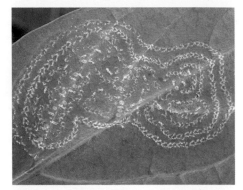

Whitefly eggs covered with wax and recently hatched nymphs.

Bemisia tabaci on *Solanum americanum* (Solanaceae).

Whitefly pupa on Asteraceae.

Male *Udamoselis* on *Ficus colubrinae* (Moraceae). It is quite large (8 mm, 5/16 in.) for a whitefly.

nymphal stage and an adult is so dramatic that biologists label the intermediate stage a *pupa*. This means that whiteflies also undergo a kind of metamorphosis, similar to that occurring in the beetles, wasps, butterflies, and flies.

Most species feed on a relatively narrow range of plants but several species feed on a wide diversity of plants, and some of these are plant pests. After insecticide applications, whiteflies demonstrate an infamous ability to rebound with a vengeance, as such spraying seems to affect their predators and parasitoids to an even greater degree. In regions of tropical America and in many other parts of the world, *Bemisia tabaci* is the most serious insect pest of tomato and several other vegetable crops, not because of its feeding per se, but rather because of the serious viral diseases it transmits. One of the few solutions is to grow young tomato plants under a fine mesh that protects them from whiteflies. If this is done for the first month and a half, the plants are then able to better endure a subsequent viral infection.

APHIDS

Aphids, though sometimes known as "plant lice," are in an entirely different superfamily (Aphidoidea) from the jumping plant lice.

Their most salient characteristic is a pair of tubes (siphunculi) that project from the rear end, and from which they discharge defensive compounds and alarm pheromones. Aphids are soft-bodied and usually green colored. Some species are black or even bright yellow. Like several other groups of Sternorrhyncha, aphids excrete honeydew (p. 87).

Aphids are notorious plant pests. They suck phloem sap (occasionally xylem sap, when they are dehydrated), sometimes transmit viral diseases to their host plants, and reproduce at a rate that would put rabbits to shame (in some cases a new generation of aphids can be produced in just five days, the shortest turnover time of all insects). They are in fact quite scarce in the forests of tropical America but can be exceedingly abundant on certain crops.

Aphid life cycles are often quite complex. In temperate regions aphids alternate between a series of asexual generations, followed by a sexual generation, the latter usually occurring before the onset of winter. In tropical America, however, most aphids appear to reproduce asexually all year round. In other words males are absent and females essentially clone themselves. Unlike sexually reproducing females, asexual females do not lay eggs but rather give birth to baby aphids, which are already pregnant the moment they are born. After first

Aphis nerii with two species of predatory lady beetle larvae (p. 124).

colonizing a plant, the first few generations of adult females are often wingless, but eventually winged forms appear in preparation for colonizing another plant.

Aphids are more diverse in northern, temperate regions than in the tropics. Although there are about 4,300 named species in the world, only a small proportion of these occur in tropical America, and of those that do occur here, very few are native species, the vast majority having been inadvertently introduced from elsewhere.

Gibbomyzus pteridophytorum on fern. This is one of the few aphids native to tropical America.

Aphis nerii (on Apocynaceae) giving birth.

Microparsus olivei on *Bocconia frutescens* (Papaveraceae).

Aphis spiraecola on *Myrica pubescens* (Myricaceae).

MEALYBUGS AND SCALE INSECTS

The superfamily Coccoidea contains very specialized plant parasites—and several of the most bizarre of all insects. The saclike adult females are wingless and often legless (or virtually so), thus making them difficult to even identify as being an insect. Mealybugs secrete a powdery wax over their bodies whereas scale insects, as the name suggests, produce a scale-like covering. Adult males of both groups are fragile insects that have legs, a single pair of wings, and one or more waxy filaments protruding from their rear end. The males never eat, live for only a short time, and as a result are less frequently seen than females.

In the males of most species the chromosomes inherited from the father are deactivated in the embryo and later eliminated from the germ line, so that only the maternal chromosomes are passed on to the sperm. In a few species, males are entirely absent and the females reproduce asexually. This brief summary, however, fails to do justice to the true diversity and complexity of genetic systems found in mealybugs and scale insects.

In both mealybugs and scale insects the female lays eggs in a cavity under her body or in a waxy covering behind her body. Upon hatching, the first nymphal stage either settles down on the same plant on which it was born or is carried by the wind (or on the body of another insect) to another plant. Subsequent nymphal stages are less mobile (mealybugs) or completely immobile (scale insects). While adult female mealybugs are capable of walking, those of scale insects spend their lives attached to a plant by their thread-like mouthparts. Unlike the females, adult males of mealybugs and scale insects have wings, which they use to fly about in search of females.

Some species feed only on specific plant species but others utilize a wide variety of host plants. Nymphs and adult females suck phloem sap and produce honeydew, except in the armored scales (Diaspididae), which feed primarily on normal plant tissue. House plants sometimes become completely covered in scale insects, probably because the parasitic wasps that normally keep their populations in check have not discovered these isolated plants. And some species are serious crop pests, the pineapple mealybug (*Dysmicoccus brevipes*) being one of them. More positively, a few species are raised commercially to produce dyes or lacquer.

Worldwide there are nearly 8,000 named species in about 30 families, though most species belong to just three families: armored scales (Diaspididae), mealybugs (Pseudococcidae), and soft scales (Coccidae).

Male scale insects such as this species (Monophlebidae) have just one pair of wings.

Female Monophlebidae. Members of this family are often larger than most other scale insects and mealybugs.

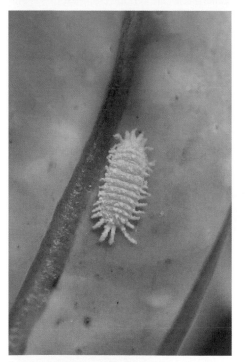

Female Ortheziidae scale insect with egg sac at rear end.

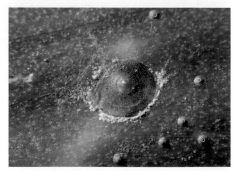

Female mealybug (Pseudococcidae) on *Drimys* (Winteraceae).

Female armored scale (Diaspididae).

DACTYLOPIUS COCCUS AND THE SPANISH EMPIRE

The *metate* (mortar) and *mano* (pestle) are used for grinding cochineal scale insects into a fine powder (stored in bowl on left). The *mano* is moved horizontally, not vertically.

For thousands of years prior to the arrival of Europeans, indigenous people of Mexico had been cultivating on prickly pear cactus the cochineal scale insect (*Dactylopius coccus*, family Dactylopiidae), selectively breeding it for its intense red color. It makes an excellent dye, as it turns out. When the conquistadors noticed cochineal in an Aztec marketplace, they sent a shipment back to Spain. Although it took decades for the Spanish to realize its value, cochineal turned out to be the most brilliant and useful red dye ever seen in the Old World. Demand soared, and the profits derived from the monopoly on this commodity, second only to those from silver, helped finance Spain's numerous wars during the 16th and 17th centuries. Spanish ships bearing cochineal from Mexico became targets for English, French, and Dutch pirates.

Throughout the 17th century most Europeans were uncertain whether cochineal came from a plant or an animal, and Spanish authorities did everything possible to conceal the source. Dutch scientist van Leeuwenhoek, the father of microbiology, eventually got it right, and, during the latter part of the 18th century, England, France, and Holland made several covert attempts to steal cochineal scale insects and transplant them to their own colonies. The cactus host was more easily transplanted than the insect, however, and it wasn't until Mexico achieved independence that Spain's monopoly was broken, at which point Guatemala and the Canary Islands began competing with Mexico as major producers. Synthetic dyes derived from coal were eventually introduced in the late 1800s, and the cochineal industry collapsed. Artisanal production—and concerns over the possible carcinogenic effects of artificial colorants—have helped maintain small-scale production. Peru is today the major producer of cochineal.

Llaveia axin (family Monophlebidae), yet another scale insect that has been used

Prickly pear cactus on which the cochineal scale insect feeds.

since pre-Columbian times, principally in the drier regions of Central America, was used to produce a kind of lacquer and also a facial paint. This giant scale insect, measuring up to 25 mm (1 in), was reared on various host trees; egg masses were inserted in plants and then protected against predatory beetles. At harvest time, the insects were placed in boiling water, and the resulting liquid was then strained through cotton cloth. True varnish comes from an Asian scale insect (*Kerria lacca*, family Kerriidae), once the main source of wood varnish until it was replaced by synthetic lacquer.

Dried cochineal scale insects.

4

BEETLES
(order Coleoptera)

Scientific lore has it that J.B.S. Haldane, the distinguished British biologist, was once asked by a theologian friend what one could conclude about the nature of the Creator from a study of his creation. Haldane is said to have answered, "An inordinate fondness for beetles." Indeed, about 400,000 species have been identified to date, with hundreds more being named every year. This current tally of named species represents about one third of all insects. The next three largest orders (wasps, moths, and flies) have approximately 160,000 named species each.

Found almost everywhere, beetles (in Spanish, *escarabajos*) have hardened front wings (known as elytra) that cover the hind wings and much of the rest of the body (beetles belong to the order Coleoptera, meaning *sheath wings* in Greek). During flight, the front wings are usually held perpendicular to the body while the hind wings do all the work. Upon landing, beetles carefully fold up their hind wings under the protective coverings provided by the front wings, thereby allowing them to squeeze into confined spaces.

Beetles—like wasps, moths, flies, and a few other insect orders—undergo complete

metamorphosis, passing from larva to pupa to adult. Beetle larvae generally have legs (unlike most fly and wasp larvae) but they lack the prolegs on the abdomen that characterize moth and butterfly caterpillars.

The larval and pupal stages of beetles are seldom seen since most species live in soil, leaf litter, dead wood, plant stems, seeds, and other concealed places. Rotting tree trunks are an especially important habitat and most of the small holes you see in wood are the work of beetles.

Beetle larvae and adults often eat the same things, unlike insects in the other orders that undergo complete metamorphosis (e.g., wasps, moths, flies), in which larvae and adults have distinct diets. Many beetle species feed on fungi or plant matter containing fungi, while others are herbivores or predators. Longhorn beetles, leaf beetles, and weevils together form an enormous branch on the evolutionary tree of beetles, and this group represents the second largest lineage of plant-feeding animals after the moths and butterflies.

Tropical America has 115 families of beetle, some of the more notable of which are discussed in this chapter.

AQUATIC BEETLES

Tropical America is home to about a dozen beetle families that inhabit a variety of freshwater bodies, from fast-flowing rivers to the water that accumulates in bromeliads. In many cases, both the larvae and adults are aquatic; in the vast majority of species, however, pupation takes place on shore. The larvae are equipped with gills and cannot leave the water. Adults, on the other hand, are capable of flying from one body of water to another.

Elmoparnus (Dryopidae). Unlike other aquatic beetles, only the adults are aquatic.

PREDACEOUS DIVING BEETLES AND WATER SCAVENGER BEETLES

Predaceous diving beetles (Dytiscidae) and many water scavenger species (Hydrophilidae) have streamlined bodies and legs adapted for swimming (note that the water scavenger beetle family also contains a good number of semiaquatic species and terrestrial species). The former move their hind legs in unison whereas the latter move them alternately. In order to breathe underwater they carry a layer of air, under the wings in the case of predaceous diving beetles, and in a dense coat of short hairs on the underside of the body in the case of water scavenger beetles. This air bubble functions as both oxygen tank and as a physical gill. The insect takes oxygen from the bubble, which in turn is able to draw oxygen from the surrounding water. The bubble slowly decreases in size, however, and eventually the beetle must come to the surface to obtain a new air supply.

Larval and adult predaceous diving beetles feed on other aquatic invertebrates. Larvae of water scavenger beetles are also predators, but adults are omnivores.

Worldwide there are more than 4,000 named species of predaceous diving beetle and about 3,500 species of water scavenger beetle.

Aggregation of water scavenger beetles (Hydrophilidae: *Anacaena*) feeding on algae-covered rock in river.

Cybister festae (Dytiscidae).

order to clean their food, whirligigs dole out their pasty secretion little by little to counteract the fish's attempt to flush away the chemicals. Species with the strongest odors are often the ones that form the largest congregations, and these groups consist of both sexes and sometimes even include more than one species. Whirligigs form temporary congregations whose concentrated defensive compounds serve to deter fish and to alert other members of the group to the presence of fish.

There are about 1,000 named species of whirligig beetle worldwide.

Larva of Dytiscidae.

Gyrinus (Gyrinidae), with arrows showing upper and lower part of eye.

WHIRLIGIG BEETLES

Shiny black whirligig beetles (Gyrinidae) can often be seen gyrating about in compact groups on the water surface. While they spend most of their time on the surface, they are also capable of diving. To facilitate life in two such very different worlds, each eye is divided into an upper and lower half, which allows the whirligig to simultaneously view the world above and the world below.

Whirligigs possess short antennae that are extremely sensitive vibration detectors. With these they can avoid collisions with fellow whirligigs, locate potential prey (insects on the water surface), and communicate during courtship. Females lay their eggs in submerged vegetation near the water's edge and males of some species set up territories in these areas. By doing this the males can intercept females as they come to lay eggs.

Whirligig beetles secrete defensive compounds from glands at their rear end to help protect them from fish predators. Since fish can swish water in and out of their mouth in

Gyrinus showing hind legs adapted for swimming.

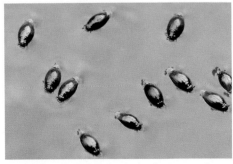

Aggregation of whirligig beetles on the water surface.

RIFFLE BEETLES AND WATER PENNY BEETLES

Riffle beetles (Elmidae) live in flowing, well oxygenated rivers. The adult beetles cannot swim but instead spend their time gripping submerged rocks, from where they feed on algae.

They can remain permanently underwater thanks to a non-collapsible layer of air that envelopes their body and extracts oxygen from the surrounding water.

The larvae of water penny beetles (Psephenidae) also live in flowing water, though adults are terrestrial. As an adaptation to life in flowing water, the larvae have extremely flattened bodies. They attach themselves tightly to submerged rocks, where they feed on algae. There are about 1,500 named species of riffle beetle and about 300 species of water penny beetle.

Riffle beetles, such as this *Disersus*, often come to night lights.

Riffle beetles (Elmidae) on the surface of a submerged rock.

Larva of water penny beetle (Psephenidae) on a submerged rock.

Ventral view of larva of water penny beetle.

TERRESTRIAL BEETLES

Although dividing beetles into aquatic and terrestrial families is somewhat helpful in organizing our thinking, it should be noted that this division does not reflect evolutionary relationships. The vast majority of beetle families are predominantly terrestrial and can be found in virtually all types of terrestrial habitats.

GROUND BEETLES

Ground beetles (Carabidae) are a diverse group of mostly predatory insects. The larvae are usually more strictly carnivorous than adults, who are often quite opportunistic in their feeding habits. In a few species, the larvae are ectoparasitoids. The larvae of bombardier beetles (*Brachinus*), for example, parasitize the pupae of water beetles, while larvae in the genus *Lebia* parasitize the pupae of leaf beetles.

Many species are dark-colored inhabitants of leaf litter and fallen tree trunks, but in tropical rainforests nearly half of the species live in the canopy, and for these species the name *ground beetle* is a misnomer. Several arboreal species hide in bromeliads or other crannies during the day, and then come out at night to hunt on the surfaces of leaves and twigs. These arboreal species often have bright metallic colors, as seen in many members of the diverse tribe Lebiini (Harpalinae).

Adult ground beetles manufacture defensive compounds that are secreted from glands at the rear end of the body. In species in the subfamily Harpalinae, these compounds consist of formic acid and hydrocarbons; bombardier beetles (Brachininae), however, spray in rapid-fire pulses a boiling-hot quinone solution.

Tiger beetles, now classified as a subfamily (Cicindelinae) of the ground beetle family, are reputed to be among the fastest sprinters in the insect world. They have bulbous eyes ideal for sighting prey.

Many species of tiger beetle inhabit the forest floor, though species in the genus *Ctenosoma* live at middle levels of the forest, where they run rapidly along small branches. Species in the genus *Oxycheila* are amphibious,

Galerita (Harpalinae). The larvae live in leaf litter and are very active.

Lebia (Harpalinae: Lebiini), a genus with hundreds of species in tropical America.

searching for aquatic prey attached to rocks in streams. Some *Cicindela* species (those in the subgenus *Opilidia*) live on sandy beaches, but they are difficult to observe due to their quick movement and camouflage coloration.

Among adult tiger beetles, many species actively seek out prey, while others are sit-and-wait predators. The larvae of many species dig burrows in the ground and wait at the entrance for passing insects, pulling their victims down into the tunnel. Larvae in the genus *Pseudoxycheila* dig their burrows in vertical clay banks while larvae that inhabit vegetation dig their burrows in rotten wood.

There are about 40,000 named species of ground beetle worldwide.

Calleida scintillans (Lebiini). Some species in this genus are predators of caterpillars.

Calophaena ligata (Harpalinae) lives inside the rolled leaves of *Calathea lutea* (Marantaceae).

Pseudoxycheila tarsalis, an essentially flightless tiger beetle occurring at mid elevations from Costa Rica to Colombia.

Cicindela (*Opilidia*) *macrocnema*, a common tiger beetle on sandy beaches, where it is well camouflaged.

Ctenostoma longipalpe (Cicindelinae). Unlike other tiger beetles, larvae in this genus live in rotting wood.

CARRION BEETLES

Only two genera of carrion beetle (Silphidae) occur in tropical America: *Oxelytrum*, a genus that appears to have originated in South America (with one species reaching Central America and Mexico), and *Nicrophorus*, a genus that originated in northern temperate regions and later (over eons) colonized tropical America.

Species in the genus *Oxelytrum* lay their eggs in the carcasses of larger mammals.

Species in the genus *Nicrophorus* (known as burying beetles) lay their eggs in the carcasses of smaller animals such as mice and small birds. Before laying their eggs, the female and male bury the carcass (which can weigh hundreds of times more than themselves); in an underground chamber, the female and male first roll the carcass into a ball and strip it of its fur or feathers. Then the female lays eggs. Both parents apply anti-microbial secretions to the dead animal in order to slow down putrefaction. Although the larvae are capable of feeding directly on the carcass, they grow faster when fed regurgitated food by their parents.

Worldwide there about 200 named species of carrion beetle, of which 15 occur in Central and South America.

Oxelytrum discicolle.

Group of *O. discicolle* on dead snake.

Nicrophorus quadrimaculatus feigning death.

N. quadrimaculatus emits a foul odor.

ROVE BEETLES

Rove beetles (Staphylinidae) have extremely short front wings and exposed abdomens that make them look a bit like earwigs. They are adept fliers; on landing, the process of tucking their hind wings underneath the stubby front wings requires a fancy folding action and acrobatic maneuvering. Rove beetles might possibly be the largest family of beetles, with weevils the only close contender.

When threatened, many rove beetles arch their rear end over their back in a menacing, scorpion-like manner. The vast majority are nonetheless harmless to humans, the main exception being members of the genus *Paederus*, which can cause severe blistering if a beetle makes contact with the skin. The responsible compound is pederin, synthesized in adult females by bacterial symbionts.

Adults and larvae of most rove beetle species eat fly larvae and other insects living in leaf litter, decomposing tree trunks, mushrooms, rotting fruits, dung, carrion, and animal nests. Several species live in termite and ant colonies, often by means of chemical camouflage. Some species live with army ants and are good mimics of their hosts, which probably helps protect them from predators—if you run with army ants and look like an army ant, you are probably fairly safe from attack. Species in the genus *Amblyopinus* inhabit the fur of field mice, where they eat ticks and lice.

Some rove beetles have deviated from the predatory habits of the majority. Species in the genus *Oxyporus*, for example, eat fungi (where they also excavate egg chambers). Similarly, species in several genera of the subfamily Osoriinae live under bark and feed on yeasts and other fungi. Species in the genus *Charoxus* enter fig fruits just after the fig wasps (p. 161) have chewed an exit hole; adults apparently feed on wasps that still remain inside, but the larvae feed on fig pollen. Those in the genus *Bledius* excavate tiny tunnels in silty sand next to water, where they feed on microscopic algae; a few species even inhabit the intertidal zone and wait out each high tide inside their burrow.

There are about 56,000 named species of rove beetle worldwide.

Paederus can cause blisters.

Paederus, with raised abdomen typical of many rove beetles.

Leistotrophus versicolor. Adults feed on flies that come to dung and carrion.

Bledius tunnels on sandy beach, with dog tracks.

SCARAB BEETLES

Members of the superfamily Scarabaeoidea are recognizable by a unique club at the tip of each antenna. This antennal club, which is used to detect odors, consists of three small plate-like segments. Species vary from minute beetles that are 2 mm (.08 in) in length to the most massive insects in the world. Some have brilliant metallic coloration but most are rather drab nocturnal beetles. Many of the latter have an auditory organ (a thin membrane) in the neck region with which they detect the ultrasonic sounds emitted by bats.

There are nearly 32,000 named species of scarab beetle worldwide.

Chrysina aurigans (Scarabaeidae: Rutelinae).

Heterosternus oberthuri. Like a few other species of Rutelini, the male has enlarged hind legs, probably used in combat with other males.

Female *H. oberthuri*.

BESS BEETLES
(superfamily Scarabaeoidea)

Bess beetles (Passalidae) are rather uniform-looking beetles that excavate tunnels in tree trunks. A female or male locates a suitable log (one that has been dead for at least a year), starts excavating, and is joined by a member of the opposite sex. The mated male and female rear their young from larvae to adulthood, a behavior that is extremely rare among beetles.

Larvae feed both on wood that has been chewed up by the parents and on the feces of adult beetles. Adults feed on their own excrement, which is usually plastered to the walls of their tunnels. Various bacteria, yeasts, and even a few protozoans are present in the hind gut of adults; some of these microbes are excreted along with the feces and continue digesting the wood in the excreted feces. In eating this microbe-rich excrement, the larvae obtain much more nitrogen than the amount present in the original cellulose, and they consequently often reach maturity in just two or three months, a record time for a large, wood-eating insect. When the larvae are ready to pupate, their parents assist them in constructing a protective case made from shredded wood and feces. Newly emerged adults often help care for the larvae.

Both adults and larvae make squeaking noises by rubbing one part of their body against another part, thereby producing about a dozen different calls. The English common name derives from the *bess...bess...bess* sound they make, or so it is said. In Mexico, their common name is *escarabajo cantor* (singing beetle).

There are about 800 named species of bess beetle worldwide.

Veturius sinuaticollis, like other bess beetles, has a gap between the wing covers and the front part of the thorax.

Larvae of bess beetles, such as these *Passalus punctatostriatus*, produce sounds by rubbing the hind leg against the base of the second leg.

DUNG BEETLES
(superfamily Scarabaeoidea)

Dung beetles are in the subfamily Scarabaeinae (family Scarabaeidae). Most species feed on excrement from a variety of mammals, especially herbivorous mammals, but some species prefer the droppings of a specific mammal species.

Dung provides a concentrated source of nutritious food for various beetle and fly species, but, as it occurs in widely scattered patches, the insects must expend energy to find it. And, because it dries out quickly and because many other insects search out the dung as food, it is essential that dung beetles locate a fresh pile of dung as quickly as possible. To do this they rely on highly sensitive antennae that can pick up a whiff of dung almost as soon as it is deposited. To ensure even more certain access to dung, a couple of species of dung beetle reside more or less permanently in the fur of sloths, and, taking matters a step further, a few species place themselves near the anus of monkeys, falling with the dung as the monkey defecates. Adult dung beetles use their mouthparts to squeeze and suck juice from the excrement; the larvae eat plant fibers and other undigested particles in the dung, and then rely on microorganisms in their digestive tract to help break down the cellulose in the plant fiber.

In most species, adults excavate tunnels in the soil below or adjacent to the dung. These tunnels lead to brood chambers, where the eggs are deposited. The female does most of the digging, while the male stocks the brood chambers with dung that he gathers during repeated trips to the surface. The dung is first formed into a ball and then placed in a chamber; finally, the female places a single egg on top of the ball so that the larva has ready access to food. Mating occurs in the tunnels; in most species, males have horns to defend against other males.

Most dung beetles have very stout bodies and thick legs adapted for burrowing. But species in the genus *Canthon* and a few other species have more flattened bodies and long slender legs. These are the dung rollers, species that laboriously roll a dung ball some distance away from the original pile. The male often begins the process by pushing the ball with his hind legs, and is soon joined by a female who assists by pulling on it. Once a suitable nesting site is located, they mate, bury the dung ball, and then the female lays an egg on it. In various species, both burrowers and rollers, the female remains and cares for the developing larvae by assuring that the dung ball does not dry out, applying antifungal secretions, and fending off competitors and predators.

Dung beetles provide valuable ecosystem services by reducing the breeding grounds of filth flies, fertilizing and aerating the soil, and by increasing the germination rate of seeds that are present in the excrement. (We know that beetles have been doing such work for a long time, and that ancient species of scarab beetles buried the dung of herbivorous dinosaurs.) Although most dung beetles feed on dung, several species variously eat dead animals, rotting fruit, fungi, or moldy leaves; the most unusual feeding habits are those of two South American species: *Zonocopris gibbicollis* rides on ground snails in order to feed on their mucus and *Deltochilum valgum* eats live millipedes!

There are more than 5,000 named species of dung beetle worldwide.

Deltochilum mexicanum, with the thin legs of a dung roller.

Dichotomius annae, with legs for digging.

Deltochilum mexicanum rolling a ball of dung.

JUNE BEETLES AND THEIR KIN
(superfamily Scarabaeoidea)

Beetles in the subfamily Melolonthinae (family Scarabaeidae) most often feed on plants, the adults usually on foliage, the larvae (white grubs) on roots. The best known members of this subfamily are the June beetles (genus *Phyllophaga*); also called May beetles, they are named for the months in which they most frequently appear. These nocturnal, stocky, brownish beetles are often seen flying clumsily into night lights. Several species of June beetle cause serious damage to crops. June beetles are restricted to the New World.

In contrast to most other Melolonthinae, species in the genus *Macrodactylus* are diurnal and brightly colored. Adults in this genus feed not only on foliage but also on flowers.

The subfamily Melolonthinae is one of the largest groups of scarab beetle, comprising about 11,000 named species worldwide.

The life cycle of *Phyllophaga* usually lasts a year and adults only appear at the beginning of the rainy season.

Phyllophaga feeding on *Trichilia* (Meliaceae).

Macrodactylus female feeding while mating; the pair can remain like this for days.

SHINING LEAF CHAFERS
(superfamily Scarabaeoidea)

The subfamily Rutelinae (family Scarabaeidae) includes some of the most spectacularly colored scarab beetles, especially those belonging to the genus *Chrysina* (previously known as *Plusiotis*). Species in this genus inhabit montane forests from the southern United States to Ecuador.

The metallic (iridescent) silver, gold, and shiny green colors are due to the microscopic structure of their cuticle rather than to pigments. The green species are among the very few insects to reflect circularly polarized light, meaning that the color does not change when viewed at different angles. The gold or silver species reflect most or all wavelengths of light simultaneously. Unfortunately, like so much that glitters, these beetles have attracted the attention of black market traffickers.

Adults of Rutelinae feed on leaves and flowers. The larvae of some species (e.g., those of *Chrysina*) feed on decomposing plant material, such as rotting tree trunks, whereas others (e.g., *Anomala*) feed on plant roots.

The subfamily Rutelinae comprises about 4,100 species worldwide.

Chrysina boucardi. To avoid predators, *Chrysina* beetles blend in with green foliage.

C. aurigans in flight.

Ventral view of *C. aurigans*.

Platycoelia humeralis.

Chrysina aurigans. Both parts of the name mean gold.

Phalangogonia sperata.

The genus Macraspis occurs from Mexico to Argentina and includes more than 40 named species.

HORNED SCARABS
(superfamily Scarabaeoidea)

Horned scarabs belong to the subfamily Dynastinae (family Scarabaeidae). Males of some species (especially those in the tribes Oryctini, Agaocephalini, and Dynastini) have very impressive horns. Especially diverse in tropical America, this subfamily includes some of the largest of all insects. Two examples are the elephant beetle (*Megasoma elephas*), which occurs from southern Mexico to Venezuela, and the Hercules beetle (*Dynastes hercules*), which occurs from southern Mexico to Bolivia. True to its name, the Hercules beetle is so strong that it is capable of escaping from a bird cage by bending the wires. The front wings of the Hercules beetle change color according to changes in humidity levels, becoming more brightly colored during the day (less humidity) and duller at night (with more humidity the spongy layer becomes impregnated with water, letting light penetrate deeper and become absorbed).

The female lays her eggs in a rotting log; shortly thereafter, or sometimes preceding the female, a male establishes his territory on that very same log. Should another male appear on the scene, the two males will joust with their enormous horns, the winner successfully using his horn to pry his rival off the log. Large males have disproportionately large horns, whereas small males (resulting from undernourished larvae) have disproportionately smaller horns; unable to compete, small males will sometimes attempt to sneak up on a female as her two large male suitors are engaged in fighting.

Generally, adults feed on overripe fruit or on sap that exudes from tree wounds. However, in species in the tribe Cyclocephalini (whose members lack horns), adults feed on flowers, especially those of aroids (Araceae). Indeed, many *Dieffenbachia*, *Philodendron*, and *Syngonium* species depend on these beetles for pollination. In many cases the flowers actually heat up, enticing the beetles with a warm, sweet smelling place to eat (the plant provides them with nutritious, sterile stamens) and to perhaps

Dynastes hercules male. The top horn arises from the thorax, the bottom one from the head.

encounter a mate. After the beetles enter the flower, the floral sheath encloses them, making them prisoners for 24 hours; once the beetles have pollinated the female flowers and become coated with pollen, the plant ensures that the beetles do not over extend their stay: In some cases, the sheath squeezes the beetles out, while in others it begins filling up with water.

A similar pollination mechanism occurs in night-flowering water lilies (Nymphaeaceae). For example, the giant Amazon water lily *Victoria amazonica* (named for Queen Victoria), which has floating leaves that grow to a diameter of 2.5 meters (8 feet) in quiet backwaters of Amazonian tributaries, is pollinated by species of *Cyclocephala*. It has been estimated that 900 species of heat-producing plants (including certain water lilies, aroids, and palms) in tropical America are pollinated by this group of beetles.

Worldwide, there are about 1,500 named species in the subfamily Dynastinae.

Dynastes hercules males can reach 17 cm (6.5 in) in length.

D. hercules females lack horns.

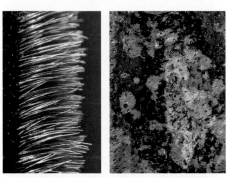

D. hercules male. Close-up of hairs on underside of horn (left); close-up of wing cover or elytron (right).

Megasoma elephas males can reach 12 cm (5 in) in length. The pale brown color is due to a coat of fine hairs.

The larvae of *Megasoma elephas*, like those of most scarab beetles, have a thick, C-shaped body.

METALLIC WOOD-BORING BEETLES

These beetles (Buprestidae), also known as jewel beetles, nearly always have at least some metallic glint on their body. Several species produce a bitter chemical (a glucoside called buprestin) that deters predators.

Some species feed on the foliage of their host plants, while other species visit flowers to feed on pollen and nectar. The most commonly noticed species are bullet-shaped, and their somewhat flattened larvae frequently feed on vascular tissue between bark and sapwood; they mostly prefer dead or dying branches, but some species (like the emerald ash borer *Agrilus planipennis* of North America) attack healthy branches and can cause serious damage. The larvae of many metallic wood-boring beetles excavate wide flat tunnels that are quite different from the more circular holes made by wood-boring beetles in other families.

With a length of 6 cm (2.4 in), the ceiba borer (*Euchroma gigantea*) is the largest species in the family. Its larvae bore into the relatively soft wood of dead balsa, kapok, and related trees in the Malvaceae family. Some indigenous people of tropical America use this gaudy insect to make jewelry and to adorn garments.

At the other end of the spectrum are the flattened, rhomboidal-shaped beetles, which are only 2 to 6 mm (.08 to .24 in) in length. The larvae of these tiny beetles are leaf miners (p. 202), and they feed on a diversity of plants, although any given species is restricted to a small number of plant species.

There are nearly 15,000 named species of metallic wood-boring beetle worldwide.

Hylaeogena, a very small Buprestidae.

Euchroma gigantea. The iridescent colors are due to the structure of the cuticle, the yellow to miniature scales.

Chrysobothris.

CLICK BEETLES

If a beetle ends up on its back, it struggles to regain an upright position. Click beetles (Elateridae) have on their thorax a hinge mechanism that stores elastic energy when it is locked; but when the hinge is released (producing a *click* sound), the beetle abruptly snaps upward in the air, with a fifty-fifty chance of landing feet first. Another function of the click mechanism might be as a defense against predators, since a beetle that snaps back and forth is more difficult to grasp. A few adult click beetles (those in the tribe Pyrophorini) have a pair of glowing "headlights" that probably serve to startle potential predators.

Adults ingest a variety of foods, depending on the species: pollen, nectar, rotting fruit, soft-bodied insects, and fungi. The larvae are elongate and their cuticle is tougher than that of most beetle grubs, hence the common name *wire worm*; a few species feed on roots

Semiotus, showing the "locked" position.

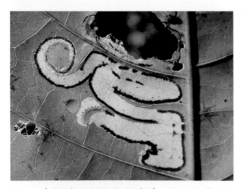

Gall formed by *Hylaeogena* on *Amphilophium* (Bignoniaceae).

Semiotus, showing the "unlocked" position.

Larva of *Brachys insignis* in leaf mine on *Genipa americana* (Rubiaceae).

of grass seedlings and sprouting seeds, but most are probably opportunistic predators of other invertebrates inhabiting leaf litter and rotting logs.

There are about 10,000 named species of click beetle worldwide. In addition, there are about 1,500 species of false click beetles (Eucnemidae) that also click.

Chalcolepidius bomplandi. Larvae in this genus feed on beetle larvae and termites in decomposing wood.

Pyrophorus showing the "headlights."

Dilobitarsus bidens.

Pyrophorus seen from below, showing two "headlights" on thorax and a small orange glow between the thorax and abdomen.

Semiotus cuspidatus. Larvae in this genus are predators that live in rotting wood.

False click beetle (Eucnemidae), with a more rounded head.

Click beetle on its back, before the jump.

Click beetle during the jump.

FIREFLIES

Fireflies (Lampyridae) are also called lightning bugs and, in Spanish, *luciérnagas*. Most species produce in organs at the tip of their abdomen a light that attracts a member of the opposite sex. Males usually fly about and flash a species-specific light pattern while females remain perched on vegetation as they assess the passing males. If a particular male catches a female's fancy, she responds with a single flash; fascinatingly, the exact duration of time between the end of the male's display and the moment when the female gives her reply lets the male know that the female was responding to him and not another. Females probably choose a male, at least partly, on the qualities of his luminous signals—certain aspects of his signal (e.g., flash duration) are sometimes correlated with how much protein he carries in his sperm package. Some firefly species merely glow rather than flash, and a few produce no light at all, relying exclusively on chemical communication (via pheromones).

Fireflies manufacture defensive toxic steroids (lucibufagins) that are similar to those of toads. These steroids are made from the cholesterol that they obtain in their food. But converting cholesterol to steroids is expensive, as cholesterol is also needed for making many other important compounds. Female fireflies in the genera *Bicellychonia* and *Photuris* forgo this expense and obtain the defensive steroids by eating the males of other firefly species, which they lure in by sending a response flash to the males' courtship display.

Firefly larvae, which can be recognized by their plate-like segments, inhabit soggy habitats and are voracious predators; they have piercing mandibles with grooves for injecting enzymes to break down their food before slurping up the partially digested brew. Some prey on snails, some on earthworms, while others are generalists. All species have a pair of lights at the rear end; somewhat confusingly, these larvae are called *glowworms*, a term that is also applied to members of a closely related family (Phengodidae, p. 119). Like adult fireflies, their larvae produce toxic steroids, and their lights probably serve to warn potential predators of their toxicity.

There are 2,200 named species of firefly worldwide.

Photuris crassa. Females of several species in this genus attract male *Photinus* with light flashes and then eat them.

Photinus. Larvae of this genus are more subterranean than those of *Photuris*.

Aspisoma. Larvae of this genus feed on snails.

Firefly larva sticking to a leaf with its rear end.

Mating *Photinus*.

Little Lights in the Vast Darkness

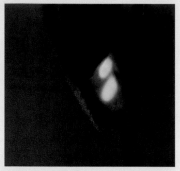

Firefly emitting light from its abdomen.

One can read by the light of *Pyrophorus* (Elateridae).

Pyrophorus (Elateridae) showing light-producing area (white area between hind legs) on underside of body.

Animals that glow in the dark do so by means of a process called *bioluminescence*. Light is produced via a complex chemical reaction that takes place within the animal. There is something about luminescent creatures that has inspired human cultures around the world. The Aztecs, for example, saw fireflies as symbols of flashes of perception, or knowledge, surrounded by a vast darkness, or ignorance.

Bioluminescence has evolved independently more than 30 times, mostly in deep-sea animals, but also in a few mushrooms, protozoans (dinoflagellates), and bacteria. Among the insects capable of producing light this way are adults and larvae of fireflies (Lampyridae), one group of click beetles (Elateridae), adult females and larvae of glowworm beetles (Phengodidae), and a few fly larvae in the family Keroplatidae (p. 277).

Most luminescent organisms produce a steady or a slowly modulated glow. Many adult fireflies, however, have the unique ability to precisely control the timing of their flashes. Evidently they do this by the following sequence of events: first the nervous system triggers the release of nitric oxide in the light producing organs; this inhibits cellular respiration, thereby resulting in the accumulation of oxygen; when the oxygen reaches a certain level, it triggers the chemical reaction that produces light.

A few adult click beetles, in the tribe Pyrophorini, also produce light. Unlike many fireflies, these click beetles produce a constant glow, from both a pair of greenish yellow (or orange-yellow) "headlights" near the front end of the body and from an orange (or green) light on the underside of the body. Indigenous peoples of the American tropics would tie a couple of adult beetles to each big toe in order to light their way at night and often smear beetle paste onto the face in order to make a more striking impression. The indefatigable explorer Alexander Humboldt noted that a dozen beetles placed in a perforated gourd make an acceptable reading lamp.

Today, biologists study bioluminescence with keen interest in applications for biotechnology.

GLOWWORM BEETLES

Glowworm beetles belong to the family Phengodidae. Males are nocturnal creatures with soft-bodies, feathery antennae, and, in most species, short front wings. Because the females and larvae live concealed in rotting vegetation, it is difficult to say whether they are nocturnal, diurnal, or both. Females resemble their worm-like larvae, and adult females and larvae both emit a yellowish-green light from a row of spots running down each side of the body. In addition, members of the subfamily Phengodinae emit light from transverse bands on the back, while those in the subfamily Mastinocerinae emit a reddish light from the head. Light production probably wards off potential predators by advertising toxicity, though it is speculated that species that possess lights on the head may use these to better see in the dark. (In some species the males also produce light.)

Males use their feathery antennae to "smell"—and locate—females. Females lay a group of eggs, and then curl their body around the eggs to protect them; as the eggs develop, they too begin producing light.

Females and larvae are predators of millipedes and other arthropods; they inject a paralyzing substance into the underside of the millipede's neck, wait for it to become immobile, and then suck out the contents, segment by segment, all the while carefully avoiding the noxious defensive glands. When finished eating, nothing remains except a pile of rings, representing the body segments of the millipede. Adult males apparently feed mostly on nectar, despite their formidable looking mandibles, which are probably used during mating. Glowworm beetles are restricted to the New World, where there are 250 named species.

Male Phengodidae.

Female Phengodidae, which resembles the larva.

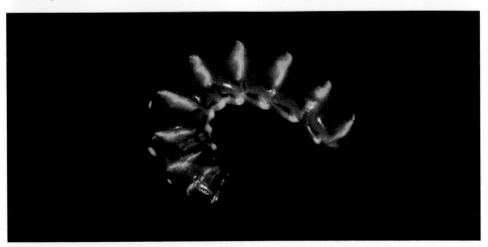

Female Phengodidae emitting light.

NET-WINGED BEETLES

Net-winged beetles (Lycidae) are soft-bodied, flattened insects whose front wings are criss-crossed by ridges, forming a netlike pattern. They harbor distasteful chemicals and advertise this fact with bright colors, especially red or yellow. Many other insects, from beetles to moths, mimic these beetles, hoping to scare off predators that mistakenly assume the mimics are also toxic.

Adults occur on flowers and leaves, where they feed on nectar and honeydew. The larvae inhabit rotting wood, or occasionally leaf litter, where they feed primarily on myxomycetes (fungus-like masses of amoebas).

There are 4,600 named species of net-winged beetle worldwide.

Mesopteron. Note the net-like pattern on the front wings.

Caenia, a genus with branched antennae.

SOLDIER BEETLES

Soldier beetles (Cantharidae) are quite similar to fireflies in several aspects (they have relatively soft bodies, for example), though they are not bioluminescent. Many species are yellow-orange and black (sometimes almost entirely bluish-black).

Adults of many species feed on pollen and nectar, but some species prey on soft-bodied insects. In one group (tribe Silini), the males appear to secrete from the first thoracic segment substances that females relish.

Adults and larvae possess paired glands on the upper part of the abdomen that secrete defensive compounds. The larvae have a velvety appearance due to a dense coat of fine, water repellent hairs; larvae of most species inhabit humid microhabitats such as leaf litter, loose soil, and decaying wood, but some larvae forage on open ground. Soldier beetle larvae are primarily predators that suck out the liquefied contents of their prey, although species in the genus *Silis* probably also eat plant matter.

There are 5,100 named species of soldier beetle worldwide.

Some *Chauliognathus* sequester alkaloids from *Senecio* flowers (Asteraceae).

Chauliognathus, showing warning coloration.

Belotus species have shortened front wings.

SAP BEETLES

Sap beetles (Nitidulidae) are rather small, drab beetles distinguished mainly by a very pronounced club on the tip of the antenna. Though many are flattened they come in a range of body shapes. On some species, the front wings are very short (which makes these beetles look similar to rove beetles); on other species, the front wings are longer and cover the abdomen.

The vast majority of sap beetles feed on a variety of fungi, including mushrooms, bracket fungi, microfungi concealed in senescent palm stalks, and fungi growing in leaf litter. As the common name suggest some sap beetles are attracted to fresh tree wounds and sap flows where they feed on yeasts and other fungi, while others feed on decaying flowers or fermenting fruits.

Several sap beetles feed on pollen and some of these serve as pollinators. For example, members of the tropical American tribe Mystropini pollinate certain palms, while others act as pollinators of aroids and Annonaceae. In some cases pollination by sap beetles is a mixed blessing for the plant, since they also lay eggs and their larvae eat the fruit.

There are 4,000 named species of sap beetle worldwide.

Conotelus (Nitidulidae) in flower of *Malvaviscus* (Malvaceae).

Larva of Nitidulidae in *Malvaviscus* flower.

PLEASING FUNGUS BEETLES

More than 20 families of beetles feed primarily on fungi (and many more feed on plant material that has been modified by fungal enzymes). While most of these fungus eating beetles are drab, members of the family Erotylidae sport delightfully gaudy colors and are thus known as the pleasing fungus beetles. In addition to being colorful, many pleasing fungus beetles are oval-shaped and very convex.

Both adults and larvae feed on fungi (mainly Basidiomycetes), especially fungi growing on dead wood, and in some species the larvae live in groups and sometimes even pupate together in tight clusters dangling from a log. In species belonging to the genus *Pselaphacus*, the female herds the larvae from one fungal patch to another under cover of darkness and then leads them to a hiding place at day break. As a result of her maternal care, the larvae reach maturity in just a week or two.

The typical pleasing fungus beetles belong to the subfamily Erotylinae, but the family Erotylidae also includes a few subfamilies that are quite different, both in appearance and biology. For example, many species in the subfamily Languriinae are elongate beetles, and either dark or metallic green; the adults often feed on pollen while the larvae of many species feed inside the stems of herbaceous plants. Species in the subfamily Pharaxonothinae are small, drab-colored beetles that are associated with cycads: the adults feed on pollen and are the main pollinators of these plants, while the larvae feed inside the male cones.

The family Erotylidae comprises about 3,500 named species worldwide.

Aegithus species graze on bracket and crust fungi growing on decomposing wood.

Iphiclus catillifer. Species in this large heterogenous group feed on fungi on decomposing wood.

Erotylidae mating on the underside of a mushroom.

Languriinae. The larvae are stem borers.

Dense aggregation of *Stenotarsus subtilis* (Endomychidae).

HANDSOME FUNGUS BEETLES

Another family that feeds mostly on fungi (Basidiomycetes) is Endomychidae. In Central America, adults of *Stenotarsus subtilis* form large aggregations of up to 70,000 individuals at the base of trees or on rock faces. Formed during the dry season, these aggregations can last for many months, during which time the beetles do not feed; apparently, by remaining in aggregations, the beetles are better able to conserve water (and perhaps amplify their chemical defenses). The beetles sometimes use the same location in consecutive years, even though the aggregation is formed by a new generation of adults each year.

There are about 1,800 species of handsome fungus beetle worldwide.

LADY BEETLES

Members of the family Coccinellidae have a variety of common names, including *lady beetle*, *ladybug*, and *ladybird*. They vary dramatically in size (from 2 to 11 mm/.08 to .43 in) and in color (from black to brightly colored).

Depending on the species, lady beetles (both adults and larvae) feed on aphids, scale insects, mealybugs, whiteflies, spider mites, or leaf beetle larvae. Because many species are fairly specialized predators, lady beetles are used in biological control more frequently than any other group of predatory insects. As a result, several species have been transported

from one country to another for the purpose of reducing local pest populations, often with great success. However, a few species of lady beetles are not predators: some species feed on powdery mildews, while those in the tribe Epilachnini feed on plant foliage, including that of beans and wild nightshades.

Many of the insect prey mentioned above (aphids, scale insects, mealybugs) excrete honeydew, which in turn attracts ants. Some of these ants protect the honeydew-excreting insects against predators such as lady beetles. This means that lady beetles have to be prepared to confront irate ants. To defend itself, the adult beetle can clamp its shiny, hemispherical body tightly against the leaf, making it difficult for the ants to get a purchase on the beetle with their mandibles. Some ladybird larvae are protected by a thick coating of wax that covers their bodies. The larvae and/or adults of several lady beetles synthesize defensive alkaloids, which are secreted from the dorsal surface of the abdomen of larvae, and from the leg joints of adults (appearing as yellow droplets). There are around 6,000 named species of lady beetle worldwide.

Harmonia axyridis, an introduced species, with fungus (Laboulbeniales) growing on the rear end of its body.

Epilachna abrupta feeding on wild nightshade (Solanaceae).

Anovia punica. Many members of the tribe Noviini feed on scale insects in the family Monophlebidae (p. 94).

Mating *Brachiacantha*

Pupa of *Harmonia axyridis*.

Larva of *Cycloneda sanguinea* feeding on somewhat toxic aphids (*Aphis nerii*).

Larva of *Epilachna abrupta*.

Larva of *Scymnus* with wax secretions (resembling a mealybug, p. 94).

DARKLING BEETLES

Darkling beetles (Tenebrionidae) are generally dark nondescript beetles, though some species (Alleculinae, Lagriinae, Epitragini) sport metallic colors and a few species even have horns.

Many species resemble ground beetles (p. 101). But unlike the predatory ground beetles, darkling beetles (both adults and larvae) feed primarily on dead plant and animal material; a few species are specialized fungus feeders. Some are pests of stored grains; grains are generally placed in a very dry location, and darkling beetles that live in stored grains are adept at conserving water in their body.

In some Latin American homes, living colonies of *Ulomoides dermestoides* (a species that originated in Asia) are maintained in jars with peanut shells. Asthma sufferers plop a couple of live beetles into yogurt—or mix them in a blender with milk—and then drink the resulting concoction as a home remedy. Although the efficacy of this treatment is still being studied, darkling beetles in general are known to secrete a wide array of chemical compounds. Some of these pungent, vile smelling defensive compounds are secreted from the rear end, which the beetle elevates before taking aim.

A closely related family (Zopheridae) includes the ironclad beetles, named for their incredibly hard bodies. They inhabit decomposing tree trunks. In the Yucatan and Guatemala, live ironclad beetles are kept as good luck charms and as reminders of the young Mayan prince who was saved from capture when the Moon God turned him into this beetle.

There are about 20,000 named species of darkling beetle worldwide.

Like bess beetles, *Phrenapates* species live in colonies in decaying wood.

Tribolium castaneum (Tenebrionidae), a pest of stored grains (shown here on rice) and grain products, hence the name "flour beetle".

Strongylium auratum. Larvae of this large genus live in decaying wood.

Larva of darkling beetle on rotten wood.

Hegemona.

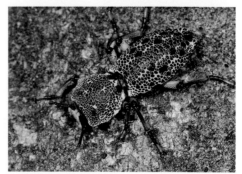

Ironclad beetles (Zopheridae) such as this *Zopherus jansoni* are flightless.

BLISTER BEETLES

In addition to their bizarre chemical properties (p. 129), blister beetles (Meloidae) have a fascinating natural history.

In most species, larvae are bee kleptoparasites; they feed on the pollen that bees store in their nest, with the result that juvenile bees die of starvation. How the larvae enter the bee's nest varies according to species. In the genera *Lytta* and *Pyrota*, the female beetle buries her eggs near the nest entrance and the larvae must work their way inside. In most other species, the female lays hundreds to thousands of eggs on flowers, and the larvae that hatch from these eggs wait for passing bees. Many of them probably die waiting, but the fortunate few who survive hop on a bee, get carried back to the nest, and then hop off. The recently hatched larvae of one species in the southwestern U.S. go one step further. They form a teaming mass that produces a scent like that of a female bee. When a duped male approaches, the beetle larvae hop aboard, ride him until he mates with a female, and then transfer to the female (males do not participate in nesting activities).

The adults in the subfamily Nemognathinae feed on pollen and nectar, while adults in the subfamily Meloinae feed on foliage. There are 3,000 named species of blister beetle worldwide.

Epicauta. Unlike other blister beetles, the larvae of this genus prey on grasshopper eggs.

SPANISH FLY

Members of the genus *Meloe* are typically flightless. Like all Meloidae they exude blister-causing droplets from their joints when disturbed.

Blister beetles can cause blisters if you handle them, but they can produce tumescence of another sort in male humans. If a concoction of pulverized beetles is taken orally, the active ingredient, cantharidin, eventually passes into the urine, causing inflammation of the genitals. It is this property that has made *Lytta vesicatoria*, a European blister beetle with the nickname *Spanish fly*, perhaps the most infamous aphrodisiac in history.

In Roman times, Nero's wife slipped it into the food of guests. The Marquis de Sade narrowly escaped the gallows for allegedly giving an overdose to a group of prostitutes. As he discovered, there is only a slight difference between an effective dose and a lethal dose. A French Foreign Legion physician reported a case in which soldiers suffering painful, prolonged erections after dining on frog legs; unbeknownst to the soldiers, the frogs had recently fed on blister beetles. Nonetheless, cantharidin does have potential medical applications; it is effective in removing warts and is known to have antitumor properties.

When threatened by a predator, blister beetles exude droplets of cantharidin, although these do not seem to deter certain spiders, mantises, and frogs. Blister beetles have relatively soft bodies and some species have shortened forewings that expose most of the abdomen, evidence that their chemical defenses are sufficiently effective to allow them to forgo the evolutionary cost of developing hard shells and longer forewings.

Adult males produce relatively large quantities of cantharidin; every time a male mates he passes (in his sperm packet) most of the cantharidin to his female mate. In some species, the female passes it along to the eggs. Larvae often produce small amounts of cantharidin.

As far as is known, only blister beetles and members of a related family (Oedemeridae) are able to synthesize cantharidin. However, a variety of other insects, including male ant-like beetles (Anthicidae), obtain the chemical, apparently from live or dead blister beetles, and use it to attract females. The latter acquire the precious chemical during copulation and use it to protect their eggs.

LONGHORN BEETLES

Members of the family Cerambycidae are equipped with antennae that are often much longer than their body. Species vary in body size from just 3 millimeters (0.1 in) to 17 centimeters (6.7 in) in the case of the Titan beetle (*Titanus giganteus*) from South America, which competes with some horned scarabs (p. 111) for being the largest beetle in the world. And these beetles are equally diverse in coloration; nocturnal species tend to be drab, while diurnal species are generally colorful. Diurnal species in the tribe Callichromatini, for example, are metallic green (and give off a pungent, vanilla-like odor).

Depending on the species, adults feed on foliage, flowers, tender bark, or seeping sap. In most species, the larvae are wood-borers in debilitated trees or shrubs, many of them feeding on the tissue beneath the bark. However, the larvae of some species feed on other plant tissues in healthy plants. For example, the larvae of species in the genus *Phaea* feed in the healthy stems of dogbanes, while those of *Tetraopes* feed on the roots of milkweeds. Females of one group (tribe Onciderini) girdle small branches in which they lay their eggs, and thus the larvae feed on tissue that was debilitated by the mother.

One of the most spectacular members of this family is the harlequin beetle (*Acrocinus longimanus*). The intricate hieroglyph-like pattern on its back affords it with camouflage on lichen-covered tree trunks, but is very conspicuous on a neutral background. Male harlequin beetles have extraordinarily long front legs, which are used for displacing other males from sites where females come to lay eggs. The larvae of harlequin beetles tunnel beneath the bark of dying or recently dead fig trees. The much smaller (2 cm/0.8 in) three-lined fig tree borer (*Neoptychodes trilineatus*) attacks living fig trees that have been weakened in some way.

There are roughly 30,000 named species of longhorn beetle worldwide.

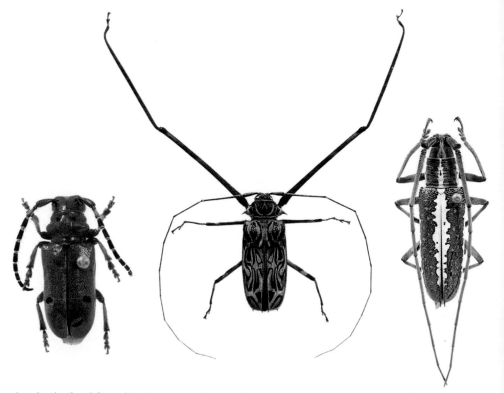

Longhorn beetles (from left to right): *Tetraopes*, male harlequin beetle (*Acrocinus longimanus*), *Neoptychodes trilineatus*.

Parandra polita belongs to the small subfamily Parandrinae, which have short antennae.

Callona rutilans, a diurnal longhorn beetle.

Scatopyrodes.

Neoptychodes cretatus.

Lagocheirus kathleenae. Larvae in this genus feed beneath the bark of dying or dead trees, often in *Bursera* and *Spondias* used as "living fence posts."

Like most longhorn beetles, *L. kathleenae* produces sounds by moving the fore thorax (blurred) over the mid thorax.

LEAF BEETLES

The members of this family (Chrysomelidae) are some of the most commonly seen beetles—the brightly colored adults often sit exposed on vegetation. However, distinguishing between these beetles (which have four tarsal segments) and the gaudy beetles in other families often requires the use of a microscope.

Virtually all leaf beetles feed on plants, both in the adult and larval stages, although the latter often feed while concealed inside plants or under soil. Larvae that feed externally generally harbor chemical defenses. Many species are gregarious, in which case they commonly feed side-by-side, forming a row (this facilitates the mechanics of chewing through tough tissue). When not feeding, gregarious larvae form a protective circle.

The sections that follow present accounts of the principal subfamilies of leaf beetle. There are roughly 33,000 named species worldwide; about half of these species are composed of skeletonizing leaf beetles and flea beetles.

SEED BEETLES
(family Chrysomelidae)

The larvae of seed beetles (subfamily Bruchinae) feed inside seeds, especially those of legumes. Adult males and females eat the pollen and nectar of various plant species; however, females lay their eggs only on specific plant species, depositing them on the surface of seeds or seed pods. Upon hatching the larva bores into the seed, where it usually encounters toxic chemical defenses. As an example, *Caryedes brasiliensis* beetle larvae bore into seeds of the *Dioclea megacarpa* vine, which contain canavanine, a non-protein amino acid. The resourceful larvae, however, are able to break down the canavanine into substances that can be used as a source of nitrogen.

Seed beetles are especially diverse in seasonally dry forests. Probably the most common species is *Acanthoscelides obtectus*, which attacks beans and is sometimes encountered as beetle bits in a plate of cooked beans.

Caryedes brasiliensis. Seed beetles are the only Chrysomelidae having clubbed antennae and notched eyes.

OVAL LEAF BEETLES
(family Chrysomelidae)

The subfamily Eumolpinae is a heterogeneous group of leaf beetles. Many species have a metallic color. Adults feed on leaves and flowers; larvae usually feed on roots. The female lays her eggs in soil, where pupation also takes place. Adults of some *Colaspis* species cause damage to bananas by gnawing on the developing peel.

Megascelis species are mostly associated with legumes (Fabaceae). The larvae feed on roots.

Colaspoides batesi on *Trichilia havanensis* (Meliaceae).

TORTOISE BEETLES
(family Chrysomelidae)

Beetles in the subfamily Cassidinae display a wide range of feeding habits and they vary in shape from parallel-sided to circular. What they have in common is a relatively small head and a small, ventrally located mouth.

Most species in the large genus *Cephaloleia* feed either on young, unfolded palm leaves or on the floral bracts or leaves of heliconias (Heliconiaceae) and prayer plants (Marantaceae). Those that feed on leaves generally do so from inside the rolled-up leaves and their larvae are extremely flattened; both adults and larvae feed by scraping the leaf surface.

Other species of Cassidinae are leaf-miners: the larvae tunnel inside the leaves of their host plant. The only other beetles having this habit are many metallic wood-boring beetles (p. 113) and a few weevils.

Tortoise beetles are among the most colorful of all beetles and at least one species, *Charidotella egregia*, has the unusual ability to turn red when it is agitated. Its hardened front wings normally reflect a metallic-gold color, which results from the mirror-like quality of the fluid inside numerous microscopic channels. When the beetle is disturbed it withdraws this fluid and the wing covers then act like a window, revealing an underlying red pigment.

The larvae of tortoise beetles feed on the foliage of their host plant and use their telescoping anus and forked tail to pile molted skins and/or feces on their backs, which serves to deter predators (the molted skins contain terpenes). Several species feed on morning glories (Convolvulaceae) and some of these have gregarious larvae that are protected by their mother, who sometimes stays with her offspring even after they have pupated.

Alurnus ornatus (Alurini). Members of this tribe feed on palm leaves and are among the largest Cassidinae.

Cephaloleia nigropicta (Cephaloleiini) in rolled leaf of *Heliconia*.

Larva of *Cephaloleia* in rolled leaf of *Heliconia*.

Sceloenopla (Sceloenoplini). Its larva is a leaf miner on *Gunnera* (Gunneraceae)

Larva of Cassidinae mining a leaf. Black spots are feces.

Ogdoecosta catenulata (Mesomphaliini).

Stolas costaricensis (Mesomphaliini).

Charidotella egregia (Cassidini) normal color.

Charidotella egregia color when disturbed.

Stolas lebasii feeds on *Mikania* (Asteraceae).

Terpsis quadrivittata (Mesomphaliini) on morning glory

Species in the genus *Chelymorpha* often have various color forms.

Coptocycla leprosa (Cassidini) on *Cordia* (Boraginaceae).

Ischnocodia annulus (Cassidini) feeds on *Cordia*.

Ventral view of tortoise beetle showing the small mouth characteristic of this subfamily.

Physonota alutacea (Physonotini) on *Cordia curassavica* (Boraginaceae).

Female *Omaspides convexicollis* (Mesomphaliini) guarding her eggs on morning glory.

Larva of *Coptocycla* (Cassidini) on *Cordia*, carrying feces at rear end.

Larva of *Charidotis* (Cassidini) on morning glory.

Larva of *Spaethiella* (Hemisphaerotini), with its basket-like fecal shield, on *Heliconia*.

CYLINDRICAL LEAF BEETLES
(family Chrysomelidae)

Larvae of these small beetles (subfamily Cryptocephalinae) use their excrement to construct portable cases around their bodies, thereby providing them with protection against predators. Adults generally eat foliage while the larvae feed on detritus in the leaf litter. Adults in the tribe Chlamisini resemble caterpillar droppings and eat the bark of living trees, whereas the larvae feed on foliage or leaf litter. Members of a closely related subfamily (Lamprosomatinae) also construct larval cases; some species of *Lamprosoma* feed on the bark of guava.

Species in the tribe Chlamisini are known as warty leaf beetles.

Cryptocephalus trizonatus, in the tribe Cryptocephalini.

Lamprosomatinae on leaf of *Hampea* (Malvaceae).

BROAD-BODIED LEAF BEETLES
(family Chrysomelidae)

Broad-bodied leaf beetles (subfamily Chrysomelinae) are some of the largest leaf beetles, measuring up to 2.5 cm (1 in) in length. Most species select a specific host plant, and pass through all life stages on the foliage of that host plant. The larvae of some species synthesize chemical defenses (terpenoids) in specialized glands; the larvae of other species sequester compounds from the plant and use them as chemical defenses. In some species, the larvae have several pairs of extendible glands along the body, tucked inside conical structures that protrude from the back, giving them a studded appearance; other species have just one pair of glands (and secrete a different set of chemicals).

Stilodes undecimlineata.

Larvae of *Stilodes undecimlineata.*

Calligrapha fulvipes feeds on mallows (Malvaceae).

Zygogramma violaceomaculata.

SKELETONIZING LEAF BEETLES
(family Chrysomelidae)

Like many other insects, a large number of skeletonizing leaf beetles (Galerucinae) devour just the tender parts of a leaf, leaving behind a skeleton of leaf veins. Adults eat the leaves and flowers of the plant, while larvae eat the roots. Species in the genera *Acalymma, Cerotoma,* and *Diabrotica* feed on beans, corn, and squash, crops that were domesticated in tropical America about 10,000 years ago. These beetles have thus been interfering with local agriculture for a very long time.

The leaves of squash and other members of the family Cucurbitaceae contain bitter terpenoids, and the adult beetles sequester these compounds in their bodies as a defense against predators.

The larvae of some skeletonizing leaf beetles feed exposed on the surface of leaves. The gregarious larvae of *Coelomera,* for example, feed on the leaves of *Cecropia*; because this plant is generally defended by ants, the beetles confine themselves to those individual plants that lack ants.

Diabrotica is a New World genus with nearly 400 named species.

Diabrotica.

Paranapiacaba.

Isotes marginella.

Isotes sexpunctata. This genus and those shown on the previous page belong to the tribe Luperini; their larvae are known as "root worms."

FLEA BEETLES
(family Chrysomelidae)

Flea beetles, members of the tribe Alticini, are small compact beetles that can leap out of sight in an instant, thanks to enlarged hind legs that contain a spring-like mechanism. Adults generally feed on foliage, leaving behind tiny circular holes that are very indicative of this group of beetles. In many species the larvae feed on roots, though some are stem-borers and others feed on foliage.

Species that live on leaf surfaces include members of the genus *Macrohaltica*, which inhabit the very large leaves of the poor man's umbrella (*Gunnera*), and members of the genus *Blepharida*, which reside on the resinous leaves of gumbo limbo trees (*Bursera*).

Disonycha reticollis. Larvae in this genus feed on foliage.

Omophoita flea beetle and *Crematogaster* ant.

Disonycha trifasciata feeds on *Byttneria* (Malvaceae)

Eggs of *Disonycha trifasciata*.

Pair of *Macrohaltica* on *Rubus* (Rosaceae).

Macrohaltica on *Gunnera* (Gunneraceae).

Larvae of *Macrohaltica crypta* on *Gunnera*.

Flea beetle on morning glory (Convolvulaceae).

Allochroma sexmaculatum on young leaf of Araceae.

WEEVILS

Weevils comprise the superfamily Curculionoidea. These insects have long snouts that might suggest—to people in a mood to anthropomorphize—a whimsical look. The mandibles are located at the tip of the snout. Most females use the snout and mandibles to drill into and pry open tough plant tissue. After penetrating the plant, the female turns around, deposits an egg in the hole, and then seals it up with a cement-like secretion. Weevil larvae lack legs and larvae of most species spend their lives burrowing in and eating plant tissue.

Weevils affect the majority of seed plant species, although each weevil species generally feeds on just a few closely related plant species. Some weevils feed in seeds, others in stems, and still others in the roots.

There are a lot of weevils in the world. While there are 62,000 named species (15% of all beetles), weevil specialists estimate that the total number in the world is probably about 220,000. The vast majority of known species belong to the family Curculionidae (51,000 species).

LEAF-ROLLING WEEVILS
(superfamily Curculionoidea)

Leaf-rolling weevils belong to the subfamily Attelabinae (Attelabidae). They are fairly host-specific; examples of host plants include Anacardiaceae, Lauraceae, and *Inga*. Females, without the aid of any adhesive such as silk, construct elaborate leaf rolls in a procedure not unlike that used to construct origami. On a young leaf, she first makes a series of strategic cuts, usually beginning by making two cuts near the base of the leaf, from each edge to the midrib. The cut parts curl up, resulting in a complex, barrel-shaped structure that usually detaches from the plant. The female lays just one egg per leaf roll and she usually does this before the roll is completely formed. When the larva hatches, it feeds on fungi that grow on the dead leaf tissue inside the leaf roll. The females of most leaf-rolling weevils infect the withering leaf tissue with a fungus, the spores of which they carry in special

pockets next to their hind legs. Other members of the family Attelabidae simply sever the sap flow in a stem, bud, or leaf, and their larvae feed on fungi that colonize the decaying tissues.

Worldwide, the family Attelabidae consists of 2,500 named species.

Omolabus conicollis constructs leaf rolls on guava (*Psidium*).

Omolabus corvinus constructs leaf rolls on *Spondias mombin*.

Leaf roll constructed by female *Omolabus* on *Spondias* (Anacardiaceae).

STRAIGHT-SNOUTED WEEVILS
(superfamily Curculionoidea)

This group of weevils belongs to the subfamily Brentinae (Brentidae). Straight-snouted weevils are characterized by their very long, slender, flattened bodies, and are generally found beneath the loose bark of dead or dying trees. The larvae tunnel in living or dead branches and trunks.

There are about 1,700 named species of straight-snouted weevil worldwide.

Straight-snouted weevil (Brentinae).

Brentus cf. *anchorago*, with an extremely long snout.

PALM WEEVILS
(superfamily Curculionoidea)

Palm weevils are in the subfamily Dryophthorinae (Curculionidae). Their antennae project out from the base of the snout, near the eyes (in most other weevils, the antennae are located part way down the snout). Most species are red and black, though some are entirely black, including the bearded weevil (*Rhinostomus barbirostris*), which is one of the most striking weevils in tropical America, due to its large size and the reddish brown hairs that cover the male's snout (which he uses to stroke the female during courtship).

Nearly all of them feed on a group of flowering plants known as the monocots (palms, bromeliads, orchids, etc.). Palm weevils are most often encountered on the trunks of recently fallen palm trees, which is where they mate and lay eggs. If two males simultaneously encounter a female, they will spar with their snouts, each attempting to dislodge the other from the vicinity of the female.

The American palm weevil (*Rhynchophorus palmarum*), a very large black weevil, is a pest of coconut and oil palms. The female places her eggs in the base of a leaf, and the larvae then bore into the stems. Occasionally the larvae transmit a round worm (nematode) that causes red ring disease (the cut trunk of diseased trees displays a red ring); this often results first in yellowing of the fronds and then rapid death of the tree. The banana weevil (*Cosmopolites sordidus*), originally from Asia, is a pest of banana corms. Several tiny, cosmopolitan weevils (genus *Sitophilus*) have larvae that feed inside stored grain, especially in rice and corn.

Indigenous people of the Amazon essentially cultivate *Rhinostomus barbirostris* and *Rhynchophorus palmarum* weevils. By cutting down certain palm trees, they provide egg-laying sites for the weevils, where they can ultimately harvest the plump larvae.

Bearded palm weevil (*Rhinostomus barbirostris*), close-up of male snout.

Bearded palm weevil (*Rhinostomus barbirostris*) male.

American palm weevil (*Rhynchophorus palmarum*).

Unlike most other weevils, broad-nosed weevils have a short snout. And females, rather than using their snout to drill an egg hole in vegetation, instead either lay their eggs on the surface of leaves or in soil. Larvae that hatch on leaves immediately drop to the ground to feed on roots. Pupation takes place in an underground chamber; on emerging from the pupal stages, the adults have cusps on their mandibles that they use to claw out of the chamber. When they exit the chamber, their cusps drop off.

Adults usually feed on foliage or flowers; both adults and larvae are often less host-specific than other weevils.

The green color on this broad-nosed weevil ("*Exophthalmus*") is due to scales.

The larva of American palm weevil emerges from the trunk and weaves a pupal chamber from plant fibers, situated at the base of the petiole or in debris at the base of the tree.

BROAD-NOSED WEEVILS
(superfamily Curculionoidea)

Broad-nosed weevils are in the subfamily Entiminae (Curculionidae). Most species are medium sized and many are brightly colored (due to scales that cover their bodies)—and often seen on plant foliage. Several species are flightless and reproduce asexually.

"*Exophthalmus*" belongs to a complex of genera that is very diverse in tropical America, especially the Caribbean.

Compsus includes over a hundred named species, mostly from South America.

Mating *Cleistolophus*.

BARK BEETLES
(superfamily Curculionoidea)

Bark beetles are in the subfamily Scolytinae (family Curculionidae). These tiny weevils have lost their snout, no longer needed since the adults bore into plants with their entire body rather than with the snout alone.

Many species, especially in northern temperate regions, bore a tunnel just under the bark of trees, where the females lay their eggs. The majority of these species feed on phloem tissue just beneath the bark and tend to attack debilitated or dying trees. Some species supplement their diet with fungi, which they sometimes carry on their body. In several beetle species, the fungus aids the beetle in killing the host tree; Dutch elm disease, for example, is caused by the fungi carried by bark beetles.

Another group of bark beetles, called ambrosia beetles, carry a fungus in a special organ on their body. The adults excavate short tunnels into the wood and cultivate the fungus there. Adults and larvae feed exclusively on this fungus; ambrosia beetles are usually less host-specific than those bark beetles that feed on the inner bark. In lowland rainforests, ambrosia beetles account for over half of all bark beetle species that occur.

However, not all bark beetles inhabit tree trunks and branches. Several species, including some members of the genus *Scolytodes*, feed inside the petioles of fallen *Cecropia* leaves. Some species, like the coffee berry borer (p. 152), eat seeds. One bark beetle, *Coccotrypes rhizophorae*, has a major effect on populations of red mangrove trees (*Rhizophora*). It burrows into the tree's propagules, young seedlings, and expanding prop roots; the burrows in the prop roots cause the roots to grow at abrupt 90° turns, giving them a step-like appearance.

Several bark beetle species inbreed, whereby sisters mate with brothers, and occasionally mothers with sons. In some species, males are produced from unfertilized eggs and females from fertilized eggs, which allows mothers to control the sex ratio of their offspring (as in wasps, ants, and bees). When there is inbreeding, it makes sense only to produce enough males to inseminate the females, with the result that female offspring sometimes outnumber males by a ratio of 30:1.

Xyleborus vochysiae. Most ambrosia beetles are generalists but this species has only been found in live trees of *Vochysia ferruginea* (Vochysiaceae).

Coccotrypes rhizophorae has an important effect on red mangroves.

Larvae of ambrosia beetles.

Bark beetle tunnels. The central tunnel was made by the adults; the numerous tunnels branching off to either side were made by the larvae.

OTHER WEEVILS
(superfamily Curculionoidea)

The remaining subfamilies of Curculionidae include the majority of weevil species. Below are just a few examples from three of the larger subfamilies.

In the subfamily Curculioninae, larvae in the tribe Anthonomini generally feed in fruit or seeds; the best known species is the cotton boll weevil (*Anthonomus grandis*), an agricultural pest. Most members of the tribe Acalyptini feed on palms or plants of the Panama hat family (Cyclanthaceae); the larvae feed on fruit, seeds, or flowers, while the adults of some species are important pollinators (one species pollinates peach palm (*pejibaye*), an important native crop). An African species (*Elaeidobius kamerunicus*) of the tribe Acalyptini was imported into Latin America to help pollinate African oil palm. In both peach palm and oil palm, the weevil larvae develop in the spent male flowers.

Other members of this tribe (some species of *Cyclanthura*) are pollinators of *Anthurium* (Araceae).

In the subfamily Baridinae, the tribe Baridini is especially diverse in tropical America and their larvae feed mostly in fruits and stems of various plants. Several of them (*Ambates, Embates, Pantoteles, Peridinetus*) are stem borers in pepper bushes (*Piper*) and the adults have the curious habit of resting inside a hole they cut in the leaf, which makes them look like a fallen flower or bird dropping.

Species in the other major tribe, Conoderini (subfamily Baridinae), can be recognized by their large, dorsally situated eyes. Compared with most other weevils, they are quite agile and some of them appear to mimic certain types of ants. The larvae of most species are wood or stem borers and some feed on ant-defended plants such as *Cecropia* (*Lissoderes, Pseudolechriops*) and swollen-thorn acacias (*Helleriella*).

Among the most colorful weevils in tropical American are species in the subfamily Molytinae that belong to the tribes Cholini and Hylobiini, most of which are medium to quite large-sized (up to 4 cm/1.6 in). Members of Cholini are associated with monocots such as bamboos, bromeliads, orchids, palms, and the ginger group (Zingiberales). In the tribe Hylobiini, three species of *Heilipus* are associated with avocado, one occurring in the stems and two in the seeds.

Cholus (Molytinae). Larvae in this genus are associated with monocots, mostly as stem borers.

Cholus (Molytinae) feeding on Asteraceae stem.

Rhyssomatus (Molytinae). The white filaments are probably wax secretions.

Heilipus areolatus (Molytinae).

Tylomus fuscomaculatus (Molytinae).

Heilipus (Molytinae). Larvae of several species in this genus feed in seeds, for example those of Lauraceae.

Cratosomus (Conoderini). Larvae of many species in this genus are borers in living tree trunks and branches.

Eurhinus magnificus (Baridinae). The larva forms galls on *Cissus* (Vitaceae).

Hoplocopturus varipes (Conoderini) on *Xanthosoma* (Araceae) leaf.

Anthonomus monostigma (Anthonomini). The larvae feed in seeds of *Miconia* (Melastomataceae).

Leaf-mining larva of Baridinae on bromeliad. Like other weevil larvae, it lacks legs.

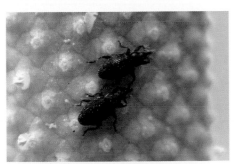

Cyclanthura (Acalyptini) on *Anthurium* flowers.

The color of this *Lixus* (Lixinae) is due to wax secretions.

Phelypera distigma (Hyperinae).

Larvae of *Phelypera distigma* feed in groups on leaves of *Guazuma ulmifolia* (Sterculiaceae).

Beetles, Coffee, and World Trade

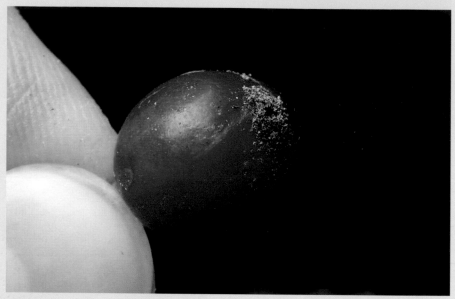

Coffee berry with hole made by the coffee berry borer.

After oil, coffee is the most traded commodity in the world, and is a vital source of foreign capital for many Latin American nations.

The coffee berry borer (*Hypothenemus hampei*) is thus of great concern to coffee growers, as it is the most devastating insect pest of this crop, worldwide. Known in Spanish as *la broca del café*, this tiny beetle, measuring 1 to 2 mm (.04 to .08 in), is originally from Africa. It first appeared in Brazil in 1913, in Guatemala in 1971, in Colombia in 1988, and in Costa Rica in 2001.

The female beetle searches out a developing coffee berry, burrows through it, then into its seed, and, during a period of several months, lays 20 to 60 eggs. She stays with her brood, never leaving the seed. Both larvae and adults feed on the developing seed, completely destroying it. Damage continues even after harvest, as the beetles continue living until the beans are put through the wet fermentation process. Like several other bark beetles, the coffee berry borer has an odd sex life; males, ten times less numerous than females (and much smaller), mate with their sisters.

Coffee bean with berry borer hole.

There are several non-insecticidal methods of controlling the coffee berry borer, including biological control (p. 166), traps, and collection of berries that remain on the trees after harvest and berries that have fallen to the ground. Of course, all methods of control add to the costs of growing coffee. And so, for farmers who grow one of the most important commodities in the world, the coffee berry borer is no trivial matter.

5

WASPS, BEES, ANTS
(order Hymenoptera)

While it may surprise some readers to learn that ants are close relatives of both wasps and bees, that is indeed the case, and winged ants (males and young queens) are easily mistaken for wasps. (Bees and ants both evolved from wasps, albeit from different groups of wasps.)

There is no obvious characteristic that uniquely defines this order. Most species have two pair of membranous wings (worker ants and a few wasps lack wings entirely), but this characteristic also occurs in some other insect orders. As in katydids and some other insect groups, most female hymenopterans have an egg-laying organ known as an ovipositor; note that in some groups, this organ has become specialized for the sole function of injecting venom, in which case it is termed a stinger (only females have stingers). In fact, wasps, bees, and ants are best known for their painful sting, although only a small minority of species is capable of stinging humans, and some species have no stinger at all.

One characteristic of the Hymenoptera is the unusual manner in which male and female offspring are produced. In most insect species (and other animals, including humans), the sex of the offspring is determined by how chromosomes are combined: a combination of two X chromosomes, for example, yields a female, while a combination of an X and a Y chromosome results in a male. In hymenopterans, however, the sex of offspring is determined by whether or not an egg is fertilized; fertilized eggs produce female offspring and unfertilized eggs produce males. Thus, the mother is able to effectively "decide" the sex of each egg she lays. After mating, the female stores sperm in a special pouch (the spermatheca) and every time she lays an egg she has the option of releasing or withholding sperm. Unmated females can reproduce, but they produce only sons.

As in other insects that undergo complete metamorphosis, the larval stages of hymenopterans consume more food than do the adult stages. While adult hymenopterans generally sip on nectar or honeydew, the diet of larvae varies according to group: most wasp larvae are carnivores (a few species feed on plants); bee larvae feed on pollen and nectar; and the diet of ant larvae varies from species to species. Generally, in wasps, bees, and ants, the mother provides food for her larvae (in most cases, the larvae are legless, maggot-like, and rather helpless). And, except for sawflies and some bees, the larvae do not defecate until they are ready to pupate, a sanitary measure made possible by their relatively nutritious diet containing little indigestible matter.

SAWFLIES

Sawflies are the oldest group of hymenopterans. They differ from other hymenopterans in having larvae that resemble the caterpillars of butterflies and moths; sawfly larvae are generally distinguished from those of butterflies and moths by having more than five pairs of prolegs (the nub-like protuberances on the abdomen). The common name *sawfly* is derived from the saw-like ovipositor that females of many species use to cut into plant tissue in order to lay eggs.

Adults are stocky, nondescript, seldom-seen insects, though in some species, the female can be seen standing guard over her egg mass. Sawfly larvae are exceptional among hymenopterans in that the majority of species have legs, and are thus mobile. Most sawfly larvae feed on the foliage of plants, either solitarily or in groups.

The three most common families in tropical America are Argidae, Pergidae, and Tenthredinidae. Among the most commonly encountered species, especially during the latter part of the rainy season, are those belonging to the genus *Perreyia* (Pergidae), whose black caterpillars form large aggregations that crawl along the ground and consume decaying leaves—a most unusual habit for a sawfly. In Brazil there are reports of cattle and pigs becoming sick after accidentally ingesting these caterpillars, which suggests that they harbor nasty chemical compounds.

Sawflies are among the few groups of organisms that do not notably increase in diversity in the tropics; in tropical America there are fewer than 1,000 named species.

Waldheimia interstitialis (Tenthredinidae) inserting eggs into leaf of *Hamelia patens* (Rubiaceae).

Sawfly larva feeding on a leaf.

Group of *Perreyia tropica* larvae crawling over the ground.

Sawfly larvae feeding together on a leaf of *Cissus* (Vitaceae).

WASPS

At least 90% of all hymenopterans are wasps, and the vast majority of wasps are parasitoids of other insects or spiders. The female parasitoid lays her egg in (or on) a host, where her larva lives in an intimate association with this host animal. In this aspect, parasitoids behave just as do parasites, but a parasitoid nearly always kills its host while a parasite generally does not (in large numbers, parasites such as bacteria and protozoans sometimes kill their host, but in most parasitoids it is just a single larva that does the killing). Although various other insects are parasitoids (e.g., tachinid flies, p. 299), there are more species of parasitoid among the wasps than in all other insect orders combined.

A smaller number of wasp species are predators, and these evolved from the parasitoid wasps. Unlike female parasitoid wasps, which generally lay their eggs on hosts in situ, a female predatory wasp first carries the targeted insect back to her nest before she lays an egg on it. While nearly all predatory wasps build nests and only one group of parasitoid wasp does (the spider wasps), these two groups are more clearly distinguished by feeding habits; parasitoid larvae feed on just a single host, but predatory larvae feed on two or more hosts.

Many large female wasps are capable of stinging if you grab them, but only the paper wasps (p. 173) are aggressive. Although these are the wasps that people fear, they represent only a very small minority of all wasps.

Paper wasp (*Agelaia*).

PARASITOID WASPS

The female parasitoid wasp flies about sipping nectar and searching for a host species (males fly about in search of females). When she finds one, she uses her ovipositor to insert an egg very precisely inside (or on) the surface of the host, and then continues looking for more hosts. Depending on the species, she can produce dozens to hundreds of eggs, though she might not be able to find enough hosts in which to lay all of her eggs during her relatively short lifetime (a few weeks to a few months). Females of a given species search in specific habitats for specific types of hosts (even when a host is located, she may reject it, because it is the wrong size, because is has already been parasitized, or for other reasons).

In species in which the female lays her egg on the surface of the host, the larva feeds externally and is called an ectoparasitoid; in species in which the female lays her egg inside the host, the larva (an endoparasitoid) feeds internally.

In most species, the female injects venom into the host as she inserts her egg (either on or in the host), though the effect of the venom varies among parasitoid species. The venom of many parasitoid wasp species (especially ecto-parasitoids) causes permanent paralysis in the host, but generally does not kill it, since a dead host would quickly rot and thus no longer provide a source of food for the larva. Permanent paralysis serves the needs of the larva so long as the host insect is concealed within, or attached to, the plant (or some other substrate), but if the host lives on the surface of foliage, paralyzing it would be unwise since a paralyzed host would probably fall to the ground and be devoured by scavengers. For this reason, many wasps (especially endoparasitoids) use a more sophisticated strategy; they insert the egg in the host, but inject venom that does not immobilize the host. Instead, the venom often alters the host's physiology to the benefit of the parasitoid by, for example, altering its immune system; making nutrients more available (e.g., rerouting nutrients from some organ to the blood, where it is becomes available for the parasitoid); or speeding up (or slowing down) the normal developmental time of the host. Regardless of the strategy employed, eventually the wasp larva kills the host, pupates (either inside or outside the host), and emerges as a new adult wasp; in some wasp species, a fully grown wasp larva emerges (to pupate on the outside) while the host is still alive. In the world of insects, *Alien* is real.

Most parasitoid wasp species lay just a single egg in the host, but a few lay several eggs. Depending on the species, the parasitoid larva feeds on the host's egg, larva, or pupa; very few species parasitize adult insects. Hyperparasitoids attack the larvae or pupae of other parasitoids.

Endoparasitoids must battle the host's immune system, which consists of blood cells capable of recognizing the alien, surrounding it, and suffocating it (a process known as encapsulation). But wasps come equipped with adaptations that allow them to avoid this fate. Some species, for example, lay their egg in the victim's nerve cord, which is isolated from the blood cells; in other species, the female injects a symbiotic virus that prevents encapsulation. This virus replicates in the reproductive tract of the female wasp, without harming her, and is passed on from generation to generation in the genome. Neither wasp nor virus can live without its partner, and each wasp has its own species of virus. Given the evolutionary arms race that goes on between host insects trying to encapsulate the invaders and parasitoids deploying mechanisms to avoid encapsulation, many endoparasitoids tend to be very host specific. Because the female wasp can only lay her eggs in a restricted set of insects, she is essentially a search and destroy machine, which makes these wasps very valuable in controlling insect pests (p. 166).

A few "parasitoids" have given up carnivory and become vegetarians. Unlike most sawfly larvae (most of which feed on foliage), the larvae of these wasps feed on the most nutritious types of plant tissue, in seeds or in galls (p. 280). Phytophagous wasps have arisen several times from parasitoid ancestors, and, quite confusingly, some species belong to families consisting primarily of parasitoids. In the case of gall wasps (family Cynipidae) and fig wasps (family Agaonidae), all members of the family are phytophagous; these families are thus treated in this section of mainly parasitoid wasps for taxonomic reasons, even though their biology is quite distinct.

In tropical America there are nearly 50 families of parasitoid wasp.

Chalcidoid wasp (Eurytomidae: *Neorileya*) examining eggs of a stink bug, into which she will lay her own eggs.

ENSIGN WASPS

On ensign wasps (Evaniidae), the abdomen is attached high up on the thorax, resembling a banner or ensign (although the abdomen is so small that it often appears to be missing). The female lays an egg in each (suitable) cockroach egg case (oothecae) that she finds. The larva begins as a parasitoid by feeding inside a single egg, but it then proceeds to consume the other eggs inside the cockroach egg case, thereby acting as a predator.

There are at least 170 named species in tropical America.

PELECINID WASPS

There are just three species in this family (Pelecinidae), one that occurs exclusively in Mexico, one in Brazil, and *Pelecinus polyturator*, whose range extends from the eastern U.S. to northern Argentina, and that occurs fairly commonly in cloud forests.

The larvae of pelecinid wasps are endoparasitoids of the larvae of scarab beetles. The

Ensign wasp belonging to the genus *Evaniella*.

female's extremely long abdomen allows her to reach down into leaf litter and loose soil in order to search for—and lay an egg in—a scarab larva. The abdomen of the male is not as long as the abdomen of the female and is slightly swollen at the tip.

Female *Pelecinus polyturator* on rotten wood, probably searching for a scarab beetle larvae (p. 112).

GALL WASPS

Most members of the superfamily Cynipoidea are parasitoids (especially of fly larvae), but gall wasps (family Cynipidae) are phytophagous, and feed by inducing galls on oak trees, either on leaves, twigs, flowers, acorns, or even roots, depending on the species. (A gall is an abnormal plant growth, with an inner layer of nutritious tissue surrounding the gall-inducing larva; p. 280.)

Although oaks occur primarily in northern temperate regions, they extend as far south as Colombia (and in Asia, down to Southeast Asia); in Central America, oaks grow mostly at higher elevations, above 1,400 meters (about 4,600 feet). A single oak species can host up to 20 or more species of gall wasp.

As is the case with gall formers in general, it is much easier to observe the galls than the actual insects that induce them, which are quite tiny.

Some species (e.g., those in the genus *Synergus*) have lost the ability to form galls on their own and instead feed inside the galls formed by other gall wasps.

There are about 1,500 named species worldwide, but relatively few of these occur in tropical America.

Female gall wasp (Cynipini).

Berry-like galls induced by *Coffeikokkos copeyensis* on oak twig. These galls eventually detach and fall to the ground, with the developing larva still inside.

Gall on oak twig, showing holes from which adult wasps emerged.

Fuzzy galls on oak leaf.

Disc-shaped gall on underside of oak leaf (6 mm or 1/4 in. in diameter).

SCELIONID WASPS

These wasps traditionally were classified in their own family, Scelionidae; today, they are placed in the family Platygastridae.

Like chalcidoid wasps, scelionid wasps are minute insects (0.5 to 12 mm/0.2 to 0.5 in) with very reduced wing venation. Females have a needle-like ovipositor that they use to insert their eggs into the eggs of other insects or spiders. In a few species, the adult female wasp hitches a ride on an adult female host, and every time the latter deposits eggs, the female wasp dismounts to parasitize (by inserting her own eggs) the freshly laid eggs of the host (a similar behavior occurs in a few chalcidoid wasps that parasitize eggs).

Each species specializes on certain types of hosts and the form of the adult body reflects the shape of the host egg from which it emerged. Thus, species with a very elongate body parasitize elongate eggs (e.g., those of katydids), whereas scelionid wasps with a very stocky body parasitize spherical eggs.

Worldwide, there are about 5,400 named species of Platygastridae.

CHALCIDOID WASPS

Chalcidoid wasps, in the superfamily Chalcidoidea, tend to be tiny, with most species less than a centimeter (0.4 in) in length; indeed the smallest winged insect in the world (1.6 mm/0.06 in), a fairyfly (family Mymaridae), belongs to this group. Most chalcidoids are therefore seldom noticed. Members of the family Chalcididae, however, are among the few species large enough to be recognized in the field. They have enlarged hind legs (which in flight hang down like those of stingless bees), and the body is usually yellow and black. Many species in this family attack the pupae of moths, butterflies, or flies.

As a group, chalcidoid wasps parasitize a very wide diversity of insects (and insect eggs) and spiders, including leaf miners, gall formers, scale insects, mealybugs, and whiteflies.

Some chalcidoids have an unusual biology. Females of the genus *Copidosoma* (Encyrtidae), for example, generally insert a single female egg and a single male egg into a host caterpillar, where each egg clones itself, the

Gryon species parasitize eggs of true bugs.

Female *Kapala* (Eucharitidae), a genus characterized by long projections at the back of the thorax.

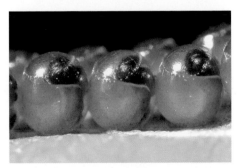

Stink bug eggs with scelionid wasps inside, ready to emerge.

Male *Kapala*, with antennae very different from those of the female.

process repeating until there are more than a thousand female and male eggs in total. A small proportion of the female larvae are soldiers that first kill off the excess males and then die before reaching adulthood. Species in the family Eucharitidae lay thousands of eggs on plants, where the larvae hatch and wait to hop onto an ant in order to be carried back to the ant's nest, where they then parasitize the ant's brood. They apparently pick up the odor of the ant colony in which they live so that their hosts are unaware of what is going on; in fact the ants care for the young parasitoids as if they were their own offspring.

Fig wasps (Agaonidae) are also members of the superfamily Chalcidoidea, but their biology is so distinct that they merit a separate description. Currently, the family Agaonidae is tentatively defined on the basis of biology rather than morphology. Although morphologically quite heterogenous, the larvae of all species are phytophagous and are associated with the

Encyrtus infelix (Encyrtidae) parasitizes soft scale insects (Coccidae).

Aphelinidae inside a whitefly nymph.

Gonatocerus (Mymaridae) laying eggs into treehopper eggs (*Bolbonota*).

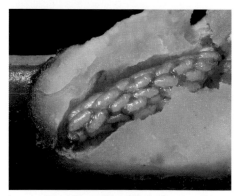

Larvae of *Copidosoma* (Encyrtidae), after parasitizing a caterpillar in a gall.

Conura (Chalcididae) laying an egg in a leaf beetle larva (Chrysomelinae, p. 139).

Larvae of *Euplectrus* (Eulophidae) feeding on a live inchworm (Geometridae).

Larva of Eulophidae feeding on a paralyzed gall midge larva (Cecidomyiidae).

Female *Torymus* (Torymidae) cleaning her wings with the hind legs.

developing fruits of fig trees. Members of the subfamily Agaoninae have very flattened heads and are the exclusive pollinators of figs, whereas species in the other subfamilies do not pollinate the plant. The biology of the non-pollinators is poorly known and the following discussion focuses on the pollinators, which are ecologically very important in tropical forests since figs are keystone species that provide sustenance for numerous other animals.

Tropical America has about 150 species of fig tree, each of which is pollinated by just one or two species of Agaoninae. Generally a given plant species depends on multiple insect species for pollination, and most insect

pollinators visit multiple species of plant, but figs are not ordinary plants. Their tiny flowers are concealed inside the green fruit, forming an enclosed inflorescence called a syconium. When the female flowers are ready to be pollinated, the trees give off an odor that attracts female fig wasps. The female wasp enters the green fruit through a natural opening (ostiole) so narrow that her antennae and wings sometimes break off when she squeezes through this hole. Once inside, she pollinates the flowers and then lays eggs in the fig ovules that are closest to the central cavity (the non-pollinating wasps lay their eggs while standing on the outer surface of the syconium). Ovules that receive eggs produce fig wasps instead of seeds, so the tree is sacrificing some of its seeds in return for the services of a very reliable pollinator. The female wasp has a vested interest in pollinating the fig's flowers, since the tree often aborts green fruits that haven't been pollinated; through pollination, then, the female ensures that the fruit in which her offspring reside will not be aborted. Once the female wasp has laid her eggs, she dies inside the fruit.

Each wasp larva induces a barely perceptible gall in the fig ovule, feeds inside this gall, pupates, and then emerges as an adult into the inner cavity of the fruit. In most cases, just a single mother wasp enters the green fruit, and the new generation of wasps she produces inside a particular fruit are therefore all siblings; because mating occurs inside the fruit, brothers mate with their sisters. In this situation, the mother generally lays many more female eggs than male eggs, since one male is capable of inseminating several females. In those cases where more than one female wasp enters the fruit (e.g., in fig species with larger sized fruits), the sex ratio is more balanced and there is less inbreeding. The blind, wingless males often assist the females in emerging from their ovules, and, after mating, they chew an exit hole through which the females depart for the outside world; the wingless males, however, shall never experience the outside world since they die inside the fruit.

At this point, just before the figs are fully ripe, the masculine flowers have matured and the female wasps, before leaving the fruit, pack pollen into special pockets on their bodies (species in the primitive genus *Tetrapus* lack special pockets and instead carry pollen passively on their bodies). When female fig wasps depart the tree in which they were born, they must locate another tree of the same species that is in just the right stage to receive them. They live for less than a week, but are capable of flying 100 kilometers (60 miles) or more when assisted by wind.

In tropical America, the superfamily Chalcidoidea comprises 20 families; worldwide there are well over 20,000 named species.

Undersides of leaves with thousands of minute fig wasps.

ICHNEUMONID WASPS

The family Ichneumonidae includes some of the largest and most easily observed parasitic wasps, though there is considerable variation in size among its members. Some species mimic paper wasps, although, unlike some members of that family, the mimics are not aggressive, stinging only if handled and with a result that is usually no more painful than a pin prick. Interestingly, even males will go through the motions of stinging, despite the fact that they lack the apparatus (ovipositor) needed to sting.

Females of some ichneumonids have a very long ovipositor, which suggests that they must drill deep into some substrate in order to reach the host insects on which they lay their eggs; some beetle larvae, for example, live deep within rotting logs. Many species

Non-pollinating fig wasp (Sycophaginae: *Idarnes*) with elongate ovipositor at rear end.

Male fig wasps (Agaoninae) lack wings.

Netelia (Tryphoninae) species search for caterpillars at night.

Mature fig fruit showing hole from which fig wasps emerged.

Interior of a mature fig fruit, with a non-pollinating fig wasp (*Idarnes*).

Polycyrtus melanoleucus (Cryptinae) parasitizes mature caterpillars.

Cocoon of Campopleginae. Members of this subfamily are endoparasitoids of caterpillars.

Larva of Polyspinctini (Pimplinae) feeding on a live spider (Araneidae)

Close-up of *Netelia* larva feeding on a live caterpillar. The black object is the empty egg shell which the wasp larva uses as an anchor.

with short ovipositors lay their eggs in young caterpillars, although the wasp larvae perhaps do not terminate their development until the caterpillar has sought out a hidden place in which to pupate, thereby providing the wasp itself with the same hidden place in which to pupate. However, many species in the subfamily Campopleginae emerge from the caterpillar before it pupates, while it is still actively feeding; in these cases, the wasp larva cannot use the host's pupation retreat for protection and must protect itself in some other manner after emerging from the host to pupate, such as, for example, by forming a cryptic cocoon that resembles a bird dropping on a leaf.

Among ichneumonid wasps that do not paralyze their host, some literally take over their host's body and modify its behavior for their own benefit; some spider parasitoids (subfamily Pimplinae), for example, cause their host to construct a type of web that it would otherwise never make.

There are nearly 25,000 named species of ichneumonid wasp worldwide.

BRACONID WASPS

Braconid wasps (Braconidae) are closely related to ichneumonids and, like them, many species parasitize caterpillars. Most braconids are smaller than the average ichneumonid and are therefore generally less noticeable. However, the cocoons of one group (Microgastrinae) are commonly seem on the remains of the host caterpillar, in some cases covering it like mass

Braconid wasp in the subfamily Microgastrinae.

of cotton; this is one of the few groups of parasitoid in which the females often lay numerous eggs in the host and hence numerous larvae later emerge to pupate. Most other braconids lay just a single egg per host.

Several species of braconid wasps, including one employed to reduce populations of the sugarcane borer (p. 166), are used in biological control. Other braconids that are important in biological control include members of the subfamily Opiinae, many of which parasitize the larvae of fruit flies, and members of the subfamily Aphidiinae, which parasitize aphids. Larvae of the latter group cause the host aphid to become dry and stiff in the later stages of parasitism. If you collect one of these aphid mummies that does not yet have an emergence hole, you can rear the adult wasp.

There are about 20,000 named species of braconid wasp in the world.

Bracon (Braconinae) species parasitize concealed larvae of moths and beetles.

Aphidiinae emerging from its mummified aphid host.

Galls on fig leaf (Ficus colubrinae) induced by Labania minuta.

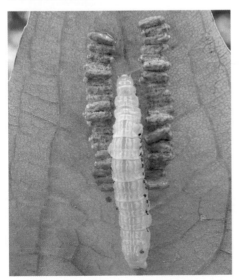

Cocoons of Glyptapanteles (Microgastrinae). Note the black scars in the dying caterpillar (Noctuidae: Pseudoplusia includens) from which the wasp larvae emerged.

Cocoons of Microgastrinae beneath its dying host, a butterfly caterpillar (Lycaenidae).host.

In all parts of the world, farmers must battle insect pests. Yet beyond the costs of purchasing and applying insecticides, these substances have unwanted side effects, including toxic residues in food, environmental contamination, and pest resistance. Agricultural experts are thus eager to find and employ alternative methods of controlling insect pests. One such alternative is biological control, in which an organism, a natural enemy of the targeted pest, is used to reduce the population of that pest.

The most effective control organisms are natural enemies that specialize on the targeted pest, and they are effective in two senses. First, they don't waste time killing insects that aren't the targeted pest—and thus they do their work more efficiently—and second, they don't end up killing insects that are beneficial to agriculture. Because many parasitic wasps are specialists on particular insect hosts, they are among the most effective biological control agents, and several species are now being used to control insect pests. Other natural enemies include predators, most of which are generalists, and microbial pathogens, many of which are specialists.

There are three strategies for using natural enemies to control insect pests. In cases where the pest originates somewhere else—and is thus able to thrive in its new environment due to an absence of natural enemies—scientists can import highly specialized natural enemies from the country where the pest originated. For example, a couple of braconid wasps have been deliberately introduced to tropical America to control the Mediterranean fruit fly (p. 291), with a certain amount of success, although more effective species are still needed. A second method involves the mass rearing of natural enemies for periodic release in the field. To cite one example, *Cotesia flavipes* (Braconidae) is being reared by the millions in various parts of tropical

Empty pupa of sugarcane borer (Pyraloidea).

America; these tiny wasps are periodically released into sugarcane fields and provide very effective control of the sugarcane borer (*Diatraea* spp.). The third method involves conserving predators and parasitoids that are naturally present in the field. By definition, natural biological control occurs all the time without human intervention, but, in disturbed habitats, pests can often gain the upper hand. In such cases, the goal is to reestablish natural control by providing natural enemies with essential resources (e.g., plants with nectar) and by removing factors detrimental to the flourishing of the control organism. Parasitic wasps are usually more vulnerable to insecticides than are pest insects and therefore the overuse of insecticides sometimes causes more problems than it solves.

CUCKOO WASPS

Cuckoo wasps (Chrysididae) are small (2 to15 mm/0.1 to 0.6 in), bright green (or blue) in color, and lack the capacity to sting because they possess a very reduced stinger). Female cuckoo wasps, similar to their bird namesake, lay their eggs in the nests of other wasps and bees. If the host nest has been sealed shut, the female uses her mandibles to chew a hole through the wall, turns around, and then, by means of a membranous tube that telescopes out of her rear end, attaches an egg to the inside of the nest wall. In some species, the female enters an open nest that is still being provisioned, where she may be violently evicted if discovered by the nest's owner. Cuckoo wasps, nonetheless, have a tough body and are persistent. Depending on the species, the larva either eats the host larva and then the paralyzed insects or pollen that the host insect provided for its own larva, or it waits for the host larva to eat all of its food and then feeds as a parasitoid on the host larva.

The family Chrysididae also includes some more drably colored wasps (Amiseginae) that parasitize the eggs of stick insect. There are about 350 named species of cuckoo wasp in tropical America.

PINCHER WASPS

Depending on the species, female pincher wasps (Dryinidae) are either winged or wingless (wingless females resemble ants). Nearly all females have pinchers on the tips of their front legs that they use to grab planthoppers (p. 74) or leafhoppers (p. 80). Once the female has "captured" a host (nymph or adult), she lays an egg on its body, usually in a membrane between two abdominal segments. The larva feeds externally and resembles a large, dark-colored sac protruding from the host's body (the size of the larva relative to the size of the host is comparable to the size of a beagle in relation to a human). When the host finally dies, the larva spins a cocoon, either on a plant or in soil, depending on the species.

There are roughly 500 named species of dryinid wasp in tropical America.

Female *Exochrysis* with membranous egg laying tube extended.

Larva of pincher wasp (black) feeding on a live planthopper nymph (Dictyopharidae).

SCOLIID WASPS

These large wasps look fearsome but are in fact innocuous. Species in this family (Scoliidae) are either all black or a combination of black and yellow (or black and orange). The female digs in soil searching for a scarab beetle larva (p. 112); on finding one, she stings it with paralyzing venom, lays an egg on it, covers it with soil, and then searches for another host. The scoliid wasp larva feeds externally on the immobilized host, and when finished consuming the host, the larva spins a cocoon in the soil and eventually emerges as a new adult.

There are about 50 named species in tropical America.

Female *Campsomeris*, with stout, spiny legs for digging.

Female *Campsomeris*. Males are thinner and have longer antennae.

VELVET ANTS

Despite the name, velvet ants (Mutillidae) are wasps not ants. Although males have wings and look like wasps, females are wingless and do resemble ants. Unlike ants, however, these females have a velvety body and many of them sport reddish orange markings on their body; these bright colors serve to warn predators of their very painful sting, which threat allows them to stroll quite casually over the ground and vegetation. Some checkered beetles (Cleridae) and several other insects mimic the colors of velvet ants to scare off potential predators.

The female velvet ant searches for nests of solitary wasps or bees that contain a mature larva or pupa; on discovering a nest, she chews a hole through the wall of the nest, deposits an egg inside, and then plugs the hole with soil or salivary secretions. Once they emerge, the velvet ant larvae feed as ectoparasitoids of the nest builder's larvae.

In tropical America there are about 1,500 named species of velvet ant.

Mating *Timulla*. This is the only tropical American genus of velvet ants in which the male carries the female in flight while mating.

Pseudomethoca chontalensis cleaning her legs, supported by just the mandibles and tip of the abdomen.

Female *Hoplomutilla xanthocerata*. Larvae in this genus parasitize bee larvae.

SPIDER WASPS

Spider wasps (Pompilidae) have the habit of flicking their wings as they walk about, which makes them easy to identify in the field. Spider wasps are parasitoids, but females of most species build nests (like predatory wasps), and they thus represent an evolutionary transition from parasitoid to predator.

In most species, the female searches for a particular species or type of spider, stings it, and then drags it back to her nest (depending on the species, the nest is either a hole in the ground, a hollow twig, or made from mud).

Pepsis wasps, known as "tarantula hawks," are large (5 cm/2 in) and have a painful sting (but they only sting if handled).

Before transporting the paralyzed spider, the females of some species (Ageniellini) remove its legs, which might otherwise impede transport. Once in her nest, she lays an egg on the paralyzed spider, scals the nest's entrance, and then ventures out to repeat the process.

Poecilopompilus carrying paralyzed spider (Araneidae: *Eriophora*).

Tachypompilus species hunt wolf spiders (and related families, p. 314) and nest in the ground.

This spider wasp (possibly *Ageniella*) walks like an ant.

Some spider wasps go hunting before they build a nest, which of course presents a slight problem, as the female must keep a watchful eye on the paralyzed spider (a tempting morsel for a marauder) as she prepares her nest.

A few species are kleptoparasites—they lay their eggs in the nests of other spider wasps. The kleptoparasite larva feeds on the paralyzed spider that was placed in the nest by the adult female of the host wasp, thus stealing the food of the host larva, which generally dies of starvation.

Among the largest wasps in the world are species in the genus *Pepsis*, whose females paralyze tarantulas in their burrows (and hence have no need to build a nest). Tropical America has about 800 named species of spider wasp.

PREDATORY WASPS

Nearly all predatory wasps build nests, except for a few kleptoparasitic species that lay their eggs in the nests of other predatory wasps. The larva of predatory wasps nearly always feed on multiple hosts (i.e., prey), unlike the larva of parasitoid wasps, which feed on a single host. The two groups are further distinguished by another aspect of their feeding habits. Almost all female predatory wasps carry host prey back to their nest before laying an egg on it; female parasitoid wasps, on the other hand, lay their egg on the host in situ, with the sole exception of the spider wasps.

After building her nest, the female flies off in search of prey (with a given species tending to focus on a specific kind of insect or spider), typically stings it with a paralyzing venom, and then brings it back to the nest, where she lays an egg on it. The prey item is generally small enough so that the female can carry it in flight; but such small prey does not provide enough food to sustain a larva all the way through to pupation, and so the female provisions each cell in the nest (where the individual larva develops) with several prey items.

Predatory wasps (Crabronidae: *Trypoxylon*), one chasing the other. Species in this genus hunt spiders.

COCKROACH WASPS, MUD DAUBERS, SAND WASPS, AND THEIR KIN

These predatory wasps (Ampulicidae, Sphecidae, Crabronidae) vary greatly in color and size (2 to 50 mm/0.1 to 2 in). Depending on the species of wasp, prey consists of spiders, crickets, cockroaches, caterpillars, flies, beetles, bees, or other kinds of insect. Upon encountering their prey, they inject it with a venom that, in most cases, causes permanent paralysis. The venom of cockroach wasps (Ampulicidae), however, has a more macabre effect; injected into the cockroach's brain, the venom converts it into a zombie, which allows the female wasp to lead the docile cockroach by its antennae to a suitable cavity and lay an egg on it. The wasp larva feeds on the listless cockroach, eventually killing it, then crawls inside the empty carcass, applies antibiotic secretions to the interior, and pupates.

The most common type of nest is either a subterranean tunnel or hollow twig, though the great variety of species within these families produces a commensurate variety of nest types. Among the most noticeable nests are those constructed from mud, sometimes globular in form (*Sceliphron*, family Sphecidae) or in a series of parallel tubes that have been likened to organ pipes (some species in the genus *Trypoxylon*, family Crabronidae). Members of both families are known as mud daubers, but it should be noted that only a minority of species in each family build mud nests.

Nest-building wasps must protect against thieves trying to break into their nest, either while the female is away on a hunting expedition or after she has finished provisioning the nest and has left her larvae to develop on their own. Such kleptoparasites include various wasps and flies that lay an egg on the paralyzed prey that the female wasp has left in her nest for the benefit of *her* offspring; the kleptoparasitic larva ends up killing the nest-builder's offspring, either directly or by eating all its food. For this reason, most nest-building wasps place a temporary plug in the entrance while they are out hunting, and all species seal the nest permanently when the nest has been fully provisioned, although these tactics are not always effective in keeping out the marauders.

Some species of Crabronidae (subtribe Bembicina) are known as sand wasps because they generally nest in sandy soil found on beaches and other places. Some resemble stinging wasps (paper wasps, p. 173), but sand wasps are not aggressive and only sting if they are handled. Many species in this group are unusual in that the mother feeds her larvae little by little (like birds do), rather than leaving the young larvae with a complete supply of paralyzed prey (as is the case in most of the other species).

In tropical America, there are about 25 named species of Ampulicidae, nearly 200 species of Sphecidae, and approximately 1,600 species of Crabronidae.

Sphex (Sphecidae) species hunt katydids and crickets, and nest in the ground.

Minute hanging nest of *Microstigmus* (Crabronidae). Some species in this genus are eusocial (p. 175).

Mud nest of *Trypoxylon* (Crabronidae), a genus with hundreds of species and diverse nesting habits.

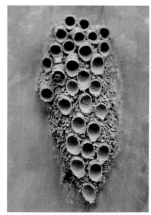

Mud nest of another species of *Trypoxylon*.

PAPER WASPS

The family Vespidae includes three quite distinct groups of wasps: the pollen wasps, the potter wasps, and the eusocial wasps. Members of the first two groups are docile and not commonly encountered; these wasps have a biology very similar to that of mud daubers and sand wasps in that each female solitarily builds and provisions her nest. Pollen wasps (subfamily Masarinae) are unique among wasps in that, like bees, they feed their larvae with pollen and nectar. Potter wasps (subfamily Eumeninae), named for the tiny mud nests that some species construct, feed their larvae with paralyzed caterpillars. In contrast to pollen wasps and potter wasps, the eusocial wasps (Vespinae and Polistinae) live in colonies consisting of workers and queens (p. 175), although these two castes are quite similar in appearance, differing primarily in behavior. These wasps, many of which aggressively defend their nests and pack a painful sting, conform to the image of wasps held in the popular imagination.

In most species of nest-building wasp, the female uses her stinger primarily to paralyze prey, whereas the eusocial wasps generally subdue their prey (mostly caterpillars) by mangling them with their mandibles. The stinger and venom in eusocial wasps have become specialized primarily for inflicting pain in any vertebrate that approaches too closely to the wasp nest. Whereas solitary wasps generally take care not to kill their prey, which otherwise would rot and prove unsuitable as food for the wasp's larva, eusocial wasps attack prey, carve off bits of meat, and return to the nest to feed the larvae little by little, thus freeing up the stinger and venom for a strictly defensive function.

The Americas are home to two subfamilies of eusocial wasps: the Vespinae (hornets and yellow jackets), whose members only extend as far south as Honduras (except for a species accidentally introduced into Chile and Argentina), and the Polistinae, which are especially diverse in tropical America, home to about 550 species.

Members of the subfamily Polistinae are known as paper wasps because they construct paper nests from masticated wood fibers. There are two types of paper wasps. Species in

Polistes erythrocephala with a masticated caterpillar, which she will bring back to the nest.

the genera *Polistes* and *Mischocyttarus* make a small simple nest consisting of just a single layer (comb) of exposed cells. These are the relatively small nests that one often sees on the ceilings of open buildings. Species in the genus *Polybia*, along with about 15 other genera (all in the tribe Epiponini), make larger nests, with one or more layers of cells (combs) usually concealed by an external paper-like covering.

Nests of the first type of paper wasp are often initiated by a lone queen, who is sometimes joined by other females who are capable of laying eggs but are prevented from doing so by the queen, the alpha female; wasps in the first generation of adults usually end up becoming workers. In *Polybia* and related genera a new nest is initiated by a swarm of workers and queens who leave their maternal colony as a group in order to locate a new nest site and begin a new colony. Nests of these so called "swarm-founding wasps" generally harbor numerous queens.

Mud nest of a potter wasp (Eumeninae).

Protopolybia on their nest on the underside of a leaf.

Mischocyttarus species, like those of *Polistes*, construct small exposed combs.

Brachygastra on their nest.

Nest of paper wasp (*Polybia*) with cover partially removed to show combs arranged one above the other.

Many animals, including a number of species of bird, live in colonies, but only rarely do colony members collaborate in rearing the young. Even less common among species that form colonies is a division of labor between a reproductive caste (queens) and a non-reproductive caste (workers). The queen is an egg-laying machine; workers build the nest, forage for food, and care for the young. This complex of characteristics is known as eusocial behavior, which, rare though it is, has evolved in more species in the order Hymenoptera than in any other group of animals. Outside of the Hymenoptera, this behavior occurs in termites—and among vertebrates it occurs only in the naked mole rat of southern Africa.

In tropical America eusocial insects include all termite species, all ant species, the paper wasps and yellow jackets, and a subset of bee species (all stingless bees and bumble bees, the honey bee, and some halictid bees). They represent far fewer than 1% of the total number of insect species, yet their combined biomass (cumulative weight of all individuals) accounts for more than 50% of the total insect biomass in lowland tropical rainforests! In addition to the sheer significance of their biomass, eusocial insects play vital ecological roles as decomposers (termites), herbivores (leafcutter ants), predators (paper wasps and many ants), and pollinators (stingless bees and the honeybee). While eusocial insects essentially own the lowland forests, they become increasingly scarce at higher altitudes.

In both hymenopterans and termites the young, newly emerged males and queens generally leave their natal colony in order to find a member of the opposite sex (usually from a different colony). Yet there are a number of differences between hymenopteran and termite colonies. Hymenopteran societies are exclusively female (except for male larvae), whereas termite colonies contain both males (workers and a single king) and females (workers and a single queen). In hymenopterans, males die right after mating with a young (newly emerged) queen; the queens, on the other hand, begin a new colony after mating, and often live for a very long time (although they usually never mate again). In termites, the king and queen begin a new colony together and continue mating throughout their long lives. The juvenile stages of hymenopterans are legless maggots, utterly dependent upon adults for their care, whereas those of termites are workers.

Within Hymenoptera itself, there is considerable variation in behavior from group to group. In bumblebees, most species of ant, and some paper wasps, a single queen initiates the colony; but in stingless bees, honey bees, and other species of paper wasp, a swarm composed of workers and one or more queens founds the new colony. While the colonies of most species contain just a single queen, in species of paper wasp that form large colonies, the colony contains numerous queens. Generally, the queen is much larger than the workers, but in paper wasps and "primitive" ants she is virtually the same size as the workers. There is also considerable variation in the reproductive potential of the workers: in a few cases some of the workers are mated females that could potentially become a queen should the resident queen die; in most cases the workers cannot mate and so the most they can do is lay unfertilized, male eggs (but the queen often eats these); in a few ants the workers lack ovaries and cannot even lay male eggs.

One explanation for why eusocial behavior is so much more common in Hymenoptera than in other groups of animals has to

Comb from paper wasp (*Polybia*) nest.

do with genetics. In hymenopterans, males are produced from unfertilized eggs and have half the number of chromosomes of females, who are produced by the union of a sperm and an egg. In most cases, the hymenopteran queen mates with just a single male, producing daughters that are all full sisters, sharing 75% of the same genes (rather than the usual 50%). Some of these daughters are new queens in potentia; whether a particular daughter becomes a worker or a new queen is usually determined by how much food she receives during the larval stage (queens get more). When workers help to rear the new queen, who is their sister, they are thus genetically better off than if they had produced their own daughters (to whom they are only 50% related).

However much genetics is capable of explaining eusocial behavior, it is quite obviously not the only explanation, as termites are also eusocial yet their genetics do not provide an explanation of why they display eusocial behavior.

One of the most interesting features of eusocial insects is the way that the members of a colony are able to cooperatively perform complex tasks (e.g., building a nest or recruiting to a food source) without receiving directions from the queen; instead, these cooperative behaviors are self-organizing. This collective intelligence ("swarm intelligence") is of considerable interest in the field of robotics.

BEES

About one hundred million years ago, certain wasps (related to those in the sand wasp group) began returning to their nest with pollen in lieu of paralyzed insects. They likely carried the pollen in their crop and then regurgitated it in the nest, as some colletid bees still do. From this inscrutable beginning emerged a very successful group of wasps that we call bees (in Spanish, *abejas*).

There are relatively few morphological differences between wasps and bees, and what few exist often require a microscope to see. Wasps, for example, have just simple hairs on their bodies, whereas bees have a combination of simple and plumose (feather-like) hairs. Most bees have some visible adaptation for transporting pollen back to the nest. For example, leafcutter bees (Megachilidae) have dense hairs on the underside of the abdomen. And many bees have hind legs modified for carrying pollen; in particular the first tarsal segment is enlarged and flattened. Four closely related tribes have a widened hind tibia (known as a corbicula): orchid bees, bumble bees, stingless bees, and honey bees.

The majority of bee species behave like the solitary wasps from which they evolved—a lone female builds a nest, usually by burrowing into the soil or a hollow twig, provisions the nest with food for the larva, lays an egg, and then repeats the process. As in the wasps, the males do very little, except mate with the females. Some bees are kleptoparasites, laying their eggs in the nests of other bees, and are very wasp-like in appearance since they have lost most of the adaptations for carrying pollen. About 10% of tropical species show eusocial behavior (p. 175), with numerous workers rearing the offspring of the queen.

Many people think that all bees are honey bees—and that all bees sting—but this is a large and diverse group, with about 21,000 species worldwide. Bees are the principal group of plant pollinators, both in the tropics and in temperate regions. While many species are generalists, gathering pollen from a wide variety of plants, other species specialize on certain flowers such as morning glories, cactus, or wild cucumbers.

Although some authors quite legitimately argue for including all bees in just one family, many entomologists recognize seven families of bees, five of which occur in tropical America. In the following accounts, we include only the two most common families, Halictidae and Apidae (orchid bees, stingless bees, and honey bees represent distinct tribes within Apidae).

Although bumblebees (Apidae: *Bombus*) are more common at high elevations, a few species inhabit the lowlands of tropical America.

HALICTID BEES

Many halictid bees (Halictidae) are metallic green, although some species, especially those occurring at higher altitudes, are predominantly black. Green species might be confused with green-colored orchid bees, which are usually larger, or cuckoo wasps, which tend to be smaller. Halictids are sometimes known as sweat bees because some are attracted to perspiration, but in tropical America this annoying habit is more common in stingless bees.

The majority of species nest in the ground, though a few species nest in dead twigs or wood. Most species are solitary; each female builds and provisions her own nest. Nonetheless, several species (both ground- and twig-nesting) are not solitary and show eusocial behavior (p. 175).

Some halictid bees illustrate the principle that eusociality is not necessarily an inviolable characteristic of a species, since eusocial behavior is sometimes present only in certain populations or during certain times of the year. *Megalopta genalis* is a facultative eusocial species in yet another way; some of its nests are eusocial (just barely so, as the nest contains

Halictid bee at a flower.

just 1 to 3 daughters), while other nests are solitary. The advantages for a daughter of remaining in the nest is that it puts her in a position to protect her developing sisters from marauding ants when the mother is away, and she also stands a chance of taking over the nest should the mother die. Apart from their facultative eusocial behavior, members of the genus *Megalopta* are unusual in two other ways; they nest in rotting sticks suspended in vegetation and they are among the very few nocturnal bees, foraging just after sunset and just before sunrise.

There are about 800 named species of halictid bee in the tropical America.

ORCHID BEES

Orchid bees (Apidae) are grouped in the tribe Euglossini. These fast-flying, medium to large (1 to 3 cm/0.4 to 1.2 in) bees come in two general forms: hairy species that resemble bumble bees and nearly hairless species that are a brilliant green, bronze, or blue. Like hummingbirds, orchid bees are restricted to the Americas and are especially common in tropical regions, where roughly ten percent of all orchid species depend on male orchid bees for pollination. Female orchid bees build resinous nests in secluded nooks and usually do so solitarily, although in a few species more than one female may collaborate in building a communal nest.

In orchid bees—indeed in all bees—females visit flowers to obtain nectar for themselves, and nectar and pollen for their offspring. In contrast, most male bees visit flowers only to supply their own energy needs, which means that they need to visit fewer flowers and, consequently, are generally less effective as pollinators.

Male orchid bees, however, are more complex. In addition to visiting flowers for a sip of nectar, they also visit specific species of orchid (and a few other flowers, rotting wood, and feces) in order to gather fragrances, which are the only reward offered by these orchids. The males spit lipids onto the odoriferous surface in order to dissolve the fragrances, mop up the resulting concoction with their front legs, and then hover for a while in order to transfer the mixture from the front legs to the middle legs, and from there to storage pockets located in the enlarged hind legs; the lipids are recycled internally from the hind legs back to the glands in the head—and the process in repeated—with the result that, over time, the perfume in the hind legs becomes very concentrated. Males of a given species have a unique blend of fragrances and they liberate these at territorial perches in order to woo females. During periodic hovering flights, the males release this perfume onto tufts of hairs on the middle legs, and the perfume is further dispersed by wingbeats.

There are about 200 named species of orchid bee in tropical America.

Female orchid bee (*Eulaema*) imbibing nectar.

Female orchid bee (*Eulaema*) with pollen (white blob) packed onto her hind tibia.

The diversity of colors shown by orchid bees in the genus *Euglossa*.

Male orchid bee (*Euglossa*) with tongue extended.

Female orchid bee (*Euglossa*).

The flattened hind tibia of female orchid bees is used to carry pollen.

The swollen hind tibia of male orchid bees is used to store fragrances.

STINGLESS BEES

Stingless bees (Apidae) are grouped in the tribe Meliponini. These are the most abundant bees in lowland regions of tropical America (and other tropical regions). Although they are generally quite docile—and lack a stinger—they are not defenseless; if provoked, they will sometimes crawl onto one's face or hair, and bite. Species in the genus *Oxytrigona* bite and simultaneously exude formic acid, which causes long-lasting blisters. Less threateningly, some small species have a craving for human sweat and can be a minor nuisance.

The reason stingless bees are generally not very aggressive is that they don't need to be; their primary defense is the inaccessibility of the nest. Most species construct their nest in hollow trees and the sole entrance is a narrow resinous tube that is guarded by a small contingent of worker bees ready to plug the portal with resin; this defense is so effective that even army ants (p. 186) seldom gain entry. Nests are constructed from a mixture of wax (secreted from their abdomen) and resins (obtained from certain plants, sometimes by scraping the leaves of the plant).

Stingless bees are the most important group of generalist pollinators in tropical lowland regions. In several species, a worker that finds a rich patch of flowers is able to guide fellow workers to those flowers, though the method of doing so varies and in many cases is still

Trigona is the largest genus of stingless bees. Note the black resin on the hind legs.

unknown. Some species deposit a scent trail (produced from labial gland secretions) from the food source some distance toward the nest. Even without actual recruitment to the food source, successful foragers of most if not all species run about rapidly (and often buzz) upon returning to the nest, which appears to stimulate other workers to leave the nest. A few species do not visit flowers; species of *Lestrimelitta*, for example, steal pollen and honey from the nests of other stingless bees, while *Trigona necrophaga* and two closely related species gather rotting meat from dead animals.

Like honey bees, stingless bees are highly eusocial, living in colonies consisting of a queen and numerous workers, whose numbers range roughly from 100 to 20,000 individuals depending on the species. There are several differences between stingless bees and honey bees, however. Stingless bees build storage containers for honey and pollen that are larger than the cells used for rearing the brood; in honey bees, the same type of cell is used for both (though different cells are used for brood and storage). In stingless bees, the cell used for rearing the brood are not reused, but rather taken down and rebuilt. Stingless bee workers completely provision a brood cell, then the queen lays an egg within it, immediately after which the workers seal it shut; in honey bees, on the other hand, the workers feed the larvae little by little. Stingless bee workers feed their queen with infertile eggs whereas honeybees feed their queen with glandular secretions. To begin a new colony, stingless bees travel back and forth, for some time, between the maternal nest and the new nest before the new colony finally becomes independent; honey bee swarms fly off from the maternal nest once and for all, and never return. In stingless bees, a new queen joins the founding swarm, whereas in honey bees it is the old queen that accompanies the swarm to the new nest. In stingless bees, the new queen generally mates with just a single male, whereas the honeybee queen is famous for her promiscuity.

There are more about 400 named species of stingless bee in tropical America.

Tetragonisca with pollen on her hind tibia.

Male *Nannotrigona* sipping sweat from finger.

Nest entrance of *Tetragonisca*.

Nest entrance of *Scaptotrigona*.

HONEY BEES

Like the orchid bees and the stingless bees, the honey bees (tribe Apini) are members of the family Apidae. Long ago, Spanish and Portuguese colonizers brought the common honey bee (*Apis mellifera*) to the New World, and in time Mexico and Cuba would become major exporters of both honey and beeswax. In an age of electricity, it's easy to forget the importance of beeswax in the production of the candles that were once one of the few sources of reading light in colonial homes.

Honey bees produce wax just as do stingless bees—by secreting it from the abdomen of worker bees—and both groups of bees use the wax to construct combs inside the nest.

Honey bees make honey by removing most of the water contained in the nectar they collect from flowers. Workers (which generally consume only a small proportion of the nectar they collect) place a drop of nectar on their tongue and then flick their tongue in and out for about twenty minutes. They then place the partially dried nectar in an empty cell in the nest, where several workers then fan their wings to dry it even further. The extremely low water content of honey (18% or less) makes it

very resistant to microbial decay. Along with pollen and glandular secretions, the honey serves as a food reserve with which to feed the bee larvae.

From colonial to recent times, bee keepers in the Americas used honey bee subspecies from Europe and the Middle East to stock their hives. These subspecies rarely ventured far from the hives provided by their human caretakers, but that changed in 1956, when 46 queens of *Apis mellifera scutellata*, a subspecies from southern Africa that is better adapted to tropical climates, were brought to Brazil. This new honey bee, sensationally dubbed the African killer bee, spread like wild fire, reaching Colombia by 1979, Costa Rica in 1981, and southern Texas in 1990. It has now largely displaced the preexisting subspecies, and today the majority of honey bees occurring in tropical America are African honey bees, except where bee keepers import queens of other subspecies (the different subspecies of honey bee look very similar to one another).

Although it took bee keepers a few years to adjust their techniques to this new honey bee, it is in fact a very good producer of honey. And though a bit more difficult to manage, it is more resistant to parasitic mites and has yet to

African honeybee (*Apis mellifera scutellata*), virtually indistinguishable from the regular honeybee.

succumb to the "colony collapse disorder" that is ravaging colonies in temperate regions.

Interestingly, the venom of this African subspecies is virtually identical to that of other honey bees. While alarmist accounts of the "killer bee" are greatly exaggerated, it is quicker to attack than its more docile cousins. But when any kind of honey bee stings, it releases a banana-scented chemical that attracts nest mates to join the attack, and thus swarm attacks are perpetrated not just by the African subspecies.

People who are allergic to bee venom are at high risk from just a single sting, not to mention multiple stings. But while most people can tolerate a certain number of stings, a swarm attack may surpass the tolerance point of even the hardiest person. That said, it should be emphasized that in tropical America more people die in traffic accidents during just a few days than from killer bee attacks during an entire year. The main concern about the African honey bee is its environmental effects; because it nests in cavities and is aggressive, the African honey bee often displaces owls, macaws, and other cavity nesting animals.

ANTS

Ants belong to the family Formicidae. All species are eusocial (p. 175); this means that they live in colonies—colony size ranges from fewer than a hundred adult ants to a couple of million, the largest colonies being those of some species of leafcutter ant and army ant—and that members of the colony belong to distinct castes: a single (in a few species more than one) egg-laying queen and a large number of workers. All workers are females; in some species, the workers are incapable of laying eggs, in others the workers lay only unfertilized male eggs (which are often eaten by the queen). Males are only produced when it is time for the colony to reproduce.

In almost all species of ant, workers show a division of labor. Generally, the task of a worker changes with age; a worker may graduate from nursemaid to forager, for example. In most species, the workers are all the same size, regardless of the task they perform. In a few species, however, workers come in varying sizes, with the size indicating an individual's function (see, for example, leafcutter ants, p. 188).

Ants belonging to the very diverse genus *Camponotus* (Formicinae).

In addition to adult ants, the colony also contains the brood (eggs, larvae, and pupae), which are produced by the queen but tended by the workers. Unlike eusocial wasps and bees, ants do not build individual cells for their offspring, but instead just pile them up inside the nest. Because the larvae are not enclosed in cells, it is much easier for the workers to move them around within the nest (to sites where the temperature is optimum for larval development) or to relocate the entire colony should the need arise. In some species (e.g., many of those living in houses) the colony simultaneously occupies several nesting sites.

New colonies are initiated via the production of new queens and males, an event that is often highly synchronized between colonies of the same species. Males come from unfertilized eggs; queens come from female larvae that have been given extra food. Upon reaching the adult stage, the males and young queens fly away to find a member of the opposite sex from another colony, and mating generally occurs outside the nest (both sexes usually mate with just one individual of the opposite sex, although young queens of army ants and leaf-cutter ants mate with numerous males). After mating, the males die whereas the newly mated queens shed their wings, burrow into the ground or a hollow plant stem (depending on the species), and begin laying eggs. (Generally, newly mated queens are not accepted back into the natal nest, except in some species that have more than one queen.) Because she is on her own, the young queen is quite vulnerable to predation, at least until she has managed to produce a new generation of adult workers in the new nest. It is still unclear how queens that live for several years manage to store sperm that was acquired during one amorous night at the beginning of their adult lives.

Worker ants require relatively little food for themselves, spending most of their time obtaining food for the larvae and queen. While

The golden carpenter ant (*Camponotus sericeiventris*) nests in tree trunks.

Male ants are sometimes difficult to distinguish from wasps. This species is a member of the subfamily Ponerinae.

Ectatomma ruidum (Ectatomminae) taking nectar from an extrafloral nectary.

Trap-jaw ant (Ponerinae: *Odontomachus*) with mandibles cocked open and ready to snap shut on an unsuspecting insect.

some ants are specialized predators, most are scavengers, diligently searching the area surrounding the nest for protein (e.g., live or dead insects) and some form of liquid sugar, such as the sugary excretions (honeydew, p. 87) of sap-sucking insects or the nectar that some plants secrete from extrafloral nectaries located on the leaves or other non-floral parts of the plant. In the canopies of tropical forests, an enormous number of ants busily harvest these sources of sugar, representing a form of indirect herbivory that is difficult to measure; ants with high carbohydrate diets generally have endosymbiotic bacteria that provide them with the nitrogen that their food lacks.

Ants have been described as "walking chemical factories" since they produce a wide array of pheromones used in communication between members of the colony. In many species, especially those with large colonies, a worker that stumbles upon a rich food source lays down a chemical trail on its way back to the nest so that other workers can readily find it, a phenomenon known as "recruitment." As with other eusocial insects, each ant colony has a unique odor, and when two ants of the same species meet, each rapidly taps the other with its antennae, like dogs smelling one another. If they belong to the same colony, all is well, but if not, a vicious fight sometimes ensues. Other insects, whether friend or foe, sometimes manage to penetrate the nest by mimicking the colony odor.

Ants are one of the most abundant groups of animals on the planet and it is estimated that their cumulative biomass is equivalent to that of humans. They are especially abundant and diverse in lowland tropical rainforests, where they play important ecological roles. Soil-nesting species, for example, churn and enrich the soil; predatory ants affect the populations of other insects (sometimes indirectly through harassment); and species that build carton nests in the canopy sow epiphytes on the surface of their nests, which benefits both the epiphytic plants and the ants (by strengthening the nest). Ants enter into a greater number of mutualistic relationships than does any other group of insects, including relationships with fungi (leafcutter ants); with insects that excrete honeydew; with plants; and with bacteria (leafcutter ants and species of ant that feed primarily on honeydew and extrafloral nectar). These are just a few examples of the many ways that ants impinge on the lives of other rainforest organisms.

In tropical America there are 13 subfamilies (out of a total of 16) of ant and about 3,000 named species.

THE BULLET ANT, TRAP-JAW ANTS, AND GIANT AMAZON ANTS

At the base of the ant evolutionary tree are seven subfamilies, six of which occur in tropical America. One of these, Martialinae, was discovered in 2000 in the Amazon rainforest, and has just a single, odd species, *Martialis heureka* (the genus name means *from Mars*). Members of these "primitive" subfamilies generally share certain characteristics: they nest in the ground; have small colonies (usually no more than a several hundred workers); are primarily predatory in their feeding habits; the queen is physically quite similar to the workers.

In the lowland rainforests of tropical America, it is not uncommon to see *Paraponera clavata* (subfamily Paraponerinae), a very large ant whose extremely painful sting gives it the common name of bullet ant. In some cases, the intense throbbing pain lasts for at least 24 hours. Visitors to the tropics

Bullet ant, with smaller leafcutter ants.

who would prefer to forego such pain are advised to take care when grabbing trees. Although bullet ants nest in the soil, they often forage in trees, hunting small insects and gathering substantial quantities of sugary liquids, which they carry as droplets suspended from opened mandibles.

The largest subfamily in this group of ants is Ponerinae, which includes the trap-jaw ants (*Anochetus* and *Odontomachus*), named for their very elongate mandibles that can snap shut in just 3 milliseconds—said to be the fastest reflex recorded in the animal kingdom. When hunting they hold the mandibles wide open at an angle of 180°, with sensory hairs in front of the mandibles serving as a trigger—and causing the mandibles to snap shut upon touching prey.

Three closely related genera of Ponerinae are restricted to tropical America. *Dinoponera* comprises 8 species; they are known as giant Amazon ants both because they are the largest

The bullet ant (*Paraponera clavata*) can reach 3 cm (1.2 in) in length.

Bullet ant stinger.

ants in the world (4 cm/1.6 in.) and because they inhabit the Amazon region. This genus is very unusual in that none of the species appear to have queens; instead it is the female alpha worker that mates and lays eggs. These ants are also unusual in that new colonies are initiated not by queens (since there are none), but rather by a reproductive worker and a contingent of non-reproductive workers, which leave their maternal colony to begin a new one. *Neoponera* comprises more than 50 behaviorally diverse species (previously placed in *Pachycondyla*), many of which are arboreal. *Simopelta* comprises 20 some species that behave like army ants.

ARMY ANTS

Army ants (Dorylinae) have extremely small eyes—and are practically blind. For the most part they don't require keen vision, however, since most species spend their lives hidden beneath the leaf litter and, like all ants, they rely on their sense of smell more than any other sense. (Among those species that hunt on the surface, the huge columns of running ants present a spectacular sight.)

Army ants are voracious predators that mostly specialize on the larvae and pupae of other ants (the white blobs they carry in their mouths), although a few species take virtually any small animal that fails to flee from their path. While army ants do sting humans, people living in rural areas sometimes welcome an army-ant invasion of their house, for when the ants depart after a few hours, the house is free of vermin.

Army ants are unusual in at least three respects. First, they always hunt in groups rather than singly; by hunting in such large numbers, they are able to overwhelm the defenses of other ant colonies. Entering the nest of another ant and plundering the offspring is no easy task, but army ants succeed by rapidly recruiting large numbers of attacking ants. Second, rather than living in a permanent nest, colonies are nomadic, living in temporary bivouacs (ants form hanging clusters that contain the queen and her brood) inside hollow trees or under logs; the colony migrates once it has

Eciton burchellii (subspecies foreli) soldier, with very long mandibles.

Army ants sometimes form bridges with their own bodies.

E. burchellii (subspecies foreli) bivouac.

E. burchellii (subspecies parvispinum) hunting.

depleted an area of food, principally the other ant colonies that had lived in the area. Third, queens are permanently wingless; thus, when a new queen leaves a colony to form a new colony, she must walk rather than fly. She is joined on her march by half the workers, with the other half choosing to stay with the maternal queen. This unusual manner of colony reproduction ensures that a new colony is never so small that it is incapable of marauding the nests of other ants. The males are also unusual in that they must locate a colony that contains a new queen (in other species of ant, the male searches for virgin queens that have left their maternal colony); upon doing so the male loses his wings and enters the colony, assuming the workers allow him to do so.

Most species of army ant hunt in long columns. But *Eciton burchellii*, perhaps the most studied of the army ants—fans out into a broad front that can be as wide as ten meters (33 ft). This species is also unusual in that about half

its prey consists of insects other than ant brood. When the colony contains hungry larvae, it moves to a new location every night, during which time the enormous queen is carefully protected by a large entourage of workers. When the larvae pupate, the queen begins laying eggs like crazy, and the colony now remains in the same bivouac each night. During the 15-day nomadic phase the colony makes daily raids but during the 20-day stationary phase raids are conducted on only about half of the days. The massive carpet raids of *E. burchellii* (and of a few similar species such as *Labidus praedator*) blanket the ground, and nearly everything in its path must flee or succumb. The fleeing cockroaches, beetles, spiders, and so forth make easy pickings for the insectivorous birds that regularly follow these army ants.

Among the species that raid in columns, *Eciton hamatum* and a few other species leave the forest floor to hunt in trees, searching for nests of arboreal ants and paper wasps.

Though humans may find it nearly impossible to rid the eaves of their home of the nests of paper wasps, a column of *Eciton hamatum* ants is so effective that the adult wasps are often forced to flee the site—and abandon their offspring to the raiders.

One species of army ant, *Nomamyrmex esenbeckii*, appears to be a specialist in attacking leafcutter ant colonies, and is the only known predator capable of killing mature colonies of *Atta* leafcutter ants. This is a formidable undertaking since leafcutters are among the few ants with colonies as large as those of army ants. When a raid begins, the leafcutters barricade the entrance holes with dirt or pieces of leaves, and send their soldiers to the front lines (where they grapple with the army ant soldiers). If the leafcutters respond quickly enough—and in sufficient numbers—their colony may have a chance of fending off the attack. Otherwise, the army ants persist and eventually, after considerable carnage on both sides, gain entrance to the nest, or at to least part of the nest.

There are about 150 named species of army ant in the Americas.

LEAFCUTTER ANTS AND OTHER FUNGUS-GROWING ANTS

Fungus-growing ants (16 related genera in the subfamily Myrmicinae) are unique among ants in that they practice agriculture, foraging for compost material that they bring into the nest as a substrate for their fungal gardens, the sole food source for the colony. The majority of species utilize caterpillar droppings, fallen flowers, and similar debris on which to grow their fungal gardens. Only two genera, *Acromyrmex* and *Atta*, cut pieces of leaves from plants to use as a substrate; these leafcutter ants are the most conspicuous fungus-growing ants, since they have much larger colonies. About the only other fungus-growing ants that make conspicuous nests are species in the genus *Apterostigma*, which make nests out of fungi on the undersides of leaves.

Bare patches of dirt—or huge mounds of excavated dirt—are the surest clues to the presence of the underground nests of leafcutter ants (although the nests of many *Acromyrmex* species are less conspicuous). The ants themselves are also easy to see; along carefully groomed trails through the forest, worker ants file one after the other, carrying pieces of leaves and other plant material that seem improbably large in relation to the size of the ant.

In *Atta* species, once workers return to the nest with pieces of leaf, smaller workers clean and scrape the leaves, then chew them into smaller pieces, adding saliva and fecal matter. The resulting sticky mass is then added to an existing fungus garden, which provides food for the colony. In a mature colony, hundreds of fungal gardens are connected by a maze of tunnels.

Leafcutter ants cultivate just a single species of fungus, *Leucoagaricus gonglyophorus*, which apparently lives only in association with these ants. Leafcutter ants take great precautions to prevent microbial contaminants from invading their fungal gardens; indeed, workers that labor in the nest chambers dedicated to the storage of garbage are never allowed to leave their posts, thus preventing them from spreading contaminants to the fungal gardens. But one of the biggest threats to the health of the fungal gardens is another fungus, *Escovopsis*, which is a parasite of *Leucoagaricus gonglyophorus*. To combat this pernicious disease, the leafcutter ants carry on their bodies a bacterium that secretes antibiotics.

Leafcutter ant crossing.

Atta cephalotes cutting a leaf.

A. cephalotes using hind leg to grasp uncut part of leaf.

A. cephalotes carrying a cut leaf, with a "hitchhiker" aboard.

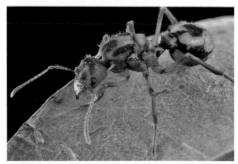

A. cephalotes carrying cut leaves.

A. cephalotes queen just after detaching her wings.

Acromyrmex queen just after detaching her wings.

Atta cephalotes queen with wings still attached.

Nest of *Apterostigma*, a fungus-growing ant that does not cut leaves.

In *Atta* species, the mature nest is an underground metropolis consisting of thousands of chambers (varying greatly in size, but many about the size of a football) and connecting tunnels, with a combined volume that is roughly equivalent to that of a large automobile. Depending on the species and soil conditions, the nest can reach a depth of 7 m (23 ft.). In such nests, the ant population can approach three or four million. Species of *Atta* are unusual among ants in that the size of the worker ant varies greatly according to caste. Soldier ants, which defend the nest against army ants and other predators, are large (and large-headed); worker ants that do the job of cutting and carrying leaves are medium-sized; another group of worker ants, the miniature workers, are generally confined to the fungal gardens within the nest.

Although usually found within the nest, the miniature workers of *Atta* do sometimes sally out. They can be seen hitchhiking a ride on a leaf fragment being carried by a medium-sized worker. One of the jobs of the miniature ant is to protect the leaf carrier from attack by phorid flies (p. 288). These gruesome flies lay an egg inside the carrier ant's head; when the maggot hatches, it eats away the contents of the head until it drops off. Little wonder that the ants panic when they detect adult flies in the vicinity. Workers that are carrying a leaf fragment in their mouth are quite defenseless since the fly lands on the leaf and inserts an egg where the mandible articulates with the head. The miniature hitchhiker ant, however, is quite proficient at chasing away flies that attempt to land on the leaf. Despite the fact that the flies are thought to be active only during the day, hitchhikers can be spotted day and night, suggesting that they have more than one function.

Leafcutter queens have one of the longest life spans of any adult insect. Depending on the species, they can live from 10 to 20 years, sometimes longer! At the beginning of her adult life, the queen mates (with several males) for the first and only time, and she continues to produce throughout her life an enormous number of eggs, all of which are fertilized with sperm she acquired at the beginning of her adult life (of course when she lays male eggs, these are unfertilized, which is the same with male eggs produced by all hymenopterans). Before a young queen departs her natal colony in order to mate and commence another colony, she places in the back of her mouth a wad of fungus, which she utilizes as seed for a new garden.

Leafcutter ants play an important ecological role that goes beyond the mere fact that they consume a huge amount of vegetation. They enrich and churn soil, thereby providing ideal sites for plant germination. And nitrogen-fixing bacteria have been discovered in fungal gardens, meaning that leafcutter ants also increase the amount of nitrogen in the soil. Fungus-growing ants, which are restricted to the New World, comprise about 250 species, most of which occur in tropical America and subtropical regions of South America.

Large mounds of dirt excavated by thousands of *Atta cephalotes* workers.

Atta bisphaerica queen in a chamber with fungal garden.

Atta cephalotes worker carrying a larva.

Excavated nest of *Atta bisphaerica* after pouring cement into the underground chambers and tunnels.

OTHER MYRMICINE ANTS

In addition to the fungus-growing ants, Myrmicinae, the largest subfamily of ants, includes a wide diversity of other ants. A trait that all Myrmicinae share (like army ants and pseudomyrmecine ants), however, is a two-segmented petiole.

Pheidole is the largest genus of ant, with more than 700 named species in tropical America, where they occur virtually everywhere, both in rainforests and homes. All of its members have a reduced stinger. Most species have two types of workers, normal workers (the majority) and large-headed workers, the latter serving to crush seeds and defend the colony. The majority are scavengers, though individual species scavenge in varying ways; some species, for example, bring fallen seeds back to their nest (at least when suitable seeds are available).

Other often seen members of this subfamily include species in the genus *Crematogaster*, which have a heart-shaped abdomen that is held elevated above the body. Most nest in dead branches, but several species use masticated plant fibers to build "carton" nests in the canopy.

Turtle ants (genus *Cephalotes*) are heavily-armored arboreal ants. Their soldiers have an enlarged, shield-like head used for blocking the nest entrance (most species nest in abandoned burrows of wood-boring beetles). If knocked off a branch, many species are able to glide (and steer) their way back to the trunk of the tree from which they fell. They feed primarily on pollen, extrafloral nectar, and honeydew, a relatively poor diet that requires supplemental nutrients that come

Turtle ant worker (*Cephalotes*). Soldiers have even larger heads, used to block the nest entrance.

Partially excavated fire ant nest with workers, larvae and pupae.

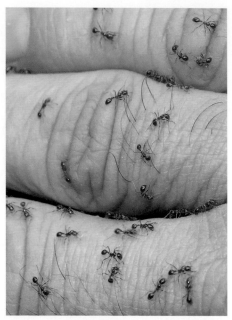

Fire ants in act of stinging.

from a diversity of endosymbiotic bacteria living in the gut of older larvae and adults. A Brazilian species (*C. specularis*) nests within the territories of a very aggressive ant (*Crematogaster ampla*) and eavesdrops on the latter's chemical recruitment trails, thereby allowing this turtle ant to locate and exploit its host's food.

Certain species in the genus *Solenopsis* are called fire ants due to their very painful sting (their venom contains alkaloids while that of most other ants contains proteins). They either nest underground (often in moist areas, under logs or rocks) or in open areas, where their excavations produce dome-shaped mounds of dirt. Fire ants are native to the Americas, though the majority of the 20 some species occur exclusively in South America. In the United States, the most notorious species is *S. invicta*, accidentally transported from northern Argentina to the southern United States in the 1930s, and known as the red imported fire ant.

The little fire ant (*Wasmannia auropunctata*) is a small (1.5 mm/0.1 in), golden-brown ant that has a surprisingly painful sting for its size. A native of tropical America, it has been

Nest of fire ant (*Solenopsis geminata*).

accidentally introduced into many other parts of the world, where it has become a serious pest, especially so on the Galapagos and other Pacific islands. Some populations of this ant reproduce in a most unusual manner. Workers (which are sterile) are produced in the normal way, from fertilized eggs, but queens are produced asexually (as a clone); males originate from fertilized eggs in which the queen's genetic material is eliminated. As a result, there is a complete separation of female and male gene pools since females produce only females and males produce only males.

PSEUDOMYRMECINE ANTS

Members of this relatively small subfamily (Pseudomyrmecinae) have elongate, slender bodies and large eyes; the majority of its species belong to the genus *Pseudomyrmex*.

Most species nest in the dead branches of a variety of trees; a few species, however, have mutualistic relationships with specific plants—or groups of plants—and occur only on those plants. One group of species in the genus *Pseudomyrmex*, for example, live on swollen-thorn acacias (*Vachellia*), from Mexico to northern Colombia. The acacias provide the ants with food bodies (Beltian bodies) produced on the tips of young leaflets, extrafloral nectar (from the leaf petioles), and a place to nest (inside the swollen thorns); in turn, the ants use their mandibles and very painful sting to aggressively defend the acacias against herbivores. They also nip off encroaching vines and make circular clearings on the ground around the base of the acacia, thereby restricting access to the plant by potential enemies (e.g., other ants) of the *Pseudomyrmex* colony; a contingent of ants is then stationed on the lower part of the acacia trunk to defend the only point of entry by (flightless) invaders.

Pseudomyrmex harvesting food bodies from leaves of a swollen-thorn acacia.

A swollen-thorn acacia showing the enlarged thorns in which *Pseudomyrmex* nest.

Swollen stem of *Cordia alliodora* (Boraginaceae) in which certain species of *Cephalotes* and *Azteca* nest.

DOLICHODERINE ANTS

Ants in the subfamily Dolichoderinae lack a stinger but are able to defend themselves by biting and by secreting defensive compounds from their rear end. If you crush one of these ants, its defensive compounds will send off a distinctive but difficult to describe odor that has been compared to the smell of both rotten cheese and anise.

Among the most common dolichoderine ants are those in the genus *Aztecta*, which are restricted to tropical America and comprise nearly 100 species. Nearly all live in trees and most of them maintain in their nests mealybugs and/or soft scales from which they obtain sugary excretions (i.e., honeydew, p. 87). *Azteca* species vary in their nesting habits— some nest in dead branches or tree cavities, others construct elaborate carton nests (primarily in South America), while still others nest in the living stems of plants with which they have an obligatory, mutualistic relationship.

For example, about a dozen *Azteca* species nest in the naturally hollow trunks of *Cecropia* trees. In addition to providing a place to nest, the plant also provides the ants with nutritious food packets (Mullerian bodies), resembling miniature rice grains, that project from felt pads at the base of each petiole; generally, you will only see Mullerian bodies on young saplings, since these lack ants (on older plants, the ants harvest the Mullerian bodies as soon as they are formed). In return, the ants protect the plant from encroaching vines and herbivores; to see this in action, slap the trunk of a *Cecropia* tree, and a teeming mass of these ants will quickly cover it. A small minority of the 60 some species of *Cecropia* harbor other types of mutualistic ants, or lack mutualistic ants entirely (especially on islands and at higher altitudes, where *Azteca* ants are absent).

Another common dolichoderine ant in tropical America is the ghost ant (*Tapinoma melanocephalum*), an extremely tiny ant that runs about erratically. This introduced species occurs only in houses and other disturbed habitats.

Azteca ants near the entrance hole in the trunk of a *Cecropia* (Urticaceae) sapling.

Azteca ants harvesting food bodies from the base of a *Cecropia* petiole.

Cut *Cecropia* stem exposing *Azteca* ants with their larvae.

Ghost ants (*Tapinoma melanocephalum*) gathered around a drop of sugar water.

FORMICINE ANTS

Like the dolichoderine ants, members of the subfamily Formicinae lack a stinger and defend themselves by biting and by spraying a chemical compound from their rear end. In this case, however, the defensive chemical compound, formic acid, lends the ants an acidic, vinegar-like smell.

The largest genus in this subfamily is *Camponotus*, with nearly 300 species in tropical America. In northern temperate regions, some species in this genus are known as carpenter ants because they burrow into wood, not to eat it but rather to build a nest. In tropical America *Camponotus* species vary considerably in their nesting habits. Some nest in preexisting cavities in dead wood while others (e.g., *C. senex* and species in the subgenus *Dendromyrmex*) use their larvae as sewing machines to join leaves together with silk; an adult ant uses its mandibles to hold a larva as it secretes silken fibers. One large, yellowish (nocturnal) species in Central America, *C. conspicuus*, has a particular fondness for nesting in electrical appliances, and the accumulation of trash that it drags inside often damages the equipment. In the Amazon *C. femoratus* is a dominant canopy ant that lives in a mutualistic association with *Crematogaster levior* (Myrmicinae); the latter builds the carton nest while the former plants epiphyte seeds in the carton (creating an "ant garden" that serves to support the nest and keep it dry) and ferociously defends the nest (most other *Camponotus* species are less aggressive).

Like most other members of the subfamily, *Camponotus* species are generalist scavengers with a diet rich in carbohydrates but poor in nitrogen. For this reason they harbor an obligate endosymbiotic bacterium (*Blochmannia*), which probably provides the ant with certain amino acids. Species in another genus of Formicinae, *Acropyga*, live underground and are totally dependent upon honeydew derived from root-sucking mealybugs (Pseudococcidae, p. 93), which they keep and tend like cattle; young queens leaving their maternal nest carry a mealybug in their mandibles with which to "seed" the new colony (the mealybugs are presumably mated females or are capable of reproducing asexually).

Camponotus harvesting honeydew from a scale insect.

6

MOTHS AND BUTTERFLIES
(order Lepidoptera)

Moths (in Spanish, *polilla*) and butterflies (in Spanish, *mariposa*) are members of the order Lepidoptera, a word derived from Greek that means *scale wing*. Indeed, their entire body is covered with minute scales, and it is the light reflected off these scales—or the pigments they contain—that give butterflies their brilliant colors (p. 243). These scales rub off easily, an attribute that probably allows many species to escape from spider webs.

Lepidopteran mouthparts consists of an elongated proboscis—coiled beneath the head when not in use—that functions somewhat like a drinking straw. The vast majority of moths and butterflies are therefore confined to a liquid diet; some species feed on nectar from flowers, for example, while others feeds on the juices oozing from rotting fruits. In some groups of moths, adults have a reduced proboscis and feed very little, if at all.

The larvae of moths and butterflies are called caterpillars. They can be distinguished from the larvae of most other insects by the presence of abdominal prolegs, which are fleshy stubs that lack the joints of true legs; most species have five pairs of prolegs (in addition to three pairs of thoracic legs). The larvae of most other insects lack prolegs (or have only one pair at the very end of the abdomen). Sawfly larvae (p. 154), which are very similar to caterpillars, usually have more than five pairs of prolegs.

As is the case with the majority of insects that undergo complete metamorphosis, moth and butterfly larvae (caterpillars) consume more food than do adults. Most caterpillars feed on plants; generally, a given species of caterpillar feeds on a restricted range of plants, many of which have chemical defenses that the caterpillar has evolved to withstand (by detoxifying the plant's chemicals) or to use to its advantage (p. 223). Caterpillars that live exposed on the surface of plants, sometimes incorporate the plants' nasty compounds into their bodies and often have bright colors to advertise their toxicity to potential predators. In many species, the defensive compounds of the caterpillar are passed on to the adult, but in other species, the compounds are eliminated during pupation.

In many species, caterpillars simply feed on plants (or plant parts) that contain non-lethal levels of defensive chemicals (or, if the plants do contain lethal levels of chemicals, the caterpillars detoxify them). Caterpillars that do not carry defensive chemicals must avoid predators by making themselves hard to find. Those that feed inside the plant's tissues (e.g., stem borers) are concealed from many predators (though not from parasitic wasps, p. 155). Other species hide during the day and feed only at night. The majority of caterpillars that live on the plant surface are cryptically colored (camouflaged) and some of these also bear false eye spots that serve to startle birds at close range.

MOTHS

At least 90% of the species in the order Lepidoptera are moths. The antennae of moths are usually either threadlike or feathery, a feature that distinguishes them from butterflies, whose antennae have swollen tips. Many male moths have noticeably branched antennae that allow them to detect very small quantities of female sex pheromones, chemical compounds that females release into the air when they are ready to mate.

Many moths are nocturnal, others are diurnal, but nearly all butterflies are diurnal; several diurnal moths are brightly colored (advertising to predators their chemical defenses), while other diurnal moths and virtually all nocturnal moths are more drably colored, often relying on camouflage to avoid predators.

Several nocturnal moths fly towards night lights, especially during a new moon, a phenomenon that is thought to result from the moths using the moon as a fixed marker while navigating. Substituting an artificial light as a marker and keeping it at a constant angle during flight, the hapless moth ends up spiraling in toward the light.

Most medium to large nocturnal moths are preyed on by insectivorous bats; to defend

Butterfly antenna (Nymphalidae: Apaturinae: *Doxocopa cyane*).

themselves, many moths have a pair of tympanic organs—located either at the back of the thorax or at the front of the abdomen—that they use to detect bat sonar. When an airborne moth detects a bat, it takes evasive maneuvers, plunging toward the ground or flying in erratic loops.

GHOST MOTHS

Ghost moths (Hepialidae) are among the most ancient of moths. Their front wings and hind wings are of similar size and shape (in most lepidopterans the front wings are larger and longer than the hind wings); in flight, the front and hind wings are not hooked together and therefore move independently of each other, whereas in most other moths the front wing and hind wing are hooked together, and move as a single unit.

At dusk, ghost moths engage in rapid flight, tracing a zigzag pattern, often near the ground. Flying females drop thousands of eggs onto the ground, near a host plant. Little is known about young caterpillars, but older caterpillars often bore into the stems of shrubs and trees that have relatively soft wood, or they feed on roots. There are approximately 600 named species worldwide.

Moth antenna (Glyphipterigidae: *Acrolepia*).

Aepytus ghost moth (Hepialidae).

Ghost moths, such as this *Druciella*, typically hang rather than perch.

Micropterigidae, the only family of Lepidoptera in tropical America having mandibles instead of a proboscis.

FUNGUS MOTHS AND BAGWORM MOTHS

The superfamily Tineoidea includes two families with species whose caterpillars protect themselves by constructing a silken tube or case around their body. In some species, the protective case is nothing more than a flimsy stationary retreat attached to the substrate on which the caterpillar feeds; in other species, the caterpillar constructs a more durable portable silk case that incorporates bits of debris.

Caterpillars of both types occur among the fungus moths (Tineidae). Caterpillars of the webbing clothes moth (*Tineola bisselliella*), for example, spin small silken feeding tunnels. They eat fur, wool, and other animal products, and are among the few insects capable of digesting keratin (the protein found in these materials). This cosmopolitan species is a pest that plagues wardrobes in many parts of the world.

Unlike most caterpillars, which feed on living plants, the tiny caterpillars of fungus moths feed on fungi, dried plant material (including cork), dried animal materials, bat guano, dry carrion, and nest debris.

In tropical America, the caterpillars of two common home invaders (*Phereoeca uterella* and *Praeacedes atomosella*) build portable cases that are flattened, oval-shaped, impregnated with sand grains or other fine particles, and

Silk cases of fungus moth larvae: *Phereoeca uterella* (left) and *Praeacedes atomosella* (right).

open at both ends. They feed on fallen hairs, dead insects, and old spider webs, becoming most noticeable when they drag their portable cases up walls in order to pupate.

The caterpillars of most bagworm moths (Psychidae) build tough, portable bag-like cases that incorporate small twigs or other bits of plant material. Only the head and legs protrude from the bag as the larva moves about, often on the underside of a leaf. Although some of the more primitive groups are omnivorous scavengers, most species feed on the foliage of trees, and are often polyphagous. Several species feed on lichens and other fungi. The caterpillars of *Perisceptis carnivora* (a species recently discovered in Panama) feed on various arthropods and then incorporate the carcasses of their prey into the larval bag.

In many species of bagworm moth, adult females are wingless; they spend their entire life within the bag created by the caterpillar. They attract males with a sex pheromone; in order to mate, the male thrusts his telescoping abdomen into the female's bag in order to reach her. In these species, the female lays her eggs within her bag; after hatching from the eggs, the larvae disperse in the wind, suspended from a silk thread, a method known as ballooning.

Worldwide, there are 2,400 named species of fungus moth and 1,350 of bagworm moth.

Larva of Psychidae that feeds on lichens.

Bagworm caterpillar (Psychidae: Oiketicinae) concealed inside its protective case. Only the legs holding onto the leaf are visible.

Bagworm moth, wingless female; unlike a larva, she has compound eyes (at left) and a scale skirt (right).

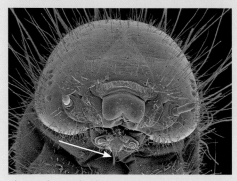

Photo taken with electron microscope of caterpillar head (Riodinidae: *Euselasia chrysippe*) showing spinneret (arrow) from which silk is secreted.

Caterpillar of small moth hanging by silk thread.

Silk is one of the strongest natural fibers that we know of. While there are noticeable differences between the various kinds of silk, all are composed of fibrous proteins. Arthropods are the only organisms known to produce silk; they store the silk-producing substances as liquid, which turns into fibers on contact with the air. Many arthropods produce silk, including arachnids, centipedes, webspinners (p. 40), barklice (p. 49), and the larvae of numerous insects (e.g., moths and butterflies).

In caterpillars, as in most insects, it is the salivary glands that produce the silk. Unlike many other insects, which use silk for a single purpose, caterpillars use silk for a variety of functions: as a means of rolling or tying leaves together into a protective shelter (a few tree-inhabiting species build communal shelters); as a lifeline that stops the fall of the insect after it jumps off a plant to escape a predator; and as material for spinning cocoons (within which the fully grown caterpillar pupates). Cocoons vary from a loose net to an extremely hard, compact case. Caterpillars that pupate in protected sites such as leaf-rolls, stems, or soil often do not spin a cocoon. Nonetheless, in most species of butterfly, even those in which the pupa is often situated on an exposed location on a plant, the caterpillar does not spin a cocoon, depending instead on other means of defense.

Silk fabric has several highly desirable properties. It is strong, light, crease-resistant, easy to dye, lustrous, and breathable. Commercial silk is obtained from cocoons of *Bombyx mori*, or silkmoth (Bombycidae), which feeds on mulberry trees and was first domesticated in China about 5,000 years ago. In ancient China the exportation of silkworms was reputedly punishable by death, a policy designed to maintain a high price on the final product, which was used as currency and traded along the Silk Road through Persia. Legend has it that in the Sixth Century two monks hid silkworm eggs in a hollow cane and smuggled them to Constantinople, thereby breaking the Chinese manufacturing monopoly and the Persian trading monopoly.

Nonetheless, the Aztecs produced silk products independently from Asian peoples, using undomesticated caterpillars. The moth *Gloveria psidii* (Lasiocampidae) and the sulfur butterfly *Eucheira socialis* (Pieridae) are two species from Mexico whose caterpillars produce silk. Rather than unwinding the fibers and then weaving them, as is done with domesticated silkworms, the Aztecs simply cut up the silken sacs and then pieced them together.

The undisputed masters of silk production are the spiders (p. 310). Unlike most insects, individual spiders produce a variety of silks, each for a specific purpose, and they produce it throughout their lives. Certain types of spider silk are renowned for their extraordinary mechanical properties (e.g., dragline silk is tougher than Kevlar), and scientists are conducting considerable research into the possibility of producing it synthetically on an industrial scale for a wide variety of potential applications. While this effort entails the latest technology, the idea of using spider silk is not new—Dr. Thomas Muffet (who perhaps inspired the nursery rhyme), a 16th century physician and naturalist with a fondness for spiders, supposedly used fresh spider webs to treat cuts.

LEAF MINER MOTHS

Many moths—and their larvae—are so small that they often go unnoticed except by specialists. Nonetheless, the tiny caterpillars of some of these moths—or at least the signs of their presence—are often quite easy to see while walking through a tropical forest, or almost any other habitat. This is because these caterpillars are leaf miners, feeding within leaves, where they make clearly visible paths across the surfaces of the leaves. Not all leaf miners are caterpillars; the larvae of some species of beetle and fly also tunnel their way through leaves, but, indeed, the vast majority of leaf miners are caterpillars of minute moths.

The most species-rich family of leaf miners is Gracillariidae, which comprises about 2,000 named species worldwide.

Phyllocnistis (Gracillariidae) comprises hundreds of species in tropical America, but very few have been named.

Glyphipterigidae. The larva of this species is a leaf miner on *Iresine* (Amaranthaceae).

Leaf-mining larva of *Phyllocnistis maxberryi* on *Gaiadendron punctatum* (Loranthaceae).

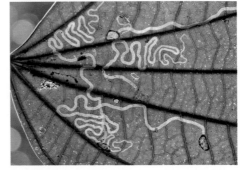

Leaf mine of a micromoth larva (probably Nepticulidae).

LEAF MINER INSECTS

The citrus leaf miner (Gracillariidae: *Phyllocnistis citrella*) is originally from Asia but was accidentally introduced into tropical America in the 1990's.

Insects and mites have evolved varied techniques to feed on plants. While many species feed externally on plants, some bore within stems or fruits, and, in yet other species, the larvae mine the interior of leaves. Such leaf miners, certain species of moth, beetle, and fly (and at least one mite species, on water hyacinth), must be tiny enough to fit within the thin leaves, through which they tunnel while eating the internal plant tissues. The leaf pattern that results from such feeding—often intricate and sometimes beautiful—varies greatly. Some leaf miners leave behind irregular blotches, others narrow meandering paths. The exact form of the leaf mine, combined with the identity of the host plant, is often a key to identifying the species that made it. Many species also have a characteristic manner of depositing their feces; some species deposit feces along the sides of the leaf mine, others down the center of the ceiling.

Some caterpillars are leaf miners only when young. On reaching a certain age, they exit the mine and begin feeding externally, often from within a sheltered site that the caterpillar itself creates; in some species, for example, the caterpillar bends (or rolls) a leaf and then ties it with silk.

Depending on the species, leaf miners either pupate within the leaf mine or exit the leaf mine in order to pupate. If you wish to attempt seeing if the leaf miner larva is still present in the mine, hold the leaf against the sun. Since a leaf that contains a leaf mine usually remains on the plant long after the culprit has left, most of the leaf mines that one commonly encounters are empty.

Nonetheless, one can often read the history of an empty leaf mine, noting where the insect entered the leaf and where it departed; you will often see an incomplete leaf mine, indicating that some tragedy struck the leaf miner before it could finish its work—many parasitic wasps, for example, specialize in attacking leaf miners.

TWIRLER MOTHS AND THEIR KIN

Walking through a rainforest, you might spot a tiny moth spinning rapidly in a circle on the upper surface of a leaf, a behavior that biologists have observed most frequently in certain species of twirler moth (family Gelechiidae) and in closely related micromoths (other families in the superfamily Gelechioidea). The function of this behavior is still unknown. In *Beltheca oni*, both sexes twirl, and they do so by anchoring one of their forelegs to the leaf surface and rotating around this pivot.

The feeding habits of the caterpillars in this group (superfamily Gelechioidea) are very diverse. Many begin their lives as leaf miners, then become leaf tiers as they grow larger. Others are stem borers, fruit borers, or gall inducers (e.g., on melastomes). Some feed on dead plant material. A few are common in kitchens, where the larvae of such species as *Sitotroga cerealella* feed on flour and other stored foods.

Worldwide there are nearly 5,000 named species of Gelechiidae.

Beltheca oni (Gelechiidae) is often found along forest edges.

Another Gelechiidae that twirls on leaves.

Gelechiidae (possibly *Strobisia*) in process of twirling.

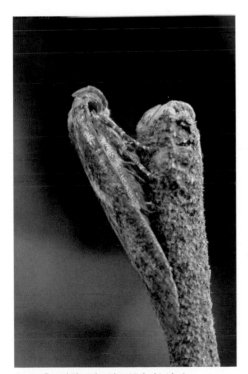

Camouflaged Blastobasidae (Gelechioidea).

Gall induced by *Mompha* (Gelechioidea: Momphidae) on *Blakea* (Melastomataceae).

Antaeotricha (Depressariidae: Stenomatinae), resembling a bird dropping (head on right).

Larva of Gelechiidae that burrows into potatoes.

FLANNEL MOTHS AND SLUG CATERPILLARS

These stout bodied, hairy moths (superfamily Zygaenoidea) usually have a very short proboscis. While the moths themselves are rather nondescript, many of their caterpillars are very distinctive and diverse (ranging from very hairy to slug-like). Many of these caterpillars are polyphagous, often feeding on old and/or tough leaves; a few species are pests of palms.

The caterpillars of flannel moths in the family Megalopygidae are extremely hairy. Do not attempt to pet them, however, for the fur conceals venomous spines that can cause lesions and even fever. This family is restricted to the New World and comprises about 250 named species, most of which occur in tropical America.

Slug caterpillars (Limacodidae) have reduced legs and prolegs; they move about on a series of sucker-like disks (lubricated by fluid silk) that are attached to the ventral surface. Many are bright green; some have tubercles with urticating spines. Among the most bizarre caterpillars of tropical America are those in the genus *Phobetron*, which, somewhat resembling a small deformed starfish, have paired arm-like projections that can be pulled off to escape a predator. The cocoons of slug caterpillars are brown, egg-shaped, and very hard (due to the presence of calcium oxalate); the cocoons have a circular escape hatch that the pupa pushes open just prior to the emergence of the adult moth. There are 1,700 named species of slug caterpillar worldwide.

Trosia nigropunctigera (Megalopygidae).

Larva of *Megalopyge* (Megalopygidae) feeding on a legume (Fabaceae).

Perola producta (Limacodidae) on twig, with front legs extended forward.

Larva of *Parasa macrodonta* (Limacodidae) with stinging spines.

Larva of *Isa diana* (Limacodidae) feeding on *Miconia calvescens* (Melastomataceae).

Larva of *Venadicodia caneti* (Limacodidae) feeding on Lauraceae.

Larva of *Acharia horrida* (Limacodidae).

Larva of *Natada* (Limacodidae) on *Miconia*.

Larva of *Phobetron hipparchia* (Limacodidae).

The caterpillars of the family Dalceridae are covered with gelatinous warts that repel ants; they move about in a manner similar to that of slug caterpillars. This family is restricted to the New World, with nearly 100 named species.

CASTNIID MOTHS

Moths in the family Castniidae are diurnal. They have clubbed antennae like those of butterflies, but tend to have thicker bodies than do butterflies. The scales on their wings are quite slippery and are easily rubbed off. The larvae are stem borers that feed on palms, bromeliads, sugarcane, bananas, and related plants.

This family has about 120 named species, most of which occur in tropical America. Castniid moths also occur in Australia and Southeast Asia.

Larva of *Perola* or *Epiperola* (Limacodidae).

Dalcerides (Dalceridae).

Telchin diva male. Castniids are day-flying moths.

Larva of *Acraga* (Dalceridae). The gel-like projections can be detached and replaced.

Telchin atymnius flies rapidly along streams and then perches.

TORTRICID MOTHS

Moths of the family Tortricidae are generally 3 cm (1.2 in) or less in wingspan; as is the case with many other families of moths, most tortricid moths are drab, though a few diurnal species are somewhat brightly attired. Many species rest with their wings held in a characteristic bell-shaped silhouette, but given their small size and drab coloration, they are fairly difficult to recognize in the field.

Tortricid caterpillars feed primarily on dicotyledonous plants. Biologists divide them into two groups.

Those in the first group (subfamilies Childanotinae and Olethreutinae) tend to be borers in stems, roots, buds, fruits, or seeds. The macadamia nut borer (*Gymnandrosoma aurantianum*) burrows into the fruits of several plants. *Cydia deshaisiana* feeds inside the seeds of *Sebastiania pavoniana*, a shrub in the spurge family (Euphorbiaceae); when the seeds fall to the ground and are heated by the sun, the caterpillar grasps the inside wall of the empty seed with its legs and snaps its body, causing the seed to hop, and giving them the popular name

Pseudatteria volcanica (Childanotinae) mimics certain Geometridae, Pyraloidea, and other Tortricidae.

Mexican jumping bean. This behavior of the caterpillar probably serves to move the seed out of the hot sun—and into a more sheltered location.

Many caterpillars in the second group (subfamily Tortricinae) are external feeders that construct leaf rolls; these species are considerably more polyphagous than members of the first group. A distinctive characteristic of caterpillars in this group is an anal fork that is used to flip excrement away from the larval shelter.

There are more than 10,000 named species of tortricid moth worldwide; if one includes the large number of unnamed species, the American tropics probably harbor more species than any other part of the world.

METALMARK MOTHS

Metalmarks (Choreutidae) are small (with a wingspan of 10 to 20 mm/0.4 to 0.8 in), diurnal moths. They are sometimes seen resting on the upper surface of leaves.

Metalmark moths in the genus *Brenthia* mimic jumping spiders (p. 316); when the moth raises its wings, the color patterns on the wings resemble the legs of jumping spiders, an illusion enhanced by the moth when it moves about in a jerky manner that mimics the movement of the spiders. Indeed, jumping spiders often react to metalmark moths as if they were jumping spiders rather than potential prey. Caterpillars in this genus spin a silken sheet over the leaf surface, beneath which they feed; they also chew into the leaf an escape hole, through which they can exit with incredible speed, the act of disappearance taking 100 milliseconds to effect! There are more than 400 named species of metalmark moth worldwide.

Tinacrucis (Tortricinae) uses scales from its abdomen to build a fence around the egg mass, to keep out ants and mites.

Hilarographa (Childanotinae) often moves backwards as if its rear end (to the right) were the head.

Tortyra. Larvae in this genus feed on fig leaves.

Casuaria (Pyralidae: Chrysauginae).

Brenthia species mimic jumping spiders (p. 316).

Diaphania (Crambidae: Spilomelinae). Larvae in this genus feed on plants in the cucumber family (Cucurbitaceae).

SNOUT MOTHS AND THEIR KIN

Members of the superfamily Pyraloidea are commonly known as snout moths due to the fact that some species have a pair of prominent palps that project in front of the head. This is one of the largest and most diverse groups of moths, with about 16,000 named species worldwide—and perhaps an equal number of unnamed species. While most snout moths are drab nocturnal creatures, some species are brightly colored (and a subset of these are diurnal).

The caterpillars of snout moths show a diversity of feeding habits. Most species feed on plants—either as stem borers (e.g., the infamous sugarcane borer), leaf tiers (which use silk to roll or fold leaves), leaf miners (p. 202), root feeders, seed feeders (including stored grains and flour), or even gall formers (p. 280). One group (subfamily Acentropinae) includes species that are adapted to living underwater in streams, where they feed on algae attached to submerged rocks. Other species feed on detritus, while a couple other species feed on the brood of paper wasps; the caterpillars of wax

Spilomelinae resembling a dead leaf.

moths eat the pollen, nectar, and wax inside honeybee hives (mostly in weakened colonies).

Within this superfamily are species in the genera *Bradypodicola* and *Cryptoses* that are called sloth moths; if you are lucky enough to see a sloth at close range, you should be able to see these little drab brown moths scuttling around in their fur; a single sloth can carry over 100 moths. When the sloth descends to the ground to defecate, the sloth moths dismount momentarily and lay eggs in the dung, which is where the sloth moth caterpillars feed and pupate.

Diaphania plumbidorsalis, with iridescent wings.

Cryptoses (Pyralidae: Chrysauginae) on a sloth.

TENT CATERPILLARS

The caterpillars of many of these moths (Lasiocampidae) live in groups that construct large sheets of silk, or tents, around the foliage of trees; these tents serve as communal shelters for the caterpillars when they are not feeding. It should be noted, however, that this is not a unique feature, as there are other families of moths in which the caterpillars build tents.

Most tent caterpillars are hairy; depending on the species, they either live in large groups

or alone (solitary species, of course, do not construct communal tents). The moths themselves are of medium size; the majority are nocturnal. Many species rest with their front wings in a roof-like position, but with the hind wings extended.

There are about 2,000 species of tent caterpillar worldwide.

Euglyphis is the largest genus of tent caterpillars, with about 350 named species in tropical America.

Larva of *Gloveria* feeding on oak (see also p. 200).

GIANT SILK MOTHS

Giant silk moths (Saturniidae) are closely related to silk moths (Bombycidae, p. 210). They have stout bodies, and wing spans that range from 4 to 15 cm (1.6 to 5.9 in).

With the exception of species in the subfamily Oxyteninae, adults lack functional mouthparts and are therefore incapable of feeding. Unlike many nocturnal moths, most giant silk moths lack "ears" for detecting bats; nevertheless their habit of flying erratically, close to the ground, and in brief spurts, probably helps them avoid predators such as bats.

Adults in the subfamily Saturniinae have on each wing a transparent window, the function of which is uncertain. They attach their cocoons to the host plant of the caterpillar; in other subfamilies, most species pupate on the ground.

Adults in the subfamily Hemileucinae have on each hind wing a large eyespot that is normally hidden when a moth is at rest; but if disturbed, the moth elevates its front wings to expose this false eye, which probably serves to startle diurnal predators by convincing them they are dealing with a much larger animal than a moth. (The false eye is surrounded by a ring, as in Saturn, hence the family name Saturniidae.)

The caterpillars of giant silk moths feed mostly on trees. They snip off pieces of older,

tannin-rich leaves, and swallow them without masticating them much at all. The caterpillars of many species in the subfamily Hemileucinae bear stinging spines. When young, Hemileucinae caterpillars are gregarious; they rest lower in the canopy during the day, and then move in columns upward to the canopy at night (their movement in procession accounts for the common name *processionary caterpillar*). As they grow older, these caterpillars become increasingly solitary.

Worldwide there are about 2,350 species of giant silk moth, and about half occur in the Americas.

Male *Copaxa syntheratoides* (Saturniinae).

Male *Rothschildia triloba* (Saturniinae).

Male *Antheraea godmani* (Saturniinae).

Female *Antheraea godmani*.

Male *Leucanella hosmera* (Hemileucinae), showing false eye spots when it opens its wings.

Epia (Bombycidae, a family related to Saturniidae) curling its wings and resembling a lichen.

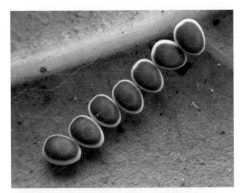

Eggs of *Copaxa* are usually laid in a single row.

Larva of *Rothschildia lebeau*.

Larva of *Molippa nibasa* (Hemileucinae) with stinging spines; it dropped to the ground and curled into a defensive position.

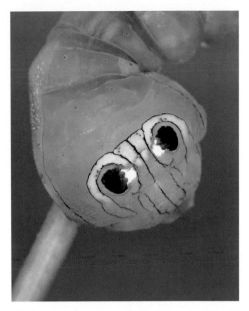

The larva of *Oxytenis modestia* (Oxyteninae) might startle a bird.

HAWK MOTHS

Hawk moths (Sphingidae) are easily recognized by a combination of features: cigar-shaped bodies, long narrow wings, and a hummingbird-like flight. They are among the fastest insects, with some species capable of flying at speeds over 50 kph (30 mph).

While most species are nocturnal, only species in the genus *Xylophanes* and some other species possess tiny structures on both sides of the mouth that allow them to detect the ultrasounds produced by bats, a key nocturnal predator. That said, hawk moths have excellent night vision, including a capacity to see colors in the dark.

Adults feed on nectar and are important pollinators of various night-blooming orchids, cacti, mimosoid legumes, and several members of the coffee family. It is estimated that 10% of the trees in some parts of tropical America depend primarily on hawk moths for pollination.

Many species have long tongues used to obtain nectar from tubular flowers; in fact some species are the longest-tongued insects in the world. In the New World, the insects with the longest tongues (much longer than the body) are two species of hawkmoth that occur throughout tropical America: in second place is *Neococytius cluentius*, with an average proboscis length of 25 cm (9.8 in), while the record holder is *Amphimoea walkeri*, with an average proboscis length of 28 cm (11.0 in).

Hawk moths often hover while sipping nectar from flowers, with some species moving rapidly from side to side while hovering, which probably makes it more difficult for predators to catch them (spiders and other predators often wait in flowers in order to ambush nectar-feeding insects).

The caterpillars can be large, with those of some species measuring up to 14 cm (5.5 in) in length). Most have a spine that projects from the rear end, which is why they are often called hornworms. The majority have cryptic (often green) coloration, and some of these (e.g., species of *Xylophanes*) also have eye-spots on the body. *Hemeroplanes triptolemus* and a few other species are snake-like in appearance, an adaptation that apparently serves to ward off predators.

When disturbed, the caterpillar retracts its head and raises the front end of its body (hence another common name, *sphinx moth*, though *sphinx caterpillar* would be more accurate) and

either adopts an intimidating pose or thrashes from side to side; as an added defense, the caterpillars regurgitate sticky material onto the assailant. Many hawk moth caterpillars feed on plants that contain alkaloids (e.g., those in the coffee family), milky latex (e.g., the dogbane family), or calcium oxalate crystals (e.g., the grape family). Caterpillars that feed on plants containing milky latex sometimes gnaw on the petiole or leaf vein before feeding in order to drain the latex from the leaf. They generally pupate in the soil without spinning a cocoon.

Worldwide, there are about 1,500 species of hawk moth.

Amphimoea walkeri showing the length of the tongue.

Xylophanes chiron. The larva feeds on Rubiaceae.

Like all hawk moths, *X. chiron* has large eyes.

Two *Xylophanes thyelia* resembling dead leaves.

Xylophanes letiranti warming up for flight.

Manduca florestan resembling lichens.

Perigonia stulta warming up for flight.

Larva of *Eumorpha satellitia*, frontal view. It feeds on plants in the grape family (Vitaceae).

Larva of *Hemeroplanes triptolemus* showing uncanny resemblance to a snake.

GEOMETER MOTHS AND INCHWORMS

Geometer moths (Geometridae) have slender bodies and broad wings. When at rest, adults hold their wings upright (in the manner of some butterflies) or outstretch them to form a semicircle. Male antennae are typically feather-like; in most species, female antennae are unadorned.

In the majority of species, the forewing and hind wing are the same color and show lines that continue from the front wing through the hind wing. Although some are brightly-colored, diurnal species, most are more dully colored and nocturnal. Among the diurnal species are the emerald moths (Geometrinae), which are light green in color; this comes from a unique pigment (geoverdin) that occurs in large concentrations only in these moths.

In the majority of species of Lepidoptera, caterpillars have five pairs of prolegs on their abdomen, but the caterpillars of geometer moths, known as inchworms (or loopers), have only two pairs. This explains in part their unique style of locomotion: they arch their body, bringing the rear end forward, place the prolegs next to the true legs, and then extend the front end forward. Most inchworms are cryptic in appearance; some resemble mossy bark or lichens, while others are adept at mimicking twigs, which they accomplish by maintaining their camouflaged body in a frozen upright posture. Inchworms generally feed externally on the foliage of vascular plants (though a few feed on lichens). They are especially diverse on understory vegetation; nearly all species in the large genus *Eois*, for example, feed on pepper bushes (*Piper*).

Geometridae is one of the largest families of Lepidoptera and currently comprises more than 23,000 named species worldwide, about a third of which occur in tropical America, where they are especially diverse in montane forests.

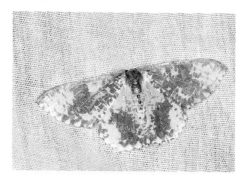
Oospila emerald moth resembling lichens.

Phrygionis polita. The larva feeds on Primulaceae trees.

Leuciris moves its wings up and down while walking.

Prochoerodes.

Rhodochlora, another genus of emerald moths.

Erateina imbibing nutrients from dung.

Epimecis perched on a tree trunk.

Larva of *Melanchroia* feeding on Phyllanthaceae.

Larva of *Thysanopyga* moving in typical inchworm manner.

Larva of *Oxydia* frozen in the position of a twig.

URANIAS

The family Uraniidae contains four subfamilies, only two of which occur in the New World: Epipleminae, the largest subfamily, with medium-sized (wing span usually 2-4 cm/0.8-1.6 in), drab, nocturnal moths; and Uraniinae, which (in the New World) includes large (wing span is 10 cm/3.9 in), colorful (with iridescent green stripes), diurnal moths that resemble swallowtail butterflies (p. 228). Of the seven genera that compose the subfamily Uraniinae, just one genus, *Urania*, occurs in the Americas. It contains (or once contained) a mere six species: *U. boisduvalii* and *U. poeyi* in Cuba; *U. sloanus* in Jamaica (sadly, now extinct); *U. fulgens* from Veracruz (Mexico) to northwestern Ecuador; and *U. brasiliensis* and *U. leilus* in South America.

Urania spp., like those in two other diurnal genera of Uraniinae—*Alcides* (in Australia and beyond) and *Chrysiridia* (in Madagascar and East Africa)—are famous for undertaking epic migratory flights. During these migrations, for a period of a week to a month, hundreds to thousands of individuals take flight, usually less than ten meters (11 yards) above the ground, traveling at about 20 kph (12 mph). These migrations apparently occur when caterpillars encounter diminishing supplies of the plants they feed on (primarily members of the genus *Omphalea*, in the family Euphorbiaceae).

Sematura (Sematuridae, a family related to Geometridae and Uraniidae).

Syngria (Epipleminae) resembling a dead leaf.

Urania fulgens on *Coccoloba tuerckheimii* (Polygonaceae).

The caterpillars of *Urania* are clad in striking colors, presumably to warn predators that they possess noxious alkaloids sequestered from their host plants. These same compounds also protect adults, which feed primarily on nectar from white flowers of legumes; males, in particular, are highly attracted to flowers of the genus *Inga*.

PROMINENT MOTHS

While most adult moths in the family Notodontidae are nocturnal and display drab colors (quite similar to the coloration of owlet moths), many members of the tropical American subfamily Dioptinae are diurnal and display brilliant warning coloration. Adults in this subfamily often mimic the warning coloration of other diurnal moths and butterflies.

The caterpillars of prominent moths are usually hairless; as a group, they show great variation, and many species have bizarre shapes. On most, the hind pair of prolegs are either much longer or much smaller than the other prolegs—and these caterpillars generally do not use this hind pair in walking. Some employ chemical defenses (e.g., formic acid) not commonly found in other caterpillars; these substances are usually emitted from glands in the neck. There are about 4,000 named species of prominent moth worldwide.

Rifargia. The larvae often feed on Anacardiaceae and Burseraceae.

Pentobesa resembling a wood chip. The larvae feed on Fabaceae.

Colax, frontal view. The larvae feed on Fabaceae.

Hardingia maximespina, frontal view.

Tithraustes (Dioptinae). The larvae feed on palms and *Heliconia*.

Larva of *Rhuda difficilis* on *Miconia* (Melastomataceae).

Larva of *Naprepa houla*, another melastome feeder.

The larva of *N. houla* turns bluish just before spinning its cocoon.

OWLET MOTHS

Members of the family Noctuidae are medium-sized moths (wingspan usually 3-8 cm/1.2-3.1 in) with robust bodies. At night they are common visitors to lights, where their eyes reflect the light, giving off a noticeable orange glow (this also occurs in other nocturnal moths, but is especially noticeable in owlet moths). While most species are drab and nocturnal, a few (e.g., members of the subfamily Agaristinae) are brightly colored and diurnal. Depending on the species, owlet moths feed on nectar, rotting fruits, or tree sap.

The caterpillars of many owlet moths feed on herbaceous plants. Indeed, several species are agricultural pests. For example, cutworms, with species in several genera, leave the soil at night to feed on low growing plants, sometimes cutting off small seedlings near ground level and leaving just a tiny stump; the fall armyworm (*Spodoptera frugiperda*) is a major pest of corn.

Caterpillars in the subfamily Plusiinae have only three pairs of prolegs (as opposed to the usual five) and move in a looping manner similar to that of inchworms (p. 215). There are about 12,000 named species of owlet moth worldwide.

Dargida owlet moth (Hadeninae). The larvae feed on grasses.

Heterochroma (Amphypirinae).

Larvae of *Peridroma semidolens* feeding on *Bocconia* (Papaveraceae).

Cutworms hide in soil during the day.

Neostictoptera (Nolidae, a family related to owlet moths). The raised abdomen resembles a broken twig.

EREBID MOTHS

Due to new understanding of the evolutionary relationships among moths, Erebidae has been recently proposed as the family name for tussock moths, tiger moths, and many species that were previously placed among the owlet moths. (The tussock moths and tiger moths were previously classified as independent families but today are placed in subfamilies within the family Erebidae.) Under this new classification, Erebidae now becomes the largest family of Lepidoptera, with 25,000 named species worldwide.

Among the most spectacular erebid moths are the white witch (*Thysania agrippina*), with the widest wingspan (up to 28 cm/12 in) of any moth or butterfly, and the black witch (*Ascalapha odorata*), not quite as large as the white, but easier to see, as it readily approaches house lights, especially during periods of rain. Caterpillars of both species feed on the leaves of woody legumes.

Another fascinating member of the family is *Letis mycerina*, a species that, during territorial flights at dusk, produces sharp cracking sounds somewhat similar to the sounds of cracker butterflies (p. 262).

Moths in the genus *Gonodonta* (40 species, all restricted to tropical America) and other genera of the tribe Calpini use their specially adapted proboscis (with tearing hooks) to pierce soft fruits (e.g., citrus) and suck up the juice.

Eulepidotis (Erebidae: Eulepidotinae).

Mellpolis fasciolaris (Erebinae)

Fernanda Retana Alvarado with *Thysania agrippina* (Erebinae).

Larva of *Antiblemma* (Eulepidotinae) on Melastomataceae and resembling a leaf vein.

Larva of *Gonodonta* (Calpinae) moving like an inchworm.

TUSSOCK MOTHS
(family Erebidae)

Tussock moths (Erebidae, subfamily Lymantriinae) are drab, nocturnal moths that do not feed as adults. The family is best known for the infamous gypsy moth (*Lymantria dispar*), which in the 19th century was introduced from Europe to North America, where it defoliates many common shade trees. Tussock moths are named for the appearance of their caterpillars, which in many species have dense tufts of hairs protruding from the body. The caterpillars of many species are polyphagous, feeding on the leaves of various tree species.

Worldwide there are over 2,500 named species of tussock moth, with only about 200 species in the New World (the vast majority occur in the Old World tropics).

Eloria (Lymantriinae) occasionally flies during the day in the forest understory.

Larva of *Thagona tibialis* (Lymantriinae) feeding on *Terminalia catappa* (Combretaceae).

Pupa of *Sarcina purpurascens* (Lymantriinae), head.

TIGER MOTHS
(family Erebidae)

Some tiger moths (Erebidae, subfamily Arctiinae) are mostly white, others have cryptic (or disruptive) patterns, while still others have flamboyant colors that warn potential predators they contain noxious chemical compounds. Caterpillars obtain these chemical defenses (mostly pyrrolizidine alkaloids; p. 224) from their host plants and pass them on to adults; in some species, the adults themselves gather the chemicals from target plants (instead of or in addition to those obtained during the larval stage). Adult males are particularly avid collectors of these chemicals since they use them not only for their own defense but also to synthesize sex pheromones and to offer as a nuptial gift during mating (the male passes some of the chemicals to the female via his sperm packet, and the female, in turn, passes some of the chemicals along to her eggs).

Species in the genus *Cosmosoma* and some other tiger moths spend nine hours copulating, during which time they are extremely vulnerable to predation; in order to deter spiders from attacking, the male discharges alkaloid impregnated fibers over the female.

In addition to relying on their chemical defenses, a number of species of tiger moth also mimic the warning coloration of other distasteful or otherwise noxious insects. Many species of *Hyalurga*, for example, closely resemble glasswing butterflies (p. 249), while members of the genus *Chetone* are very similar to passion-vine butterflies (p. 258). Numerous tiger moths show a very precise mimicry of wasps—in their color pattern, narrow transparent wings (and the way in which they fold them), and even by having waists similar to those of wasps.

Many tiger moths, especially the less colorful species, fly at night, when bats are active. Unlike other moths, tiger moths generally do not undertake evasive maneuvers when they detect bat ultrasounds, but instead emit a burst of ultrasonic clicks that apparently serves to warn bats that they contain distasteful compounds.

In most species, the caterpillars are hairy. The hairiest of the caterpillars are known as woolly bears. While some feed on just a few types of plants, many are quite polyphagous; in many cases, however, even the polyphagous species are discriminating, seeking out only plants that contain alkaloids. Caterpillars in the subfamily Lithosiinae feed on blue-green algae (cyanobacteria) and lichens.

Worldwide there are about 11,000 named species of tiger moth. There are more species in tropical America than in any other part of the world.

Argyroeides notha (Arctiinae). The pale green area at the base of the abdomen gives it the look of a wasp waist.

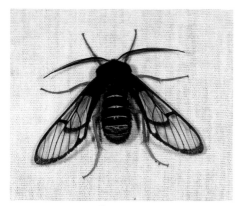

Homoeocera in normal resting position.

Same moth performing its warning display.

Hipocrita drucei, like several other tiger moths, emits a foul froth when disturbed.

A wasp-mimicking *Myrmecopsis*.

Hipocrita drucei is one of the most dazzling moths attracted to lights.

Utetheisa, a day-flying tiger moth.

Halysidota, a night-flying tiger moth.

Xanthomis grandis, day-flying but attracted to lights.

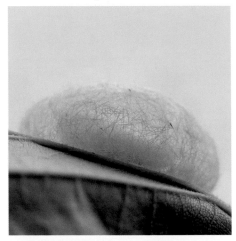
Cocoons of Arctiinae are loosely woven.

Larva *Turuptiana obliqua* feeding on Lauraceae.

Processionary larvae of *Lampruna rosea.*

PLANTS, DEFOLIATORS, AND CHEMISTRY

Heliconius hecale (Nymphalidae: Heliconiinae).

Insects and fungi are the prime enemies of plants, whose most effective defense is often chemicals that botanists call secondary metabolites. In what might be described as a coevolutionary arms race between plants and insects, a given plant species evolves new or more potent chemical defenses against an insect pest, while that insect, in turn, evolves adaptations that allow it to detoxify, excrete, or somehow avoid those very chemical defenses, and so the cycle continues.

This is why most phytophagous insects feed on just a few types of plants—they have evolved to withstand the chemical defenses of their host plants but not those of other plants. Insects that specialize on a specific kind of host plant often locate that plant by means of the odors that its chemical compounds produce, while insects that do not specialize on that plant are often deterred by those very same odors. We use some of these compounds as spices or drugs, though a compound that functions in a salutary fashion when taken in small doses can sometimes be toxic at larger doses. Humans have bred crops such as lettuce to reduce their levels of chemical defenses, thereby making such crops both tastier and more vulnerable to attack by insects.

In many cases—in some species of caterpillar and leaf beetle, for example—instead of using enzymes to detoxify a plant's nasty chemicals, the insect uses those chemicals to defend itself against predators. Other insect species are capable of internally manufacturing their own chemical defenses. When insects contain defensive compounds in their bodies, whether those compounds are sequestered from a plant or manufactured (or both), they nearly always advertise their toxicity with bright colors, a phenomenon known as aposematic coloration. Interestingly, since animals generally contain fewer types of secondary metabolites than do plants, insects that feed on plants have had to evolve more enzymes for detoxifying nasty chemicals than have the carnivorous insects that prey on other insects, with the result that insect crop pests tend to be more resistant to insecticides than are the carnivorous insects that feed on them.

The diversity of secondary metabolites is staggering. Often, a given secondary metabolite occurs in just a few closely related species of plant, though a given plant species generally contains many different kinds of secondary metabolites—since a combination generally provides a more effective defense. There are three principal categories of these defensive compounds in plants.

Nitrogen-containing metabolites. These include alkaloids, cyanogenic glycosides, and glucosinolates. More than 12,000 kinds of alkaloid have been described, including, for example, those contained in caffeine, cocaine, morphine, nicotine, quinine, and strychnine. Many have a bitter taste. Alkaloids occur in many species of plant, indeed in at least 30% of all flowering-plant species. Cyanogenic glycosides release hydrogen cyanide when brought into contact with certain enzymes; they are produced by a wide variety of plants (e.g., cassava, which is why these tubers must be ground and washed before being eaten) as well as by millipedes and various lepidopterans. Glucosinolates, which contain both sulfur and nitrogen, are responsible (at least in part) for the flavor of cabbage, cauliflower, mustard, and other members of the family Brassicaceae.

Terpenoids. Also known as isoprenoids, terpenoids make up the largest group of secondary metabolites—with more than 30,000 kinds identified to date. They are components of the scents given off by eucalyptus and mint—and of the taste of both cinnamon and ginger. They also include steroids such as cardiac glycosides (or cardenolides), named for

their ability to cause cardiac arrest in humans. Some terpenoids that are given off by a plant only after it has been attacked by an insect herbivore serve to attract predators and parasitoids that prey on the attacking insect. Many insects synthesize their own defensive terpenoids, including termites, fireflies, blister beetles, and broad-bodied leaf beetles.

Aromatic compounds. These compounds, also known as phenolics, contain one or more rings of carbon atoms (benzene rings); the term *aromatic* in this case is a technical term from chemistry and does not imply anything about odor. Common examples include capsaicin, which makes chili peppers hot, and tannins, which cause the puckery sensation when consuming unripe fruits or red wine. Tannins from oak trees are used in tanning animal hides into leather, a process that binds proteins together, making them less vulnerable to decomposition. By the same process, when an animal consumes plants that contain large quantities of tannins, proteins in the food are often rendered less digestible. Another group of aromatic compounds is the flavonoids, which several butterflies sequester from plants and use as pigments in their wings.

Table 6-1. Examples of some secondary metabolites

	Plants containing them	Insects that sequester them
Nitrogen-containing metabolites		
Cycasin	Cycads	J: blues (*Eumaeus*)
Pyrrolizidine alkaloids	Aster family (Eupatoriae, Senecioneae), borage family (several genera), a few others	J & A: tiger moths A: leaf beetles; milkweed butterflies, glasswings
Tropane alkaloids	Nightshade family	J: hawk moths and glasswings
Cyanogenic glycosides	Many plants	J: passion-vine butterflies
Glucosinolates	Cabbage family	J & A: harlequin bug
Terpenoids		
Cardiac glycosides	Dogbane family	J & A: seed bugs J: tiger moths, milkweed butterflies
Cucurbitacins	Cucumber family	A: skeletonizing leaf beetles
Iridoid glycosides	Many plants in the order Lamiales (verbenas, etc.)	J: geometer, owlet & tiger moths; buckeye & checkerspot butterflies
Aromatic compounds		
Aristolochic acids	Pipevine family	J: swallowtails (Troidini)

J = juvenile stages (larvae or nymphs), A = adults (not indicated are some cases where adults do not sequester, but receive metabolites from the larvae).

BUTTERFLIES

Butterflies have clubbed antennae, while the vast majority of moths do not. Nearly all butterflies are diurnal, while many moths are nocturnal. (A fascinating exception to this general description is one small family of butterflies that occurs in tropical America. Members of the family Hedylidae, also known as American moth-butterflies, lack clubbed antennae and are nocturnal.)

In the popular mind, greater appreciation is shown toward butterflies than moths, perhaps because many people mistakenly believe butterflies are all prettier than moths.

In point of fact, butterflies tend to fall into one of two general categories, one of which consists of species that are somewhat drab, have relatively stout bodies, and escape birds and other predators through fast, erratic flight. In the second group are butterflies that have bright colors, use chemical defenses to warn off potential predators, and fly rather slowly.

Brightly colored species of butterfly often mimic one another (both in color pattern and wing motion), and are thus often difficult to distinguish. But why do the species mimic one another? If, for example, five colorful species in a given geographical area are all the same pattern, it becomes much easier for predators to learn to recognize which species are unpalatable. Imagine if the five species each displayed a different color pattern; in such a case, predators would have to learn five distinct lessons about what butterflies to avoid eating. In some species, males and females have different coloration, and it is only the females that are mimics; in other species, populations in one place mimic the coloration of species x while populations in another region mimic the coloration of species y; in a few species, females mimic a different species than do males.

The adults of some butterfly species feed on nectar from flowers, while adults of other species feed on the juices from rotting fruits. (Nectar-feeders generally live for only about a week, but juice-feeders often live up to three weeks.) Young males of many butterfly species engage in an activity called mud-puddling, in

Colobura dirce (Nymphalidae: Nymphalinae), underside.

which they aggregate along the margins of mud puddles or rivers (or on soil where a mammal has urinated) in order to suck up sodium (and/or other nutrients such as amino acids). In some species, sodium is passed along to the female in the male's sperm package (sodium is often scarce in the diet of caterpillars, so adults need to obtain an additional supply). In their search for salt, several species sometimes even take a sip from the eyes of crocodiles and turtles.

Tropical America is the richest region on earth for butterflies; there are about 7,000 species in the region, representing roughly 40% of the world's butterflies.

SKIPPERS

Skippers (Hesperiidae) are named for their rapid, erratic flight. A hooked club at the tip of the antenna distinguishes them from all other butterflies. Other distinguishing characteristics include a broad head with widely separated antennae, a stout body, and relatively small wings. In many species, the wings are brown with a few small spots of transparent sheen (the spots are often located on the forewings); several species show brighter colors, of a metallic hue in some cases. It must be noted that it is often extremely difficult to distinguish one species from another. Skipper caterpillars generally have a large head that is joined to a narrow neck.

Depending on the species, adults feed on floral nectar or bird droppings, darting from place to place in search of food. Males of many species pugnaciously defend territories.

Skippers are currently placed in seven subfamilies, three of which are restricted to the Old World. Tropical America has four subfamilies with collectively more than 2,000 species: Eudaminae (400 species), Pyrginae (700 species), Heteropterinae (140 species), and Hesperiinae (nearly 1,000 species). Species in the first three subfamilies generally perch with both sets of wings held vertically (like many other butterflies), while those in Hesperiinae perch with their hind wings spread horizontally and their forewings only half opened. Caterpillars in the subfamilies Eudaminae and Pyrginae feed on dicots; these caterpillars often make two transverse cuts into a leaf blade, fold over the flap, and then tie it down with silk. Caterpillars in the subfamilies Heteropterinae and Hesperiinae feed on grasses and other monocots, often constructing leaf rolls and incorporating a white powdery wax (produced by abdominal glands) into the inner walls of the shelter.

Astraptes alardus latia (Eudaminae).

Male *Urbanus dorantes* (Eudaminae).

Jemadia scomber (Pyrginae).

Yanguna spatiosa (Pyrginae).

Male *Bolla cupreiceps* (Pyrginae).

Male *Pyrgus adepta* (Pyrginae).

Creonpyge creon (Pyrginae) on *Rondeletia* (Rubiaceae).

Dalla eryonas (Heteropterinae) males drinking from mud.

Male *Anthoptus epictetus* (Hesperiinae).

Synale cynaxa (Hesperiinae) on *Impatiens*.

Mating *Cynea cynea* (Hesperiinae).

Vettius aurelius (Hesperiinae).

Leaf-tying larva of *Vettius coryna conka* (Hesperiinae) on grass.

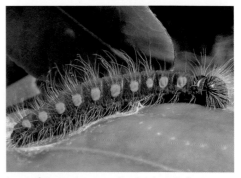

Larva of *Creonpyge creon* (Pyrginae) on Araliaceae.

Larva of *Nascus* (Eudaminae) on Sapindaceae; folded leaf has been opened.

Larva of *Astraptes* (Eudaminae) on *Erythrina* (Fabaceae).

SWALLOWTAILS

Swallowtails (Papilionidae) can be distinguished from other butterflies by the habit of fluttering their wings as they imbibe nectar from flowers. Many swallowtails have a tail-like structure on each hind wing, which lends the family its common name (this feature is not limited to swallowtails, however). The family Papilionidae is divided into three subfamilies: Baroniinae (with just a single species, *Baronia brevicornis*, which occurs in central and southern Mexico), Parnassiinae (in northern temperate regions), and Papilioninae, which occurs worldwide and contains about 500 species, 140 of which occur in the Americas. Papilioninae comprises one small Old World tribe and three large, cosmopolitan tribes: Troidini, Leptocircini, and Papilionini.

The tribe Troidini (50 species in the Americas) includes the spectacular birdwings of Southeast Asia, the largest butterflies in the world. Species in tropical America are smaller, and the majority of species belong to the genera *Battus* and *Parides*. Adults of these two genera lack tails and are mostly black, although most species of *Parides* have a red patch on each hind wing. Members of the tribe Troidini are often known as the pipevine swallowtails because their caterpillars feed on pipevines (Aristolochiaceae), from which they sequester defensive compounds that are passed on to the adults.

Members of the tribe Leptocircini (about 40 species in the Americas) are characterized by their rather short, upturned antennae. They include two groups: The genus *Mimoides*, in which adults and caterpillars of many species mimic *Parides* (tribe Troidini), and in which the adults of some other species mimic passion-vine butterflies (p. 258); and the genera *Eurytides*, *Protesilaus*, and *Protographium*, in which adults are clad in yellow and black, have a long tail on each hind wing, and tend to be strong-flying inhabitants of open areas. The caterpillars in this tribe feed primarily on plants in the soursop family (Annonaceae).

The tribe Papilionini contains a couple of small genera in southern Asia and the cosmopolitan genus *Papilio*, which includes about a third of all swallowtail species (50 species in the Americas). Adults of many species are yellow and black (with tails), others are mostly dark (with or without tails), and a few are mimics of *Parides*. The caterpillars of most species feed on plants in the citrus family (Rutaceae), but some feed on other plant families (e.g., Apiaceae, Lauraceae, and Piperaceae). *Papilio* caterpillars are cryptic from a distance (some species resemble bird or lizard excrement), but at close range, warning coloration can be detected on some species.

The caterpillars of swallowtails have a two-pronged eversible gland (osmeterium) at the front end of the thorax. When disturbed, the caterpillar extrudes its osmeterium, from which it liberates defensive compounds. These chemicals are mostly synthesized by the caterpillar, and consist primarily of terpenoids in Troidini and aliphatic acids and esters in Leptocircini. Caterpillars of Papilionini produce terpenoids (which deter ants) throughout most of their life, but in the last larval stage they switch to rancid-smelling methylbutyric and isobutyric acids (which deter many vertebrates). The caterpillars themselves may be solitary or gregarious (depending on the species). Swallowtail pupae are angular, often resembling wood fragments, and are fastened to the plant by a silken thread that loops around the middle of the body.

Papilio (*Heraclides*) *anchisiades* drinking from puddle while ejecting copious streams of water from its rear end.

Papilio (*Heraclides*) *thoas*.

Papilio (*Heraclides*) *cresphontes*.

Papilio (*Pterourus*) *garamas*.

Parides iphidamas (Troidini).

Last larval stage of *Papilio* (*Heraclides*) *thoas* on *Piper*.

The same larva as on left, but with osmeterium everted.

Larva of *Papilio* (*Pterourus*) *menatius cleotas* on avocado.

Larva of *P. menatius*, frontal view showing false eyes on thorax.

Eggs of *Battus polydamas* (Troidini).

Last larval stage of *B. polydamas*, on Aristolochiaceae.

Pupa of *B. polydamas*.

SULFURS, WHITES, AND OTHER PIERIDS

Butterflies in the family Pieridae come in many colors. Some species are mostly yellow (the sulfurs), a few are mostly white (the whites), and several come in yet other colors. Adults in this family feed on nectar. Some species undergo local migrations (probably in search of host plants for the caterpillars).

The family Pieridae is divided into four subfamilies, three of which occur in the Americas, where there are about 320 species in total. Although the subfamilies are distinguished on the basis of adult morphology, they also differ in terms of the plants that their caterpillars prefer.

Adults in the subfamily Dismorphiinae (about 50 species in tropical America) fly slowly. The color varies from species to species; many mimic the appearance of passion-vine butterflies (p. 258) or glasswings (p. 249). Their caterpillars feed on legumes (especially those in the genus *Inga*).

Adults in the subfamily Coliadinae (about 70 species in tropical America), called sulfurs, are fast-flying fairly large, and usually yellow. In many species, the upper-wing surface of males reflects ultraviolet light, which is invisible to the human eye but visible to female butterflies, who use the quality of the reflected light in choosing a mate. Their caterpillars feed predominantly on legumes.

Adults in the subfamily Pierinae (about 200 species in tropical America) vary in color: some are predominantly white, others have a checkered pattern, while still others mimic the warning colors of other butterflies. Caterpillars of the whites feed mainly on plants in the cabbage family (Brassicaceae) and on related plants such as those in the caper family (Capparidaceae), whereas other members of the subfamily feed on mistletoes (Loranthaceae, Viscaceae, and other families).

The caterpillars of Pieridae are frequently green. Their head capsule has a granulate surface. The pupae are cryptic and have a girdle of silk that holds them to their host plant.

Dismorphia theucharila (Dismorphiinae), a mimic of a glasswing.

Lieinix (Dismorphiinae).

Leodonta tellane chiriquensis (Pierinae) recently emerged from pupa.

Male *Catasticta eurigania straminea* (Pierinae).

C. cerberus.

Male *C. teutila* recently emerged from pupa.

Male *Melete leucanthe* (Pierinae) procuring nutrients such as sodium from sandy river bank.

Various species of sulfur butterflies engaged in "mud-puddling."

Pupa of *Catasticta flisa*.

Pupa of *Ascia monuste* (Pierinae) in urban area.

Larvae (young last stage) of *Catasticta sisamunus*.

Phoebis sennae (Coliadinae).

Pupa of *P. sennae*.

Eggs of *Hesperocharis costaricensis* (Pierinae).

Head of last larval stage of *H. costaricensis*.

Larva of *H. costaricensis* preparing to pupate, with a silken thread holding it to the plant.

Pupae of *H. costaricensis* (arrows) resemble leaves.

Adult *H. costaricensis*.

BLUES AND HAIRSTREAKS

Members of the family Lycaenidae are sometimes called gossamer wings. Further adding to their delicate appearance, many species are quite small (wingspan usually less than 5 cm/2 in). The family is divided into 8 subfamilies, but only 4 of these occur in the Americas, and only 2 in tropical America: Polyommatinae (blues) and Theclinae (hairstreaks). In tropical America, these two subfamilies are represented by about 50 and 1,000 species, respectively.

The common names given to these butterflies are confusing. While virtually all blues are blue on the upper side of the wing, so too are many hairstreaks (species that are not blue are grayish brown). The term *hairstreak* refers to the hair-like tail on each hind wing; but, this tail is absent in some hairstreaks (these are sometimes called elfins) and present in a few blues!

When perched with their wings in a vertical position, species with these hair-like tails move their hind wings so that they resemble a pair of antennae; in addition, many species also have an eye spot on the undersurface of each hind wing, all of which creates the illusion of a false head. This serves to divert an attack by predators that rely on sight to hunt, as they focus on the false head rather than the true head.

The stout caterpillars of blues and hairstreaks have a head that can be retracted into the thorax. And most species have access to yet another defensive strategy: they produce abdominal secretions that deter aggression by ants (or, in some species, that attract ants for protection). The caterpillars often feed on flowers and/or fruits, and many species are quite polyphagous. A few are pests of fruits. Some species of *Strymon*, for example, feed on pineapple, while *Oenomaus ortygnus* feeds on soursop (Annonaceae). Many species (*Calycopis* and related genera) feed on dead flowers and leaves on the forest floor. While most caterpillars probably rely on cryptic coloration rather than chemical defenses, the reddish caterpillars of *Eumaeus* feed on the leaves of cycads and sequester the plant's chemical defenses for their own defense.

Celastrina gozora females are not very blue.

Echinargus isola (Polyommatinae).

Cupido (*Everes*) (Polyommatinae) drinking from mud.

Female *Calycopis* (Theclinae).

Iaspis andersoni (Theclinae) moving its false antennae.

Kisutam syllis (Theclinae) drinking from moist towel. This genus is related to *Calycopis*.

Female *Theritas hemon* (Theclinae).

Panthiades bathildis (Theclinae).

Arawacus togarna (Theclinae).

Eumaeus godarti (Theclinae).

Larvae of *Eumaeus godarti* feeding on cycad.

Male *Theritas mavors* (Theclinae).

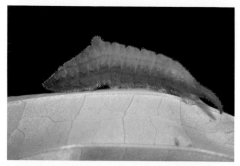

Larva of *T. mavors* on *Miconia* (Melastomataceae).

Larva of *Rekoa palegon* (Theclinae) feeding on flower buds of Asteraceae.

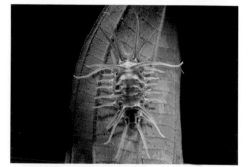

Pupa of *Thecla* (Theclinae).

Larva of *Thereus lausus* (Theclinae) feeding on *Struthanthus* (Loranthaceae).

Pupa of *Rekoa marius* on *Cassia* (Fabaceae).

Egg of *Laothus oceia* (Theclinae).

METALMARK BUTTERFLIES

Metalmarks (Riodinidae) are small to medium sized butterflies that vary greatly in color. Mimicry is probably more common in Riodinidae (especially in females) than in any other butterfly family, at least in terms of the diversity of lepidopterans that its species mimic. On males, the two front legs are very small, which obligates them to walk on just four legs rather than the full six as in virtually all other insects (the brush-footed butterflies are the only other lepidopterans to walk on just four legs, but in this group it is both sexes that do so). Worldwide there are about 1,600 species of metalmark, 98% of which occur exclusively in tropical America, many in the Amazon basin, where individual species often have quite restricted geographic distributions.

Adults usually feed on floral nectar. Species in the genus *Eurybia* are unusual among butterflies in that their tongues are much longer than their body, which means they can access the nectar in long tubular flowers (though with the consequent tradeoff that it takes more time for them to obtain the nectar). Males of many metalmarks feed on carrion (this is especially true of species with relatively small wings).

There are 3 subfamilies of metalmarks, 2 of which occur in the Americas: Euselasiinae and Riodininae (the latter includes the vast majority of the species). Adults of Euselasiinae perch with their wings held together above their bodies (like many other butterflies), whereas adult Riodininae perch with their wings extended horizontally at the sides of their body, or partially open, often while on the undersides of leaves.

The caterpillars of Euselasiinae often feed, and travel, in groups, whereas caterpillars of Riodininae are usually solitary. They feed on a wide diversity of plants, although each species usually restricts itself to a limited set of host plants.

Though these caterpillars are hairy, Eurybiini and Nymphidiini (two tribes of the subfamily Riodininae) are not; caterpillars in these two tribes produce sugary secretions (from the abdomen) that are eagerly lapped up by ants. Because the ants protect the caterpillars, the latter announce the availability of sugar by means of substrate-borne vibrations, and some species also attract ants by producing chemical compounds that apparently mimic ant-alarm pheromones. Caterpillars in the tribe Eurybiini feed on flowers of the prayer-plant family (Marantaceae) or gingers (Zingiberaceae). Caterpillars of many Nymphidiini are associated with plants that have extra-floral nectaries. The caterpillars drink the nectar (and eat the leaves) of these plants; some species are monophagous, others are polyphagous and often associate with particular ant species rather than particular plants. The caterpillars of a few species such as *Setabis lagus* (Nymphidiini) are carnivorous and feed on treehopper nymphs that arboreal ants keep as a source of honeydew.

Rhetus arcius (Riodininae) occurs in a wide range of habitats. The larva feeds on Combretaceae and Euphorbiaceae.

Emesis lucinda (Riodininae).

Rhetus dysoni (Riodininae).

Mesosemia telegone (Riodininae).

Semomesia croesus (Riodininae).

Female *Detritivora hermodora* (Riodininae). The genus name refers to the larvae, which feed on decaying leaves.

Larva belonging to the subtribe Lemoniadina (Riodininae: Nymphidiini), feeding on Fabaceae.

Theope virgilius (Riodininae: Nymphidiini).

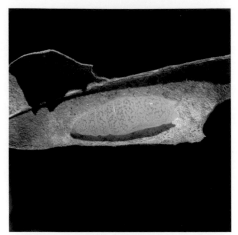

Pupa of *Corrachia leucoplaga* (Euselasiinae).

Male *C. leucoplaga.*

Young larvae of *Euselasia chrysippe* on *Miconia* (Melastomataceae). They all molt together.

Male *E. bettina.* The front legs are not used for perching.

Larva of *Anteros kupris* (Riodininae), with white and red balloon setae (possibly for defense) covering the head.

A. kupris, underside.

Female *Calephelis iris* (Riodininae).

Egg of *Calephelis*.

Larva of *Calephelis* on Asteraceae.

Male *Chalodeta lypera* (Riodininae).

Chalodeta lypera, frontal view.

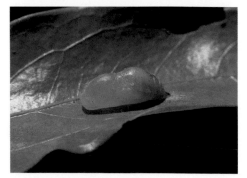

Pupa of *Chalodeta lypera* on *Zygia longifolia* (Fabaceae).

Catocyclotis adelina (Riodininae: Nymphidiini) males perch on tree trunks, dash out, chase one another, and fly between trees in a roller coaster pattern.

C. adelina females lays eggs on plants where *Myrmelachista* ants are present.

Larva of *C. adelina* with *Myrmelachista* ants.

Larva of *C. adelina* with *Myrmelachista* ant.

The Colors of Butterflies
and other Insects

Example of pigment color, *Siproeta stelenes biplagiata* (Nymphalidae: Nymphalinae).

Example of structural color, *Morpho helenor narcissus* (Nymphalidae: Satyrinae).

Cultures past and present have commemorated the beauty of butterflies in works of art. Images of butterflies adorn frescoes in the Aztec temple of Tlalocán in Teotihuacán, Mexico. The Aztecs believed that butterflies were the reincarnated souls of dead warriors. And people in the Amazonian regions of Peru once believed that the morpho butterfly was a spirit intent on luring people into forests, where they would become lost forever. But it was not only native people who fell under the spell of the butterfly's jewel-like appearance; Sir Walter Raleigh himself reputedly presented a morpho specimen to Queen Elizabeth, who adorned her hair with it before appearing at a ball.

Butterflies—and many moths—display an incredible diversity of color patterns. Even within a species, moreover, the coloration of males can be quite distinct from that of females. And, to further complicate matters, the upper surface and lower surface of the wing often show different patterns. The wing membrane itself is usually colorless; it is the miniature scales that cover the membrane that give the wing its color, and the overall color pattern is essentially a finely tiled mosaic. Each scale is a cuticular outgrowth from a single cell.

The growth of the wing commences in the early stages of the caterpillar, which possesses tiny wing buds within its body; in the final caterpillar stage, the wings develop venation and begin to develop patterns. The scales themselves develop during the pupal stage. The diversity and complexity of butterfly scales makes them among the most complicated extracellular structures manufactured by a single cell (colored scales also occur in several other groups of insects, including springtails and weevils).

In general terms, insect colors are produced in one of two ways, either by pigment or structure, though in many cases color results from the interaction of the two. Pigments selectively absorb certain wavelengths of light and reflect a given color, which is what we see. Structural colors, on the other hand, are produced by the physical interaction between light and nanostructures on the surface of the insect (e.g., wing scales). Pigments, actually chemical compounds, commonly produce reds, oranges, yellows, and black; structures produce all iridescent colors and most blues and greens. Some of the chemical compounds that compose pigments—flavonoids, for example—are derived from diet, while in other cases the insects manufacture the chemical compounds.

Research on butterfly colors has produced knowledge that might end up leading to revolutionary technological innovations. The nanostructures that produce iridescent colors, for example, are of considerable interest to scientists and engineers studying photonic crystals, which affect the motion of photons (light) in a way similar to how semiconductor crystals affect the motion of electrons in a circuit. Photonic crystals may someday enable engineers to make optical computers, whose architecture would be vastly superior to that of today's computers.

BRUSH-FOOTED BUTTERFLIES

All members of this diverse family (Nymphalidae) walk on just four legs; the front legs are small, retracted against the body, and often hairy (or brush-like, hence the common name).

Nymphalidae is the largest family of butterflies, with 12 subfamilies and more than 6,000 named species worldwide. Ten subfamilies—and roughly 2,000 species—occur in tropical America; the following accounts discuss all of these subfamilies except for Libytheinae and Apaturinae, two small families whose species are less commonly seen.

Siproeta epaphus (Nymphalinae: Victorini; p. 267). The larva feeds on leaves of Acanthaceae.

Colobura dirce (Nymphalinae: Nymphalini; p. 267), underside. The larva feeds on leaves of *Cecropia* (Urticaceae).

Myscelia cyaniris (Biblidinae: Epicaliini; p. 262).

Temenis laothoe (Biblidinae: Epiphilini).

Anartia fatima (Nymphalinae: Victorini).

Baeotus beotus (Nymphalinae: Coeini; p. 267).

Historis acheronta (Nymphalinae: Coeini).

Larvae of *Historis* feed on *Cecropia* (Urticaceae).

MILKWEED BUTTERFLIES
(family Nymphalidae)

The subfamily Danainae consists of 3 tribes, 2 of which occur in the New World: milkweed butterflies (Danaini) and glasswings (p. 249).

Milkweed butterflies are primarily an Old World (Indo-Australian) group with a total of some 160 species, only 12 of which occur in the New World. Seven species of *Anetia* and *Lycorea* (closely related genera) occur only in tropical America; the genus *Danaus* (with species in both the Old World and the New) is represented by 5 species in the New World, one of which, *Danaus plexippus*, is the monarch butterfly, among the most charismatic of all butterflies, and famous for its epic migrations (p. 247).

Like male glasswings (p. 249), the males of a given species of milkweed butterfly visits a specific species of plant (usually species in the aster and borage families) to obtain pyrrolizidine alkaloids, which they use for their own defense; also to synthesize sex pheromones; and as a nuptial gift during mating (by passing some of the alkaloids along to the female in the sperm packet). On adults, the bright orange-brown color, outlined with black, warns predators of their chemical defenses, although some species of bird are somehow able to tolerate these nasty compounds.

The caterpillars are also conspicuously colored, featuring colored rings around the velvety body; their body also shows one or more pairs of long, soft tentacular structures. Caterpillars of *Danaus* feed primarily on milkweeds and other members of the dogbane family (Apocynaceae), from which they sequester cardiac glycosides (p. 224), and perhaps certain alkaloids, as a defense against predators; at least some of these chemical defenses are passed on to the adult stage. Less is known about the other tropical American genera, except that caterpillars of *Lycorea* feed on figs and plants in the papaya family.

Danaus plexippus egg.

D. plexippus larva.

D. plexippus pupa.

D. plexippus laying eggs.

INSECT MIGRATION

Monarch butterflies aggregated at overwintering site in Mexico. They fly on warm days.

Only a minority of insect species migrate. Unlike vertebrates that migrate, insects generally do not make a round-trip journey; instead, the return trip is made by individuals of a subsequent generation. Nonetheless, in some species, the same individual does make a round-trip; in western Central America, for example, some true bugs, beetles, paper wasps, and butterflies migrate to higher altitudes (e.g., up the sides of volcanoes) during the dry season, and then return to the Pacific lowlands during the rainy season in order to reproduce.

The monarch butterfly, *Danaus plexippus*, undertakes a massive long-distance migration that biologists have studied in greater depth than any other insect migration. Interestingly, not all populations of this species migrate, including populations in Central America and northern South America. It is only two distinct sets of populations in the United States—one west of the Rocky Mountains and the other east of the Rocky Mountains—that migrate. Monarch populations to the west of the Rocky Mountains make a yearly journey to two areas in central California, Pacific Grove and nearby Santa Cruz, both located on the coast. But it is the populations living to the east of the Rocky Mountains that make the epic journey.

In early September, monarch populations in eastern North America begin flying south, traversing about 50 km (30 mi) per day for 75 days (a total distance of 3,600 km/2250 mi), reaching Michoacán in central Mexico by mid-November. They winter in Mexico in immense aggregations of up to 50 million individuals on fir trees at very specific sites—nine small mountain tops, located 70 to 170 km (40 to 100 mi) west of Mexico City. Conservation of these mountain forests is obviously critical to the survival of monarch populations. In mid-March these same butterflies fly north, arriving in early April at Texas and Louisiana, where they lay eggs on milkweeds and then die. Adults of the new generation continue northward (some going as far as southern Canada), laying eggs along the way. Depending on the temperature, two or three additional short-lived generations are produced. The final generation of the summer makes the return trip south to pass the winter in Mexico and therefore has a much longer adult life span, seven months as opposed to just a month or two for the preceding generations.

Several other species of moth and butterfly that occur in tropical America also migrate. These include hawk moths (p. 211), *Urania* moths (p. 215), swallowtails (p. 228), sulfurs (p. 231), and daggerwings (p. 265). In many cases, these migrations are related to the availability of host plants on which to lay eggs; in some places, during specific times of the year (e.g., in dry forests, during the dry season), as host plants die off, the insects are forced to migrate. In tropical America, some species of dragonfly also migrate, though the details of such migrations remain sketchy.

The insect migrants with which we are most familiar fly close to the ground, but some insects (e.g., certain moths) take advantage of airstreams hundreds of meters above the ground, where wind speeds greatly exceed the insects' flying speed. These species adopt flight headings that correct for crosswind drift; they do so by using the sun or moon as a compass, and somehow also use the earth's magnetic field to orient themselves over long distances. Diurnal insects flying closer to the ground can also use landscape features to gauge their speed and drift.

GLASSWINGS
(family Nymphalidae)

Glasswings (Ithomiini) belong to the same subfamily as do milkweed butterflies. They come in a variety of color patterns. Some species are clear-winged (hence the common name), some a translucent amber, and others are tiger-striped (like many passion-vine butterflies, p. 258). They generally have elongate wings, a slender abdomen, relatively small eyes, and a weak antennal club. Their flight is slow and fluttery, with deep wingbeats.

In contrast to the closely related milkweed butterflies, glasswings occur exclusively in tropical America, where there are about 350 species. Denizens of the forest understory, they are especially diverse at middle altitudes. Many species migrate up and down the mountains (apparently in response to changing seasons).

Both sexes feed on flower nectar; females of certain species also feed on fresh bird droppings in order to obtain additional nitrogen for making eggs. Like the milkweed butterflies (p. 246), males obtain alkaloids (pyrrolizidines) from particular plants (e.g., borages and certain species in the aster family); these substances provide protection against predators and are also used to produce a sexual attractant that is disseminated from tufts of hairs on the front margin of the hind wing. When releasing these attractants, males often form groups (leks) that sometimes include males of other glasswing species. The defensive compounds are passed on to females during mating, and the females probably choose males on the basis of their chemical content.

In most species of glasswing, the caterpillars feed on nightshades (Solanaceae), though caterpillars of a few species feed on other plants such as dogbanes and gesneriads. On the evolutionary tree, caterpillars in the basal (oldest) lineages obtain defensive chemicals (alkaloids) from the host plant and advertise their toxicity by displaying warning colors (the caterpillars do not pass their chemicals on to the adults). In species that evolved more recently, however, the caterpillars are palatable and so attempt to evade predators through camouflaged coloration. The pupae of many glasswings are silvery, golden, or green with iridescent spots.

Dircenna klugii on *Ageratum* (Asteraceae) flowers.

Ithomia heraldica feeding on decaying substance on leaf.

Pteronymia simplex on *Ageratum* flowers

Ithomia patilla perched in light gap.

Male *Ithomia heraldica* exposing hair-like scales from which sexual attractant is disseminated.

Pupa of *Mechanitis menapis saturata*.

Larvae of *Mechanitis polymnia* on *Solanum*.

Pupa of Ithomiini.

LEAF BUTTERFLIES
(family Nymphalidae)

When perched with their wings held vertically, most leaf butterflies (Charaxinae) resemble a dead leaf (hence the name); in addition, members of the tribe Anaeini have a short, broad tail on each hind wing that resembles a leaf petiole. Other members of the subfamily resemble the surface of wood when perching. While all species have cryptic coloration on the underside of the wings, the coloration of the upperside varies from species to species; many have reddish-orange or iridescent blue-green bands, but a few are more drably colored. This primarily Old World subfamily consists of nearly 400 species, of which about 110 occur in tropical America.

Leaf butterflies fly quickly and erratically, which appears to be their main defense against predators. Adults of many species spend most of their time in the canopy, but occasionally come down to the forest floor to feed on rotting fruit, dung, or carrion—the diet of all species in the subfamily, as reflected in their short, stout proboscis. Males are sometimes attracted to the perspiration of sweaty hikers.

From species to species, the type of plants that the caterpillars feed on varies widely.

Male *Zaretis ellops*, with underside resembling dead leaf.

Male *Archaeoprepona amphimachus*.

Memphis proserpina.

Male *Agrias amydon philatelica*.

Prepona praeneste isabelae.

Larva of *Hypna clytemnestra* feeding on *Croton* (Euphorbiaceae).

Larva of *H. clytemnestra* preparing to pupate.

Pupa of *H. clytemnestra*.

H. clytemnestra, underside.

Larva of *Zaretis ellops*, frontal view.

The larva of *Archaeoprepona demophoon* feeds on Lauraceae and moves like a dead leaf in the wind.

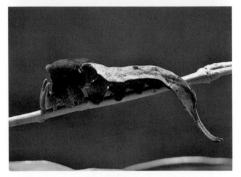
Larval head of *Memphis pithyusa* in rolled leaf of *Croton* (Euphorbiaceae).

A collection of *Agrias aedon narcissus* (Charaxinae).

SATYRS
(family Nymphalidae)

The subfamily Satyrinae, closely related to the leaf butterflies, is the largest subfamily of brush-footed butterflies, with approximately 2,500 species worldwide, about half of which occur in tropical America. Members of three tribes—Melanitini, Haeterini, and Satyrini— are collectively known as satyrs. Tropical America is home to about 1,100 species of satyr, most of which are in the tribe Satyrini.

The subfamily Satyrinae also includes the owl butterflies (p. 254) and morphos (p. 256).

Named for the woodland man-goat of Greek mythology, most satyrs are creatures of the understory, flitting just above the ground in a characteristic bouncy pattern of flight, as if being jerked up and down by a string. On

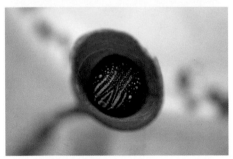
Manataria maculata (Satyrinae: Melanitini).

Cithaerias pireta (Satyrinae: Haeterini).

the vast majority of species, small eye spots appear on the underside of the wings, but above, the wings are dull brown; a few species, including some members of the small tribe Haeterini (which occurs exclusively in tropical America), have a very distinct appearance, showing transparent forewings and bright colors on the edges of the hind wings.

Most species have a swollen vein at the base of each forewing, which functions as an ear and is perhaps used to detect the sounds of approaching birds. *Manataria maculata* (the only New World genus in the tribe Melanitini) flies at dusk and at dawn, and (unlike almost all other butterflies) is capable of detecting the ultrasounds emitted by bats during echolocation.

Satyrs usually feed on rotting fruit, fungi, and other decaying matter, but some also feed on nectar from flowers.

The caterpillars have a forked tail. They feed mostly on grasses, including bamboos. The high silica content of grasses wears out the mandibles of herbivores and impairs absorption of nitrogen, but satyrs appear to have overcome these problems, having coevolved with grasses.

Both adults and caterpillars generally lack chemical defenses.

Oxeoschistus tauropolis (Satyrini).

Hermeuptychia harmonia (Satyrini).

Oressinoma typhia (Satyrini).

Euptychia westwoodi (Satyrini). The larva feeds on spikemosses (*Selaginella*).

Eggs of *Pronophila timanthes*, with larval heads visible.

Larva of *P. timanthes*, frontal view, on bamboo.

P. timanthes recently emerged from pupa.

Pronophila timanthes (Satyrini).

OWL BUTTERFLIES
(family Nymphalidae)

Satyrs (p. 252), owl butterflies, and morphos (p. 256) all belong to the subfamily Satyrinae. Like the morphos, to which they are closely related, owl butterflies (Brassolini) are restricted to tropical America. There are about 100 species of owl butterfly. Most owl butterflies have two or three large eye spots on the underside of each hind wing; in many species, at least one of these spots resembles the eye of an owl, which lends this family its common name.

Owl butterflies have a very erratic pattern of flight. They are most active at dusk and dawn, during which time the males of some species assemble at certain sites (e.g., forest edges) to defend mating territories. Like morphos, owl butterflies do not feed on nectar, but rather on rotting fruits or fungi (species in the genus *Brassolis* have a reduced proboscis and do not feed at all).

The caterpillars of most species have 2-3 pairs of horns on the head and a forked tail. The caterpillars feed at night on monocots; the principal host plants are palms and grasses (especially bamboos), but several species of owl moth in the genera *Caligo* and *Opsiphanes* have expanded their diet to include heliconias (and banana) and members of the prayer-plant family (Marantaceae). Species in the genus *Dynastor* are unique in that they feed on bromeliads. *Brassolis* species are unusual in that hundreds of individuals rest during the day in a large silken bag (made by sewing palm leaflets together).

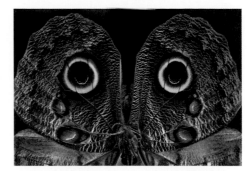

Around midnight sleeping *Caligo* species turn upside down.

Caligo telamonius showing false ("owl") eye.

Caligo atreus.

Larval head of *Opsiphanes bogotanus.*

Larvae of *Brassolis isthmia* crawling up palm tree.

Pupa of *Caligo telamonius.*

MORPHOS
(family Nymphalidae)

Morphos, like satyrs (p. 252) and owl butterflies (p. 254), all belong to the subfamily Satyrinae.

Although morphos are fairly diverse in appearance, on all species, the undersides of the wings are a cryptic brown color, often overlaid with a variety of eye spots. When perched, with the wings held upright, many species resemble a dead leaf.

Morphos belong to the tribe Morphini, with 42 species in 3 genera, all of which are restricted to tropical America. The majority of the species (29) are in the genus *Morpho*; several of the butterflies in this genus are large (with wingspans up to 16 cm/6.3 in) iridescent-blue butterflies—probably the most iconic insects of tropical America, adorning countless tourist brochures. However, not all species of *Morpho* are large (nor spectacularly beautiful); some are brownish and a few are pure white. And even in the case of blue species, the females are generally less colorful.

Species in the other two genera, *Antirrhea* (11 species) and *Caerois* (2 species), vary in color but are characterized by angular hind wings with a short tail.

All morpho butterflies feed on fermented fruits, sap flows from wounds in trees, and fungi. They appear to rely on erratic flight behavior to escape from predators. Among the blue *Morpho* species, males are frequently seen patrolling up and down forest edges (e.g., along rivers), whereas females tend to stay in the forest; males in one group of species (the *hecuba*-group) spend most of their time in the canopy and have shifted to a gliding type of flight (as opposed to the normal flapping flight).

Caterpillars of Morphini are multicolored and have prominent tufts of urticating hairs (which is unusual for butterfly caterpillars) and an oversized hairy head. Those of *Antirrhea*, *Caerois*, and the basal (most ancestral) species of *Morpho* feed on monocots (e.g., palms), whereas the caterpillars of most *Morpho* species feed on legumes and various other dicots.

Antirrhea pterocopha. Species in this genus inhabit the deep shade of the forest understory, where they feed on rotting palm fruits.

Antirrhea philoctetes tomasia.

Larva of *Morpho menelaus* on Fabaceae.

Morpho cypris.

M. cypris.

M. helenor narcissus.

Head of young larva of *Morpho helenor narcissus.*

M. helenor receiving evening sun on underside of wings.

PASSION-VINE BUTTERFLIES
(family Nymphalidae)

Worldwide, the subfamily Heliconiinae consists of about 500 species, currently divided into four tribes. Three of the tribes occur in the New World.

Most species of Argynnini (known as fritillaries) are restricted to northern temperate regions. Twelve species do occur in tropical America, however, the majority found in the high Andes. The caterpillars of species in this region feed on plants in the violet family and on passion vines.

Acraeini is centered in Africa, but there are some 50 species in tropical America (again, mostly in the Andes). Caterpillars of the most "primitive" species feed on passion vines, but all species in the Americas feed on plants in the aster family.

Heliconiini is restricted to the New World, primarily in tropical America, and comprises 70 some species. Their caterpillars feed on passion vines, and although passion vines are probably the ancestral host plants for the entire subfamily, the name *passion-vine butterfly* is

Heliconius erato.

Larvae of *Dione juno* showing defensive behavior.

reserved for the tribe Heliconiini. The following account is devoted to this group, which has contributed so much to our understanding of evolutionary processes in the tropics.

Passion-vine butterflies are characterized by elongate front wings, large eyes, and long antennae. The "primitive" genera are relatively palatable and fast flying, whereas the more recent genera (*Eueides*, *Neruda*, *Heliconius*) are relatively unpalatable and slow flying. The latter display bright warning coloration and belong to mimicry rings that include both passion-vine butterflies and certain glasswings. Passion-vine butterflies often congregate at night, using the same roost night after night; species that mimic each other roost in similar habitats, sometimes in the very same aggregation.

Passion-vine butterflies consume nectar from various plants, but unlike other butterflies, species of *Heliconius* also feed on pollen; indeed, they occupy home ranges based largely on the presence of pollen plants. Some species in this genus harvest pollen primarily from plants in the cucumber family (*Gurania* and *Psiguria*) whereas other species obtain most of their pollen from other plants such as species in the genus *Lantana* and in the coffee family. The pollen grains stick to the outside of the butterfly's proboscis, where, over a period of several hours, they are ruptured by means of salivary enzymes and by the coiling-uncoiling movements of the proboscis. The amino acids from the pollen are then ingested and eventually utilized in egg production, prolonging adult life span (up to three or more months), and in synthesizing chemical defenses. Although *Heliconius* butterflies destroy most of the pollen they collect, they are nonetheless the principal pollinators of several of the plants they visit.

Males of many *Heliconius* species defend territories and court the females that enter these areas; in some species (the *erato* and *sara-sapho* groups) the males search larval host plants for mature pupae, sit on the pupae a day before emergence, and mate with the female as she begins to emerge from the pupal case. Depending on the species, females lay eggs on tendrils, stipules, leaf tips, or leaves of passion vines (Passifloraceae); other species lay egg masses on older leaves (the caterpillars hatching from these eggs live in groups).

The caterpillars have six rows of spines; they tend to be cryptically colored when young and brightly colored when mature. In some species the caterpillars feed on just one or two species of passion vine, while others feed on several. The host plants have evolved various means of protecting themselves from the caterpillars. Some plants produce false eggs on their stipules, thereby dissuading female butterflies from laying eggs; most passion vines have extrafloral nectaries that attract ants, which attack the eggs; probably all passion vines produce defensive chemicals, including alkaloids and cyanogenic glycosides (which produce cyanide). Nevertheless, the caterpillars are capable of detoxifying the plant's chemical defenses, and in most species both caterpillars and adults manufacture their own cyanogenic glycosides as a defense against predators. However, some species of *Heliconius* (the *sara-sapho* group) have lost this ability and have become chemically dependent upon particular species of passion vines, from which they sequester certain types of cyanogenic glycosides and then pass them on to the adult stage. In all species, the female transfers cyanogenic glycosides to the eggs and, at least in some species, she receives a supplemental supply of these chemical defenses from the male's spermatophore (which also supplies her with amino acids).

Heliconius charithonia visiting Rubiaceae flowers.

Pupa of *Dione moneta*.

Egg (yellow) of *Heliconius charithonia* on passion vine tendril.

Larva of *Dione moneta*.

D. moneta imbibing nutrients from leaf litter.

Egg of *Dryas iulia moderata*.

D. iulia moderata.

D. iulia moderata at a water faucet.

Heliconius erato petiverana procuring nutrients from moist soil.

H. erato petiverana sleeping aggregation.

H. erato petiverana feeding on *Lantana* flower.

SISTERS
(family Nymphalidae)

The subfamily Limenitidinae (closely related to the passion-vine butterflies) contains more than 800 species worldwide, only 10% of which occur in tropical America. In North America, this group includes four species in the genus *Limenitis* (viceroy and admiral butterflies), some of which extend south to central Mexico. In most of tropical America, Limenitidinae is represented exclusively by the genus *Adelpha*, which has more than 80 species (known as sisters), several of which reach the southern United States.

Species of *Adelpha* are quite similar in appearance—the upper side of the wings is dark brown, with a subapical orange patch on the forewing; many species also have a white band that traverses both wings (or just the hind wing). Adults are fast fliers, which is probably their main defense against predators. They feed on rotting fruits, though high altitude species also imbibe nectar.

In about 50% of the species in the genus *Adelpha*, caterpillars feed on plants in the coffee family (Rubiaceae); the other caterpillars consume a variety other plants. All caterpillars in this genus rely on camouflage for protection, resembling, variously, bird droppings, a piece of dead leaf, or a small moss-covered twig; when not feeding, some species rest among bits of frass, silk, or dead leaf. The pupae vary in coloration from a combination of brown and green to a mirror-like chrome colour (depending on the species).

Adelpha leucophthalma.

A. serpa celerio.

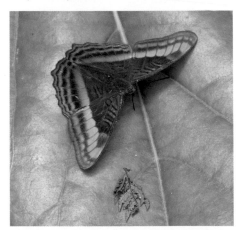

Adelpha tracta. The larvae fold leaf tips of *Viburnum* (Adoxaceae).

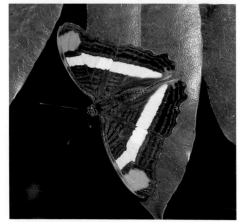

Like other *Adelpha*, *A. fessonia* shows continuous "fly and perch" behavior along forest edges.

The larva of *Adelpha serpa celerio* resembles moss.

Pupa of *A. serpa celerio*.

CRACKERS, NUMBER BUTTERFLIES, AND THEIR KIN
(family Nymphalidae)

Members of subfamily Biblidinae are so diverse that it is difficult to describe a set of characteristics shared by all species, and as they spend most of their time in the rainforest canopy, one seldom sees many of them.

The subfamily is composed of six tribes, all of which are found tropical America. Worldwide, there are 350 species, about 250 of them occurring in this region: Biblidini (6 species), Epicaliini (70 species), Ageroniini (28 species), Eubagini (40 species), Epiphilini (35 species), and Callicorini (72 species). Ageroniini, Eubagini, Epiphilini, and Callicorini are restricted to tropical America.

Adults generally feed on rotting fruits, fermented sap from wounds on tree trunks, and/or mammal dung.

The caterpillars of most species in Biblidini, Epicaliini, Ageroniini, and Eubagini feed on plants in the spurge family (Euphorbiaceae); caterpillars in Epiphilini and Callicorini feed on plants in the soapberry family (Sapindaceae).

The twenty species of *Hamadryas* (tribe Ageroniini) are commonly known as either cracker butterflies (for the sounds that males make) or calico butterflies (for the speckled colors of many species). The sounds are produced by a sudden buckling of one of the wing veins, though not all species produce sounds audible to humans; while females are also capable of clicking, they rarely do so.

Male crackers perch on tree trunks with their head downward and their wings extended flat against the surface, their colors merging with the mottled bark. When another individual of the same species flies into his territory, the male takes flight and emits a series of cracking or clicking sounds; if the intruder is another male, the resident male chases him away, but if it's a female he courts her. Males are so territorial that they occasionally chase away other insects.

Dynamine, the only genus in the tribe Eubagini, consists of small butterflies known as sailors, many of which are metallic blue. Their caterpillars are unusual in that they feed on developing leaves or flowers (some species bore into the developing ovary).

Species of *Callicore* and *Diaethria* (tribe Callicorini) are known as number butterflies or numberwings, names derived from patterns on the underside of the hind wings that resemble the numbers 66, 69, 88, or 89. Some species occur near rural dwellings (and even enter them), and more than one *campesino* has been known to buy a lottery ticket containing a number seen on a passing butterfly.

Hamadryas amphinome mexicana, a cracker butterfly.

Like other crackers, *H. februa ferentina* defends territories near a food source (fermented fruits).

Hamadryas guatemalena.

Nica flavilla canthara (Epiphilini).

Diaethria eupepla, a number butterfly appearing to bear the number 88.

Diaethria astala sipping from a concrete wall.

Larva of Hamadryas amphinome.

Pupa of Hamadryas hanging from tree trunk.

DAGGERWINGS
(family Nymphalidae)

Daggerwings (subfamily Cyrestinae) take their name from the long tail on each hind wing. This subfamily contains three genera and some 40 species worldwide. A single genus (*Marpesia*) occurs in the Americas, which is home to17 species. Adults feed on floral nectar, while caterpillars feed on plants in the fig family (Moraceae).

These butterflies are unusual in that they are very long-lived (over 5 months) and congregate at nocturnal roosting sites.

Marpesia marcella valetta obtaining nutrients from mud.

Marpesia zerynthia dentigera drinking on a concrete bridge.

Marpesia merops.

M. merops. The larvae feed on *Brosimum* (Moraceae).

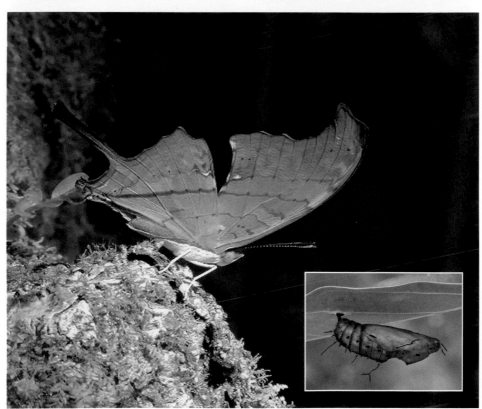

Marpesia petreus, with inset showing pupa on fig leaf (*Ficus*).

BEAUTIES, PAINTED LADIES, CHECKERSPOTS, AND THEIR KIN
(family Nymphalidae)

Members of subfamily Nymphalinae have a plethora of common names, which suggests their great diversity. The subfamily consists of 6 tribes and approximately 500 species; 5 tribes and about 170 species occur in tropical America.

The tribe Coeini, with just two genera, is confined to tropical America. There are 4 species of Baeotus (known as beauties) and 2 of Historis. Adults feed on rotting fruit and dung.

The tribe Nymphalini comprises roughly 100 species worldwide; about 25 species occur in tropical America. The adults of most species feed primarily on rotting fruits and dung.

The caterpillars of Colobura and Tigridia feed on cecropias and defend themselves against the ants that occupy these trees (p. 194) by secreting a repellent from a gland in front of the thorax and (when they are young) by resting on top of their excrement. The caterpillars of Smyrna and Hypanartia feed on plants in the nettle family.

Painted ladies (genus Vanessa) are distinct from other members of the tribe in several ways: adults feed principally on nectar; caterpillars feed on a wide range of plants rather than a single family or group of families; and some species migrate long distances and thus have a wide distribution. Indeed, Vanessa cardui is one of the most widespread butterflies in the world—nearly cosmopolitan—though found only at high altitudes in tropical America.

The tribe Victorini comprises just 10 species, all restricted to tropical America. Adults feed on floral nectar as well as rotting fruits, carrion, and dung. Caterpillars feed primarily on plants in the acanthus family.

The tribe Junonini is primarily an African tribe; the New World has just one introduced species (Hypolimnas misippus) and 4 species in the genus Junonia, whose members are known as buckeyes for the eye spots on the upper surface of the wings. Adults feed on nectar; caterpillars feed on plants from the verbena, acanthus, and other families that contain iridoid glycosides, which they sequester but do not transfer to the adult stage.

Checkerspots (tribe Melitaeini) occur in northern temperate regions as well as in tropical America; there is evidence to suggest that about 34 million years ago some members of this group (the ancestors of the subtribe Phyciodina) left North America, passed through the Greater Antilles, and colonized South America, long before the creation of the Panamanian land bridge.

Checkerspots comprise about 250 total species, nearly half of which occur in tropical America. Their antennal club is flattened at the tip, and they generally come in one of two color patters; some species are checkered; other mimic passion-vine butterflies (p. 258) or glasswings (p. 249).

Adult checkerspots feed on nectar, while the caterpillars of most species feed on plants in the acanthus family or aster family. The caterpillars of many species feed in groups. In at least one tropical American species (Chlosyne theona) that feeds on a different set of plants (Indian paintbrush), the caterpillars sequester iridoid glycosides from the plant and use them for their own defense but also pass them on to the adult stage.

A buckeye, *Junonia evarete* (Junonini).

Female *Eresia ithomioides alsina* (Melitaeini). Speci in this genus form mimicry rings with Ithomiini Heliconiini.

A checkerspot, *Anthanassa ardys* (Melitaeini).

Tegosa anieta (Melitaeini) males gathered at side of stream to imbibe nutrients.

Chlosyne janais (Melitaeini).

C. janais.

rva of *C. janais* feeding on Acanthaceae.

Pupa of *C. janais.*

7

FLIES AND THEIR KIN
(order Diptera)

Gnats, midges, mosquitos, and flies (in Spanish, *moscas*), collectively often referred to as flies, belong to the order Diptera, which means *two wings*; the second pair of wings, reduced to tiny stubs, are difficult to see. The muscles that power the first pair of wings are huge, occupying nearly the entire volume within the thorax, and allow them to flap their wings at an astonishing 200 to 1,000 times per second! Males of many species of fly form aerial aggregations with the purpose of attracting females.

Flies (like beetles, wasps, and moths and butterflies) undergo complete metamorphosis; the eggs hatch into larvae (or maggots) that feed, molt a few times, transform themselves into pupae, and then finally become adults. The legless maggots inhabit a greater diversity of habitats—and consume a greater diversity of foods—than any other group of insect. Over half of all aquatic insect species are flies.

The larvae of many species in many fly families (especially in Tachinidae, p. 299) are parasitoids of other insects. The larvae of bot flies (p. 301) and the screwworm (p. 297), to cite a few instances, burrow into vertebrate flesh. Other larvae feed on plants, boring into stems, leaves, fruits or seeds—and no other group of insects has so many species that induce plant galls (p. 280). Many fly larvae nourish themselves in less colorful fashion,

feeding on fungi or filtering out microbes from rotting vegetation, excrement, or carrion. Upon completing their development, the larvae pupate; in the most ancestral families (e.g., crane flies, mosquitos, fungus gnats, etc.), the larvae form a naked pupa, but in the majority of families the larvae of the last larval stage pupate inside a puparium, a hardened exoskeleton (the cuticle of the larva) that looks like a brown seed.

Lacking masticatory mandibles, adult flies generally shun solid food, instead using specially modified mouth parts to imbibe nectar, blood, and other liquid foods. Various species of fly, for example, sip nectar from flowers (and some of these species are important pollinators). Among the species that suck blood from vertebrates are the sand flies, mosquitos, black flies, biting midges, deer flies, and the stable fly; it is only the females that suck blood (the males generally feed on nectar), except in the case of the stable fly, in which both females and males suck blood. Females in these groups require the amino acids found in blood in order to produce eggs. Members of one genus in the family Chloropidae, and another genus in the family Drosophilidae, collectively often referred to as gnats, do not suck blood but instead seek out secretions from around the eyes and nose, an annoying experience undergone by many a hiker.

CRANE FLIES

While some fly specialists divide crane flies into four families, others place them in a single family: Tipulidae. Crane flies have long, fragile legs and tend to resemble mosquitos, although they are generally larger; species of crane fly range from 3 to 36 mm (0.1 to 1.4 in) in length, while mosquitos measure from 3 to 8 mm (0.1 to 0.3 in). Adults of many species apparently do not feed, although adults of some species are

Nephrotoma. The larvae in this genus live in soil.

Sigmatomera seguyi. The larvae in this genus live in water in tree holes.

Crane fly resting on a spider web. It has lost a couple of legs, though this is common in crane flies.

known to feed on nectar. Compared with many other flies, crane flies are rather feeble fliers.

Crane flies are commonly found resting on shaded vegetation, though some species hang from spider webs—and are able to do so without becoming entangled. When disturbed, resting crane flies move rapidly up and down, which makes them difficult to see. A few species reside beside groups of daddy-long-legs (p. 305), possibly because the latter have chemical defenses.

The larvae of crane flies live in a wide variety of habitats; depending on the species, larvae can occur in well oxygenated streams, wet mud (or sand) at the edges of water, water-filled tree holes (or bromeliads), intertidal zones, moist cushions of mosses or liverworts, decaying wood, decaying leaves, and soil. Most species feed on decomposing plant material, but some are predators and others feed on mosses.

While most crane fly larvae are difficult to see without digging into the substrate in which they live, those of *Geranomyia recondita* live on the surface of leaves in the forest understory. The larva of this species secretes a translucent, gelatinous mass around its body; it feeds on liverworts, algae, and lichens growing on the leaf surface.

Worldwide there are more than 17,000 named species of crane fly, about 4,000 of which occur in tropical America.

MOTH FLIES AND SAND FLIES

Moth flies and sand flies are members of the family Psychodidae, a group of small (body length = 1-5 mm / 0.04-0.2 in) flies whose bodies are densely covered with hairs. Moth flies (subfamily Psychodinae) tend to be slightly larger than sand flies; at rest they hold their wings in the form of an inverted V (roof-like). Sand flies (subfamily Phlebotominae) are tiny and hold their wings straight up, at a right angle to the body. Worldwide the family Psychodidae consists of about 3,000 named species, of which roughly 500 occur in tropical America.

The larvae of moth flies are generally aquatic or semiaquatic. The flattened larvae of *Maruina*, for example, live on submerged rocks in streams. The larvae of a few species (*Clogmia albipunctata* and *Psychoda* spp.) live in drain traps and septic tanks (the adults of these species are the most commonly seen species of moth fly since they are common inhabitants of houses). The larvae feed on decaying organic material. Unlike sand flies, adult moth flies do not bite.

Sand fly larvae live in soil, where they feed on decaying organic material.

Unlike adult female moth flies, adult female sand flies do bite, that is, suck blood (from vertebrates), mostly in the evening. In tropical America, various species of *Lutzomyia* are the vectors of about 20 species of protozoan (genus *Leishmania*) that cause cutaneous leishmaniasis, which results in non-healing ulcers or crusted nodules on the skin. On rare occasions the disease causes destruction of mucosal tissue, resulting in disfiguration of the face. In most of their range, these parasites depend on wild animals such as marsupials, sloths, and rodents to maintain the disease cycle, but in parts of South America, the disease is becoming urbanized and the parasites have begun to depend on humans.

Diagnosis of cutaneous leishmaniasis is often a challenge. There is no vaccine, and treatment, generally injections of antimony (a metalloid that can have severe side effects), never fully eliminates the parasites from the body. The good news is that alternative treatments are becoming available.

A more serious human disease is visceral leishmaniasis (also known as kala-azar), which is caused by the protozoan *Leishmania infantum*, which, in the New World, is transmitted by *Lutzomyia longipalpis*. The natural reservoirs of this parasite are mammals in the dog family, including domestic dogs. The protozoans migrate to the internal organs and, if left untreated, almost always cause death. Various drug treatments are available, and still others are under investigation.

In the Andes of Colombia, Ecuador, and Peru, certain species of *Lutzomyia* transmit the bacterium *Bartonella bacilliformis*. About three weeks after being bitten, humans will experience a massive invasion of their red blood cells. This phase of the disease, known as Carrion's disease (or Oroya fever), lasts 3 to 4 weeks and is potentially life-threatening. In the succeeding chronic phase of the disease, known as *Verruga peruana* or (Peruvian wart), the bacteria invade cells lining the inner surface of the blood vessels, resulting in benign skin eruptions with nodules. Fortunately, antibiotics can treat both phases of the disease.

Clogmia albipunctata, a common moth fly in bathrooms.

MOSQUITOS

The English word *mosquito* (derived from the Spanish and Portuguese for "little fly") refers specifically to the family Culicidae. It is only the females that suck blood (males feed on nectar), though in a few species of mosquito, neither the females nor males suck blood. Attracted to emissions of carbon dioxide and body odors, females generally seek a blood meal before each bout of egg laying. Some species feed exclusively on bird blood, others on the blood of reptiles, amphibians, or mammals.

Each species has its own daily schedule of activity. For example, *Culex quinquefasciatus*, a species that is widely distributed around the world, is active in the middle of the night,

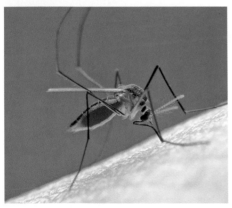
Several species of *Haemagogus* are vectors of jungle yellow fever, which often infects monkeys.

whereas species in the genus *Aedes* are usually active during the day.

Mating often occurs near water from which adult mosquitos are emerging from the pupal stage. The two sexes recognize one another by the whining sound generated by their rapidly beating wings. Males have more sensitive hearing, but each sex alters its wing-beat frequency in response to the flight tone of the other so that their frequencies become synchronized. In other words, the courting couple flies in tune.

Eggs are laid individually (as in the genera *Aedes* and *Anopheles*) or clustered (as in the genus *Culex*), on the surface of the water; along the margins of quiet pools; in tree holes, bromeliads, leaf axils, crab holes; and on or near other bodies of water, depending on the species.

The larvae of most species obtain oxygen from the surface of the water, either directly (genus *Anopheles*) or via a siphon. They feed by filtering out small organic particles, either at the surface (*Anopheles*) or below water, which they reach by wriggling their bodies. In a few species, the larvae feed on other mosquito larvae. The non-feeding pupae are mobile, moving toward the water surface, where they obtain oxygen.

Worldwide there are about 3,700 named species of mosquito; only 20% include human blood in their diet, and only a few of these are disease vectors (p. 273).

Wyeomyia feeding on human blood. Larvae of this genus occur in water-filled bromeliads.

Larva of a mosquito (*Culiseta*).

MOSQUITOS AND THE
DISEASES THEY CARRY

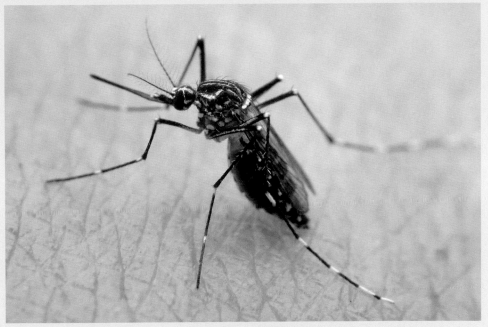

The principal vector of dengue, *Aedes aegypti*, distinguished by the white stripes on the top of its thorax.

Before drawing blood, the female mosquito injects saliva (into capillaries), which prevents coagulation. The itching that you experience is due not to the insertion of the proboscis but rather to the saliva itself. The momentary discomfort of a mosquito "bite" is nothing, however, as compared to the effects of the many microbes that mosquitos are known to transport, and indeed mosquitos transmit more human diseases than any other group of insects. These include elephantiasis and dog heartworm (caused by the round worms carried by some mosquitos); malaria (a protozoa); and a host of viruses, including yellow fever, dengue fever, encephalitis (of various types), and the West Nile Virus.

Worldwide, malaria claims the lives of more children than any other infectious disease—and it is the most significant of the parasitic diseases that afflict humans. About 90% of all malaria deaths occur in Africa, where *Plasmodium falciparum*, the most virulent of the four main protozoa that cause human malaria, is so common.

In *Daisy Miller*, as in real life, people once believed that the cause of malaria, or Roman fever, was pestilent air. It was not be until the final years of the 19th century that scientists doing research on *Anopheles* mosquitos would confirm that protozoan parasites, carried by the mosquitos, were the transmitters of the disease. Not all species in the genus *Anopheles* are potential carriers of the protozoa, though all members of the genus appear to stand on their head when they bite.

All four of the principal protozoans that cause human malaria were introduced into the New World, probably having been carried to the region in the bodies of visiting Europeans (the most widespread species in the Americas is *Plasmodium vivax*). As ill fortune would have it, several New World species of mosquito in the genus *Anopheles* proved capable of transmitting the malaria protozoans. On one voyage of plunder to the West Indies, Francis Drake was forced to halt the journey after hundreds of his men died from malaria. Europeans visiting the New World took note of reports of a Spanish countess who, while staying in Peru, cured herself of malaria by drinking a concoction made from the bark of the native cinchona tree, which contains quinine. With demand on the rise for the various plant species in the genus *Cinchona*, countries outside the region attempted to import seedlings to set up commercial production, a plan viewed unfavorably by Andean nations intent on monopoly. In spite of stiff penalties imposed on people caught exporting the plant, cinchona trees would eventually also be grown outside the region.

The 1950s saw the launch of a largely successful campaign to eradicate malaria worldwide, mostly by utilizing recently discovered insecticides to spray the walls and ceilings where mosquitos tend to rest. Nonetheless, malaria has made a comeback in many parts of the world, in part because some mosquitos have evolved resistance to the very insecticides meant to kill them.

Native to Africa, the yellow fever virus—and its main mosquito vector, *Aedes aegypti*—probably reached the Americas on slave ships. This disease so decimated Napoleon's troops in Haiti that it probably played a role in convincing the French to withdraw from the New World. In the wake of the Spanish American War in 1898, U.S. forces occupied Cuba, a country especially plagued by yellow fever. It was here that Army surgeon Walter Reed led a team that confirmed the idea of the Cuban doctor, Carlos Finlay, that yellow fever is transmitted by mosquitos. There is now a vaccine for yellow fever, though it is beyond the limited means of many poor people in Africa.

Today, the most serious mosquito borne disease in Latin America is dengue fever, which causes headaches and aching joints, and sometimes worse. There are four strains (serotypes) of the virus. A first infection is usually not serious, but if infected a second time—with a different strain—the result is potentially life-threatening. Dengue fever probably originated in Southeast Asia, but became a worldwide problem only in the later part of the 20th century, due to increasing international travel and growth of human populations. There is no vaccine for dengue and its principal mosquito vector, *Aedes aegypti*, is very difficult to control since it breeds in water collected in discarded beverage containers, water storage containers, and the like—and requires only a week or two to complete its life cycle. A second vector of dengue, the Asian tiger mosquito (*Aedes albopictus*), was unintentionally introduced into Brazil and Texas in the 1980s and has subsequently become quite widespread.

Larva of *Aedes aegypti*.

BLACK FLIES

Black flies (Simuliidae) are quite small and often go unnoticed as they suck your blood, but their bite soon leaves an itching, swollen area with a red dot in the middle. Black flies rarely venture indoors and only bite during the day.

It is only the females that feed on blood, in most cases mammal blood, although females of a few species exclusively suck blood from birds. Several species transmit a tiny round worm that causes river blindness (onchocerciasis), an African disease carried by slaves to the New World, where river blindness occurs only in a few isolated locations in southern Mexico, Guatemala, and northern South America.

Females either lay their eggs in batches on objects in (or along) streams or, while flying, drop their eggs into the water a few at a time. The larva spins a silken thread that it uses as an anchor while it drifts downstream, eventually finding a suitable object on which to settle, at which point it spins a patch of silk to attach itself to the object; if necessary it can drift to a new location, again using a line of silk. The larvae normally remain within 30 cm (12 in) of the surface of the water. They use sticky, foldable fans surrounding their mouth to filter out fine organic particles (e.g., algae and bacteria) from the flowing water. The larvae excrete large quantities of fecal pellets, which sink to the bottom, providing food for other aquatic invertebrates—and black flies therefore play an important ecological role by capturing nutrients from the flowing water that would otherwise be carried downstream. The larvae pupate under water and the adults emerge in a bubble of air.

Worldwide there are more than 2,000 named species of black fly, roughly 30% of which bite humans.

BITING MIDGES

Biting midges (Ceratopogonidae), also called no-see-ums in North America, are so small that the human eye can barely perceive them, and they are able to pass right through most mosquito netting.

Worldwide there are some 6,000 named species of biting midge. The females of some species suck blood from vertebrates. Only about 1% of the species bite humans, and these belong to just two genera: *Leptoconops* and *Culicoides* (note that only a minority of species in the latter genus feed on humans). Members of *Leptoconops* are generally restricted to coastal beaches and bite during the daytime, usually on the lower legs. At least some species bury themselves under a thin layer of sand when they are resting. *Culicoides* on the other hand are more widely distributed and can bite at any time, though they tend to be crepuscular or nocturnal. The aptly named *Culicoides furens* is very abundant in mangroves and can make life in some coastal areas very miserable.

The females of many species of biting midge feed on other insects. Some are

Female *Simulium quadrivittatum* feasting on human blood.

Female *Forcipomyia* imbibing blood from a swallowtail caterpillar (*Battus polydamas*).

essentially ectoparasites, feeding for extended periods of time on the blood of large insects such as dragonflies, katydids, stick insects, crane flies, butterflies, and caterpillars. In other species of biting midge, the female preys on male non-biting midges as they form swarms. In a few species, the female feeds on males of her own species—as she is mating with him!

Adult females also feed on nectar, whereas adult males feed exclusively on nectar; some species are important pollinators of cacao and, possibly mango. The larvae are mostly predators or detritivores that live in semiaquatic and aquatic habitats of all types, including mud, wet sand, treeholes, bromeliads, lakes, streams, and decaying vegetation.

Female *Forcipomyia* sucking blood from the antenna of a stick insect (Pseudophasmatidae: *Stratocles unicolor*).

Close-up of *Forcipomyia* on stick insect antenna, with an extremely engorged abdomen.

NON-BITING MIDGES

Non-biting midges (Chironomidae), also called water midges, look like mosquitos but lack the blood-sucking proboscis. They inhabit every conceivable kind of freshwater habitat, from rivers and lakes to the water that accumulates in tree holes and bromeliads; a few species inhabit marine habitats.

This is the most diverse of the families of aquatic insect; at a single location, it is often possible to find nearly a hundred species in a particular site. Moreover, they are sometimes very abundant, reaching densities up to 50,000 individuals per square meter. In the San Juan River, which separates Nicaragua from Costa Rica, certain species sometimes emerge as adults in such astronomical numbers that the surrounding areas are inundated with clouds of midges, a nuisance to local inhabitants, though causing no direct harm. Since many species are attracted to lights, midges occasionally interfere with nighttime events.

Water midges spend most of their lives as larvae in water. Adults, which live for a very short period of time, generally do not feed at all. The larvae of most species feed on detritus

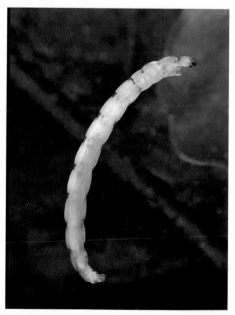

Larva of Chironomidae. This species is unusual in that it induces galls on *Iresine diffusa* (Amaranthaceae) petioles.

or small algae, though some larvae are carnivores; among the more innovative species are larvae that feed on submerged vegetation, sometimes as leaf miners or even as underwater gall-formers (on Podostemaceae).

In many species, the larvae use silk to build tube-like shelters, which in some cases are attached to other aquatic animals, including sea turtles! In a few species that live in oxygen-poor water, the larvae have hemoglobin, which gives them a bright red color.

Worldwide there are more than 7,000 named species of water midge, though the actual number of total species is probably closer to 20,000. Several thousand species occur in tropical America.

Males of non-biting midges usually have very feathery antennae. They are often mistaken for mosquitos.

FUNGUS GNATS

The members of six families of fly go by the name *fungus gnat*, a name derived from the fact that the larvae in these families feed mostly on fungi; the two largest families are Mycetophilidae and Sciaridae, respectively comprising about 4,500 and 2,500 named species worldwide.

Adults are nondescript mosquito-like flies that live for about a week. Males of one species, *Bradysia floribunda* (Sciaridae), are the only known pollinators of the orchid *Lepanthes glicensteinii*. This orchid induces the male flies to pollinate it through deception; its flowers, which smell like female fungus gnats, trick the sexually aroused males into copulating with the flowers, during which each flower attaches a pollinarium on the copulating fly. Many other species in this large genus of orchid are probably pollinated by similar means, deceiving yet other species of fungus gnat. (For other examples of orchids using deception in order to induce flies to pollinate them, see page 299).

Members of the family Keroplatidae (with about 1,000 species worldwide) are called predaceous fungus gnats. While the larvae of most fungus gnats feed exclusively on fungi, larvae in this family have a more diverse diet. Living under rotting logs, where they use salivary secretions to construct mucous webs, the larvae feed on fungal spores or small insects that become trapped in the webs. Usually, a larva rests at the top of the web, hauling up, in fishing-line fashion, a food-laden string of mucous, at once consuming prey and eating (and recycling) the mucous. The larvae of a few species are luminescent; at night, groups of glowing maggots can sometimes be seen crawling along the ground.

Hybosciara gigantea (Sciaridae). The larvae live in compact masses on the ground.

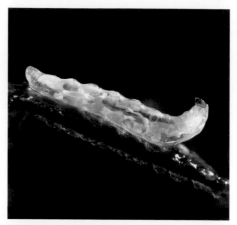

Larva of *Bradysia* (Sciaridae) from leaf mine on fern.

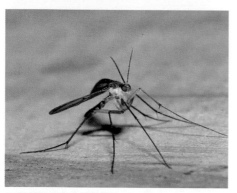

Keroplatidae at night light.

Larva of Keroplatidae in its mucous web.

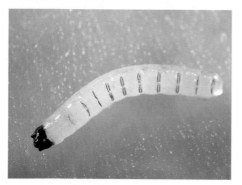

Larva of Mycetophilidae from wild mushroom.

Leaf-mining larva of Mycetophilidae on *Tropaeolum*.

GALL MIDGES

As tiny as gnats—or sometimes even tinier—gall midges (Cecidomyiidae) are rarely noticed, though evidence of their presence abounds. The larvae of this abundant and diverse family form galls (p. 280) on a variety plants, including ferns; indeed, most species of flowering plant probably host one or more gall midge species. Some gall midges are associated with crop plants; for example, there are gall midges that inhabit the leaves of cassava, the stems of both sweet and hot peppers, and the fruits of avocado. Given that most species are very host specific, and that most flowering plants probably have gall midges, the number of species that potentially occur in tropical America is staggering, possibly in the tens of thousands; although we don't even have a good estimate of how many species there are, it is almost certain that very few of them have scientific names.

In most species, adults live for a very short time, feeding on nectar (if they feed at all). Females attract males with a sex pheromone, mating ensues, and the female searches for a suitable host plant on which (or in which) to lay her eggs. On hatching, the larva forms a gall on the leaves, stems, flowers, fruits, or roots of the host plant, depending on the species. Usually, a single larva lives within the gall, though in some cases several larvae occupy the same gall. The larva feeds on nutritive tissue inside the gall.

In most tropical species, the larva pupates within the gall. Escape from the gall is done in one of two ways. Either the pupa exits through a tunnel previously cut by the larva (just prior to pupation) with a spatula-shaped dermal structure on the prothorax or the pupa itself cuts its way out, using hard, sharp cephalic structures powered by peristaltic abdominal motions.

While most species form galls, the larvae of some species live as inquilines in the galls of other gall midges. And some species have no relationship at all to galls, feeding instead in buds or flowers, or, in the case of the most ancestral groups, feeding on fungi (as do their closest relatives, the fungus gnats). The larvae of some groups of gall midges are predators of mites and small insects, or endoparasitoids of aphids and jumping plant lice.

Gall midges (Cecidomyiidae) have very reduced wing venation.

The tiny, fragile gall midges probably comprise more species than any other insect family.

Pupae of Cecidomyiidae often have horns, used to cut through plant tissue when the adult is ready to emerge.

Larvae of Cecidomyiidae are characterized by a dark rod (sternal spatula) at the anterior end (top).

Cecidomyiid gall on Piperaceae.

There are myriad ways in which insects feed on plants, but the most sophisticated manner involves an insect inducing a plant to form a gall, a growth made of plant tissue that the insect eats. The stimulus for gall formation comes from insect secretions such as larval saliva, but the exact mechanism by which these secretions manipulate the plant is poorly understood. Somehow the insect deceives the plant into sending the gall nutrients—as if it were a seed—and plants heavily infested with galls have fewer reserves to dedicate to their own growth and reproduction. Insects are not the only organisms that create galls; some mites, nematodes, and fungi are also gall makers.

Unlike a tumor, a gall stops growing at some point, reaching a characteristic size and shape determined by the organism responsible for inducing it. Indeed, galls come in a dazzling variety of shapes and sizes, and they can be found leaves, stems, flowers, fruits, or roots, depending on the species. Some galls are quite beautiful, and various indigenous peoples in tropical America use them to make necklaces.

The insect larva benefits from the gall in two ways. First, it lives protected within the gall; second, it feeds on the surrounding gall tissue, which the plant enriches with amino acids and other essential nutrients (for this reason the tissue immediately surrounding the insect is known as nutritive tissue).

The normal chemical defenses that plants use against plant-feeding insects appear not to work against gall formers, although some plants can defend against galls in other ways. Some plants, for example, fail to react to the stimulus for gall formation; in other plants, the cells immediately surrounding the larva are sometimes made to die, thus robbing the larva of food; and in a few species, the plant aborts leaves that are heavily infested with galls.

Many groups of insects induce plant galls. Among these are the flies (gall midges, fruit flies, and a few others), jumping plant lice (p. 89), wasps (p. 158), the caterpillars of various moths, and a few beetles. About 75% of all insect-induced plant galls in tropical America are caused by gall midges. The only group of gall formers that rivals the gall midges in diversity are the mites (p. 324), which generally form smaller, simpler galls.

Each species of gall former is usually restricted to just a single plant species or to several related species. Thus, by identifying the host plant and noting the form of the gall, one can often distinguish the species of gall former. It should be noted that not all individuals of a particular plant species harbor galls since some individuals are more susceptible to gall formation than others.

Cecidomyiid gall on cassava (*Manihot esculenta*).

Cecidomyiid gall on *Otopappus* (Asteraceae).

Cecidomyiid gall on *Picramnia* (Simaroubaceae).

Cecidomyiid gall on *Manekia naranjoana* (Piperaceae), with empty pupae in exit holes.

Cecidomyiid galls on *Psychotria* (Rubiaceae).

Cecidomyiid galls on *Mikania* (Asteraceae).

DEER FLIES AND HORSE FLIES

Deer flies and horse flies are both in the family Tabanidae; the term *deer fly* is often used to describe smaller species, *horse fly* for larger ones, but the biology of both small and large species is quite similar. Females of most species feed on blood, delivering quite a painful bite (they slash the skin with a pair of blades) before sponging up the blood from the wound. Some have the irritating habit of buzzing around one's head before landing, while others surreptitiously attack the legs. Most are active only during the day, but a few species are nocturnal or crepuscular.

Not all species bite humans, though females of most species do feed on the blood of other mammals; some species feed on large birds or reptiles (even caimans are noticeably annoyed by these vicious biters). Females of a few species shun blood, feeding instead on nectar; all males feed exclusively on nectar.

Whether in search of food, a mate, or a place to lay eggs, deer flies and horse flies probably depend heavily on visual cues, though biologists poorly understand exactly how they see the world through their iridescent eyes.

What little we know about their mating behavior suggests that it is quite variable. Males of some species apparently aggregate in small swarms in the tops of trees, in others the males hover individually, while the males of a few species wait for females on a perch and then dart out after them. Females usually deposit eggs in layered clusters of 100 to 1,000, on plants or other objects overlying shallow water or wet soil (including the sediments in tree holes), with each species laying eggs in a particular spot. On hatching, the larvae drop to the substrate (in those species whose larvae inhabit floating vegetation, the eggs are laid directly on the vegetation). The larvae of most tabanids are voracious predators, using their fanglike mandibles to paralyze and liquefy other invertebrates.

There are about 4,500 named species worldwide, of which at least 1,000 occur in tropical America.

Fidena trapidoi. At least one species in this genus sucks blood from caimans.

Deer flies, such as this *Catachlorops*, have forked antennae. Note also the dagger-like mouthparts.

TIMBER FLIES

Measuring up to 4.5 cm (nearly 2 in), timber flies (Pantophthalmidae) are the largest flies in the Americas.

Development from egg to adult can take a couple of years. Adults, not seen very often, are thought to live for just a short period of time (2 to 15 days). Depending on the species, the larvae excavate horizontal tunnels in either living or dead trees; as they carve their tunnels, the sound of their rasping can sometimes be heard from several meters away. Larvae apparently feed in their burrows on fermenting sap.

There are only 20 species of timber fly, all occurring exclusively in the American tropics.

Pantophthalmus planiventris.

Pantophthalmus bellardii.

SOLDIER FLIES

Soldier flies (Stratiomyidae) show considerable variation in size, shape, and color. Despite this diversity, however, a feature common to nearly all species is a small enclosed cell in the middle of each wing, a trait difficult to see in the field. As do flower flies (p. 288), many species of soldier fly mimic the colors of wasps and visit flowers.

Males of some species congregate in order to attract females, whether in mating swarms, or in leks on vegetation; within these congregations they compete for females.

Larvae of most soldier flies are unusual in that their cuticle is much harder than that of most other insect larvae (this is because it contains calcium carbonate crystals). Larvae of species in the subfamily Stratiomyinae are freshwater filter feeders, while those in the subfamily Pachygastrinae search out prey beneath bark on dead trees. Species in other subfamilies feed on decomposing plant material; some members of the genus *Merosargus*,

Ptecticus species resemble wasps, especially in flight.

Hermetia illucens (see larvae on following page).

Larvae of *Hermetia illucens*.

for example, commonly feed on detritus in the floral bracts of *Heliconia*, while some species in the genus *Plecticus* feed on the fallen flowers of monkey pot trees (*Lecythis*).

An especially interesting species is *Hermetia illucens*, a large (2 cm/0.8 in), black fly whose appearance—and habit of making buzzing noises—fools some people into thinking it is a wasp, though it is harmless. Native to the Americas, but now cosmopolitan in distribution, biologists are studying this species as a possible means of managing the size of manure piles in feedlots. The larvae reduce manure accumulation by 50%, while the prepupae, easy to harvest, can be used as high-protein animal feed.

Worldwide there are about 3,000 named species of soldier fly; they are especially diverse in tropical America.

BEE FLIES

Members of this family (Bombyliidae) are extremely variable. Many species are densely covered with fine hairs (like bees), but some species are nearly hairless. Males and females feed on nectar; females also eat pollen, a component of their diet necessary for nourishing the eggs developing in their reproductive tract. Females gather pollen with modified hairs on their front tarsi, and then scrape off the pollen onto their mouthparts.

In most species, the female bee fly has a sand chamber at the tip of her abdomen that she fills with sand grains (or similar particles)

used to coat her eggs to prevent desiccation. The female searches for the entrances to wasp nests and other potential host sites, and, while hovering in flight, flicks her coated eggs toward the location. On hatching, it's up to the larva to penetrate the nest and locate a host.

The larvae of most bee flies are ectoparasitoids of insect larvae or pupae, especially those of solitary nest-building wasps and bees, but also those of tiger beetles, darkling beetles, and longhorn beetles; some are endoparasitoids of caterpillars. The larvae of a few species are predators of grasshopper eggs.

Among the most common bee flies are those belonging to the genus *Anthrax*, whose members show wings with black bands. They are often seen slowly hovering just above the ground, or perched with outstretched wings. Some species in this genus attack wasp and bee nests, while others attack the larvae of tiger beetles (p. 101).

Bee flies are less diverse in tropical rainforests than in arid regions. Worldwide there are more than 5,000 named species of bee fly, at least 500 of which occur in tropical America.

Bee flies, such as this *Chrysanthrax*, often perch on the ground with outstretched wings.

ROBBER FLIES

Nearly all species of robber fly (Asilidae) have widely separated eyes and a depression on top of the head. They vary greatly in other respects, however, in shape, color, and size (3 to 40 mm/0.1 to 1.6 in). A few species mimic orchid bees and are quite spectacular to behold; yet other species blend in artfully with the substrate on which they perch.

Males and females are rapacious predators of flying insects, especially of adult flies, beetles, wasps, or bees, depending on the species of robber fly. They generally watch for potential prey from a perch, with each species favoring a perch of a particular height and kind (branch, twig tip, leaf, etc.); some species prefer to perch in the sun, others in the shade. From this perch they dart out after potential prey, capture it with their legs, stab it with their dagger-like proboscis, and return to their perch. They inject the prey with saliva, which paralyzes it and liquefies the internal tissues, allowing the robber fly to suck it dry. Species in the subfamily Leptogastrinae (which are very slender robber flies) do not sally forth from a perch, but rather fly about slowly and swoop down on a stationary insect or spider.

A given species generally lays its eggs in one of three ways. Some females drop their eggs on the ground, others shove them into soil or rotten wood, and a third group lays a chalky eggs mass on vegetation (on hatching, the larvae fall from the vegetation onto the ground). The larvae burrow through soil or rotten wood in search of the juvenile stages of other insects and, on encountering this prey, they begin to feed. A single robber fly larva may consume from one to more prey, and they do so by injecting toxins and enzymes, and then sucking up the contents of the prey. Many are ectoparasitoids on the larvae of scarab beetles (p. 12); *Eccritosia zamon*, for example, is a beach-inhabiting species that appears to feed on scarab beetle larvae (Trogidae), which themselves feed on damaged sea turtle eggs.

Worldwide there are approximately 7,500 named species of robber fly.

Adult *Mallophora* capture bees, while the larvae feed on scarab beetle larvae.

Diogmites species perch on low shrubs or on the ground.

Lampria species perch on sunlit leaves.

DANCE FLIES

In the American tropics, there are three families of dance fly (a fourth family occurs in northern Mexico). The two largest families are Empididae and Hybotidae (the other family is Brachystomatidae); in total these three families comprise more than 5,000 named species worldwide. They are relatively small flies (1 to 12 mm/0.4-0.5 in), and usually have a proboscis (the length varies from species to species).

In most dance flies, both males and females are predators of other insects; depending on the species, they search for prey either in flight or while running about on the surface of plants (or on exposed rocks in streams). They generally use their front legs to grasp and hold prey as they feed on it (in members of the subfamily Hemerodromiinae, family Empididae, the front legs are enlarged—as in mantises). Mating generally occurs on the ground or on vegetation.

Members of the subfamily Empidinae (family Empididae) deviate from the general description in two respects. First, the females in this subfamily have lost their ability to capture prey; instead, they feed on nectar and prey items that males give them during courtship.

In some species of Empidinae the male even gift wrap such nuptial gifts in silk (which is secreted by the tarsi on the front legs) before giving it to the female; in yet other species the male's nuptial gift is nothing more than silk-wrapped debris or a frothy balloon consisting of anal secretions. The second way in which members of the subfamily Empidinae differ from other dance flies is that mating occurs in aerial swarms, with individual flies moving in synch, so that the swarm as a whole bounces up and down (hence the common name, dance flies). In most species the aerial swarms are formed by males, and females enter a swarm to choose a mate based on the male's nuptial gift and his body ornamentation. However, in other species the sex roles are reversed, with females forming swarms and males enter the swarm with their nuptial gift; in these cases, the females compete for access to a male (and his gift) and it is the male that chooses the female, based on her ornamentation (e.g., wing coloration, scales on the legs, inflatable sacs on the abdomen).

The larvae of dance flies are generally found in moist soil, rotten wood, dung, or in a variety of aquatic habitats. Like the adults, they are predators, feeding on fly larvae and other arthropods.

Empididae on *Maianthemum gigas* flower (Asparagaceae).

Hybotidae on rolled-up fern frond.

LONG-LEGGED FLIES

Closely related to dance flies, long-legged flies (Dolichopodidae) are among the few flies that can often be recognized in the field. These slender, shiny-green flies usually measure from 1 to 9 mm (less than 0.5 in) in length. Depending on the species, they occur on the surface of leaves, tree trunks, mud flats, or exposed rocks in streams—either standing alert or running about.

Adults feed on small, soft-bodied invertebrates such as midges, bark lice, and mites, enveloping prey in their spongy proboscis, tearing them open with their minute teeth, and absorbing their bodily fluids. Many species also imbibe honeydew and some have elongate mouthparts for reaching nectar in flowers. Males are generally endowed with enlarged genitalia slung beneath their abdomen and many also sport enlarged hairs, silver spots, and other adornments with which to court the female. Most males in the subfamily Dolichopodinae engage in mating dances;

Condylostylus is common along trails and forest edges.

Aggregation of *Symbolia* on a leaf along a stream.

and, in some species of the subfamily, males also defend territories. After copulating, some males in the subfamily Hydrophorinae hold on to the female with their front legs in order to prevent her from mating with other males.

The larvae live in wet leaf litter, mud, moss, algal mats, sap oozing from trees, tree holes, and under bark. They are thought to be mostly predators or scavengers. Just before pupating, the larva spins a protective cocoon in which to pupate, incorporating soil particles and other debris.

Worldwide there are some 7,500 named species of long-legged fly.

PHORID FLIES

Also known as scuttle flies and humpbacked flies, members of this family (Phoridae) are very small, measuring from 1 to 8 mm (0.04 to 0.3 in) in length. Nonetheless, they are exceedingly abundant—and identifiable by the way they run about in jerky zigzags.

Megaselia scalaris, a very common species throughout the world, probably has the record among insects for the most diverse larval diet, which includes decomposing vegetables, carrion, injured insects and lizards, shoe polish containing wax, and, rarely, human wounds.

The larvae of many species eat decaying organic material, but phorid fly larvae show a wide range of feeding strategies. Some species are fungivores; a few are herbivores (e.g., in flowers of pipevine, family Aristolochiaceae); a few are predators; some live as kleptoparasites in ant or termite nests, eating the food intended for the ants and termites themselves; and some species parasitize millipedes, fireflies, ants, or bees, depending on the species. Of the parasitic species, some go after injured hosts while others attack healthy hosts.

Among the most interesting phorid flies in tropical America are those that attack leafcutter ants. Some of these, known as "ant-decapitating flies," land on the leaf fragment being carried by an ant, lay an egg (through the mandibular suture) in the ant's head capsule, and the resulting larva feeds inside the ant's head until it drops off. For protection, the leafcutter ants often have much smaller hitchhiker ants (that ride on the cut pieces of leaves) to drive

Neodohrniphora curvinervis attacks outbound leafcutter ants (*Atta*) and lays its egg in the back of the head.

Apocephalus ritualis lays its egg in the mandibular articulation of leafcutter ants.

away flies attempting to attack the larger ants (p. 189). However, this particular defense is probably less effective against other species that attack leafcutter ants in other parts of the body. Numerous phorids attack other species of ant, often harassing them to the point where the ants' ability to forage is greatly impaired. Those that parasitize stingless bees attack while their hosts are on flowers, near the nest, or in mating aggregations.

Worldwide there are nearly 5,000 named species of phorid fly, though the actual figure is surely many times that number.

FLOWER FLIES

Flower flies (Syrphidae), also known as hover flies, are extremely agile fliers, capable of hovering in place and, in a split second, darting backward or sideways (males search for females by hovering in sunlit areas of the forest, perching on leaves or twigs, patrolling an area, or some combination of these tactics). While many flies mimic wasps or bees to gain protection from predators, this family has the greatest number of examples.

Given that most flower flies feed on nectar and/or pollen, they are probably important

Allograpta centropogonis flower fly. Larvae of this species are leaf miners on *Centropogon* and *Lobelia* (Campanulaceae).

pollinators, though tropical species have not been well studied. Some species feed on honeydew and pollen present on the surface of leaves.

The larvae are very diverse. Those in the subfamily Microdontinae (one of three subfamilies) inhabit the nests of ants, where they prey on the ant brood. Larvae in the subfamily Eristalinae feed on decaying plant material, excrement, and other detritus. This subfamily includes the genus *Ornidia*, whose metallic green adults are very common throughout tropical America; it also includes the so-called rat-tailed maggots (tribe Eristalini), which inhabit stagnant water or wet organic material and breathe through a long, retractile tube that functions as a snorkel. Larvae in the third subfamily (Syrphinae) are unusual among flies in that they often live exposed on plants, where the greenish or translucent pink maggots generally prey on aphids, scale insects, and other small insects. In tropical America, some larvae in this subfamily do things that are virtually unheard of in temperate regions: they mine leaves (of *Centropogon*); feed on pollen inside flowers (Indian paint brush, *Castilleja*); and use the honeydew of whitefly nymphs (p. 90) to attract adult flies, which they then ensnare with a sticky secretion and consume. Worldwide there are more than 6,000 named species of flower fly.

Palpada. Larvae in this genus feed on decaying vegetation.

A bee-mimicking species of *Palpada*.

Mating *Toxomerus marginatus* (male on top).

Larva of a predatory Syrphidae.

Pupa of a predatory Syrphidae.

STILT-LEGGED FLIES

The aptly named stilt-legged flies (Micropezidae) are fairly easy to see in the understory of forests in tropical regions of the Americas. Adults spend a lot of time on the upper surface of leaves, where they feed on honeydew, fallen fruit, and bird droppings. Some species are characterized by a white band near the tip of each front leg; as these flies raise up the two legs in front of their body, the motion mimics the movement of the white-banded antennae of many wasps. In one such species, *Ptilosphen viriolatus*, small groups of adults sleep together on understory foliage and then mate between five and seven o'clock in the morning. During mating, the male drools on the sides of the female's head and the latter eagerly laps it up; the secretion is so attractive that other females sometimes barge in on the free treat.

The larvae of stilt-legged flies feed on rotting vegetation; very little is known about their biology. Worldwide there are about 600 named species of stilt-legged fly.

An ant-mimicking species of *Paragrallomyia*.

FRUIT FLIES

Fruit flies (Tephritidae) generally have patterned wings, though they are not the only flies with this characteristic. Adults feed on plant exudates, bird droppings, honeydew, and microbes found on plant surfaces.

During courtship, males display or move their wings in a variety of patterns, secrete sexual-attractant chemicals, and produce close-range sounds inaudible to human ears.

Females of many fruit flies lay their eggs in firm, fleshy fruits (as opposed to the rotting fruits preferred by pomace flies, p. 193,

Grallipeza feeding on dung.

Female *Micropeza* feeding on bird droppings while mating.

The papaya fruit fly (*Toxotrypana curvicauda*), often confused for a wasp.

Tomoplagia (Tephritidae). The larvae of this genus live in flower heads of Asteraceae.

organ sometimes harbors several fly species, an individual plant might host a dozen similar-looking species of fly.

Worldwide there are nearly 5,000 named species of fruit fly.

PYRGOTID FLIES

Unlike the vast majority of other flies, pyrgotid flies (Pyrgotidae) are nocturnal, as evidenced by the pale bodies of adult flies; creatures of the night have no need to invest in bright colors.

Pyrgotid flies are unusual in at least one other respect. The female attacks scarab beetles in flight, inserting her egg in the beetle's abdomen, where her larva feeds as an endoparasitoid, eventually killing the host. Females presumably attack beetles as they fly, which is when the abdomen of the beetle is most exposed (at rest, beetles cover their abdomen with their sturdy front wings).

Worldwide there are about 350 named species of Pyrgotidae.

and many other fly families); their larvae feed within the fruits, and the nutritional value of this food source is probably enhanced by the presence of endosymbiotic bacteria (the bacteria have been found in all life stages of fruit flies). In tropical America several species of *Anastrepha* cause serious damage to a number of fruit crops; one serious pest species, the Mediterranean fruit fly (*Ceratitis capitata*), was introduced to tropical America in the 1950s. One of the most gaudy members of this family is the papaya fruit fly (*Toxotrypana curvicauda*), which resembles a wasp.

Not all fruit fly larvae feed on fruits, however. Many feed in the flower heads of plants of the aster family (Asteraceae); other species induce plant galls, mostly stem galls on plants in the aster family. Both flower feeders and gall-formers generally pupate within the plant (frugivorous species usually pupate in the soil). Yet other species are leaf or stem borers. Larvae of the genus *Blepharoneura*, one of the most ancestral groups of fruit flies, are all associated with plants in the cucumber family (Cucurbitaceae); and individual fly species in this genus often specialize not only on a particular plant species, but also a particular organ of the plant (such as male flowers, female flowers, and seeds). As each

Leptopyrgota.

Sphecomyiella.

BLACK SCAVENGER FLIES

Black scavenger flies (Sepsidae) are small insects with a spherical head. Most species are black, with transparent wings, though a few species have a black spot near the tip of each wing. When they perch or walk, they often simultaneously move their wings slowly forward and backward, a behavior that probably warns predators of the flies' inherently unpleasant taste. The larvae inhabit excrement or carrion and adult males and females often occur near fresh cattle dung. (The eggs have a long tail that the female carefully lays out on the surface of the substrate, allowing the developing embryo to obtain oxygen despite being immersed in the anoxic environment of wet manure or excrement.)

Females accept or reject a particular male on the basis of his ability to stimulate her. The males have highly modified front legs that are used to grasp the bases of the female's wings, and in each species the males have a different set of spines, bumps and grooves that serve to stimulate the female. Males of some species have species-specific, moveable lobes on the lower surface of the abdomen, which have evolved several times in this family. Males also squeeze the female in specific ways with clasper-like portions of their genitalia and they use their legs to brush chemical signals onto her wings or antennae. Worldwide there are about 350 named species of black scavenger fly.

Archisepsis diversiformis on dung.

Close-up showing structural color patterns of wings caused by a phenomenon known as "thin film interference".

LEAF MINER FLIES

While the vast majority of leaf-mining insects are caterpillars of micromoths (p. 202), the larvae of various fly families are also leaf miners, and the most prominent of these families is the family Agromyzidae.

Adult male and female leaf miner flies are tiny and seldom seen, though the leaf mines that their larvae create are quite a common sight in tropical forests, especially on herbaceous plants. The larvae of a given species tend to make a characteristic type of mine. Some mines are linear, others are blotch-like; the location of the mine on the leaf itself also varies from species to species. The larvae of some species burrow in stems or in seeds of plants such as those in the aster family (Asteraceae). In general, the larvae of leaf miner flies live inside the leaf anywhere from one week to several months; they pupate either inside the leaf mine or on the ground.

The larvae of *Melanagromyza rosales*, a species of fly that occurs in Central America, create leaf mines that are sometimes more than a meter long on *Bromelia pinguin*, a terrestrial bromeliad. Native to the Americas, *Liriomyza huidobrensis*, *L. sativae*, and *L. trifolii* have been inadvertently introduced into several countries in other parts of the world; their larvae cause serious damage to a host of vegetable crops. Sometimes, the use of insecticides actually swells the populations of these insects. Applied either against the leaf miners themselves or other insect pests, the insecticides are often more effective in killing parasitic wasps, which usually keep leaf miner populations in check.

There are more than 3,000 named species of leaf miner fly worldwide.

Liriomyza, a genus that contains some notorious crop pests.

Larva of *Nemorimyza* in leaf mine on *Crassocephalum crepidioides* (Asteraceae).

originated in Hawaii (a hotbed of *Drosophila* evolution), and from there colonized the rest of the world. The larvae of *Cladochaeta* inhabit the spittle masses of some spittlebugs (p. 79), apparently parasitizing the spittlebug without killing it. Larvae of *Hirtodrosophila batracida* feed on the eggs of Central American tree frogs and glass frogs, and are often the primary cause of egg mortality. Worldwide there are more than 4,000 named species of pomace fly.

Larva of *Cladochaeta* on spittlebug nymph (*Clastoptera*).

Drosophila willistoni on banana.

POMACE FLIES

Of all the pomace flies (Drosophilidae), the most renowned is *Drosophila melanogaster*, the fruit fly of genetics laboratories, which, because it is easy to rear and because it has chromosomes that are easy to study, is one of the most thoroughly studied of all organisms. Somewhat confusingly, members of another family, Tephritidae (p. 290), are collectively named *fruit flies*, though they are quite distinct from *Drosophila melanogaster* and other members of its family.

Pomace flies are the tiny flies commonly seen hovering above discarded fruit—the word *pomace* is a botanical term for crushed fruit pulp. Because rotting fruit smells something like vinegar, these flies are also called vinegar flies.

The larvae of most pomace flies feed on yeasts and bacteria found in fermenting plant material. Many larvae feed in living or fallen flowers; in some species, the larvae are quite host specific. Species of *Scaptomyza* are leaf miners on herbaceous plants growing at higher altitudes; this genus is unusual in that it appears to have

Drosophilidae on fallen mango.

BAT FLIES

Bat flies are members of two closely related families, Streblidae and Nycteribiidae. The adults spend almost their entire life in the fur and wing membranes of bats, feeding on their hosts' blood.

Bat flies are host specific, with a given species of bat fly generally living on a specific species of bat. An individual bat sometimes harbors as many as five species of fly; moreover, a bat that harbors a large number of individuals of one species of fly often harbors large populations of other fly species (but note that different fly species tend to occupy different parts of the bat's body). In general, bats that roost in caves or hollow trees have higher parasite loads than those that roost on foliage. Although bat flies do not cause lesions on the bat's skin, heavy parasitism can cause the bat to lose weight.

The adult female produces one egg at a time; the egg remains within her uterus, and the larva that hatches from the egg also remains in the uterus as it completes its development, feeding on secretions provided by the mother. When the larva matures, the mother temporarily leaves her host bat to deposit the fully grown larva in the vicinity of her host's roosting area. Once deposited, the larva immediately pupates and in 3 to 4 weeks an adult fly emerges, and then locates and colonizes a host bat.

In the Americas, Streblidae and Nycteribiidae comprise about 160 and 50 species, respectively.

Louse flies (Hippoboscidae) and tsetse flies (Glossinidae) are two families closely related to the bat flies. Like the latter, both families share the unusual habit of bypassing a free-living larval stage. Adult louse flies generally parasitize birds, though one species lives on deer. Tsetse flies, which are restricted to Africa, are much better fliers than louse flies and bat flies; they feed on various large mammals and reptiles, and are the infamous vectors of African sleeping sickness.

Streblidae on bat (*Carollia*).

Two Streblidae on underside of bat wing (*Carollia*).

MUSCID FLIES

The family Muscidae is best known for the house fly (*Musca domestica*), a native of the central Asian steppes that has followed humans and their garbage around the world. But this family also includes about 900 native species in tropical America. Depending on the species, adults feed on decomposing organic material, plant or animal exudates, other invertebrates or, less commonly, the blood of vertebrates. The diet of the larva also varies widely between species, with some species feeding on decaying matter, while others feed on animals, either as predators or as parasites. The larvae of the fifty species in the genus *Philornis* are found in birds' nests, where they feed on organic material or on the blood of nestlings (the larvae often burrow just under the skin of the nestlings). The larvae of some species in the subfamily Coenosiinae inhabit freshwater, where they prey on the larvae of moth flies, black flies, non-biting midges, and other small invertebrates.

In the stable fly (*Stomoxys calcitrans*), another Old World species introduced in the Americas, both females and males suck blood from warm-blooded vertebrates (in other biting flies, only the female sucks blood). Females require more blood than do males, generally feeding 2 to 3 times per day (each session lasting a couple of minutes) but sometimes feeding more often if some of their feeding sessions have been interrupted. While the bite of the stable fly is merely painful to humans, it is a serious problem for cattle, who lose weight when they devote too much time to warding off flies—and too little time eating. Cattle often bunch together with their heads facing inward in order to protect their front legs, a favored feeding site for the fly. The larvae of stable flies live in moist decomposing plant material such as that found in the soil of pineapple plantations (fruit companies in tropical America are currently trying to resolve this problem). The cycle from egg to adult takes 2 to 3 weeks; adults live for about the same amount of time.

Stable flies on a cow.

BLOW FLIES

Many members of the family Calliphoridae are brightly colored flies sometimes referred to as blue bottle flies or green bottle flies. Females lay their eggs in fresh excrement or the dead bodies of vertebrate animals or other kinds of decomposing materials. The larvae, or maggots, that inhabit carcasses initially feed on the fluid serum and bacteria found between the muscle fibers of the dead animal; as they do so, they excrete enzymes that break down the tissue to a gooey mass, on which they then feed by rasping and filtering.

Because each species prefers a cadaver of a certain stage of decomposition in which to lay eggs—and because each species needs a certain amount of time to develop from egg to adult—forensic entomologists use blow flies to provide evidence about the time of death of human bodies. Traditional methods of estimating the postmortem interval are generally unreliable after 72 hours, and it is only blow flies (and a few other insects) that can provide more accurate estimates. Moreover, since each insect species lives in a certain habitat, the community of insects inhabiting a cadaver sometimes indicate whether the cadaver has been moved (if the insects in a cadaver do not correspond to those in the surrounding area, the body has been moved).

In a more positive light, perhaps, some physicians and veterinarians are beginning to take seriously the use of "maggot therapy" to clean intractable wounds that do not heal.

The stable fly (*Stomoxys calcitrans*) has an elongate proboscis, unlike the house fly.

The familiar house fly (*Musca domestica*).

Live, disinfected larvae (usually of *Lucilia sericata*) are introduced into the wound to eat away the dead tissue, kill off bacteria, and stimulate wound healing. Most species that feed in wounds restrict their activities to the wounded tissue, but the screwworm (*Cochliomyia hominivorax*) proceeds to burrow into healthy tissue (p. 297).

Blow flies are more diverse in the Old World than in the New World; of the 1,500 named species worldwide, about 250 occur in tropical America.

Chrysomya. Three species of this Old World genus have been accidentally introduced into tropical America.

Mesembrinella, previously part of Calliphoridae, now placed in a separate family (Mesembrinellidae) that occurs exclusively in tropical America.

STERILE MALES AND THE
WAR AGAINST SCREWWORMS

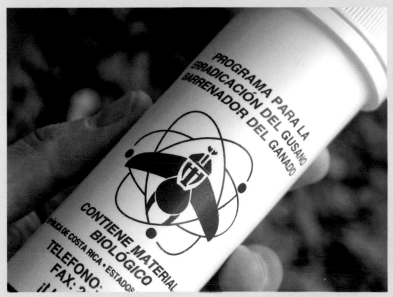

Container for sterilized males of the screwworm.

One of the most serious pests of livestock in tropical America is the screwworm, *Cochliomyia hominivorax*, a member of the blow fly family. The females lay their eggs in the open wounds of birds and mammals, including humans; after hatching from the eggs, the larvae (maggots) burrow into and feed on healthy tissue around the wound. Infected animals become jittery and make frantic attempts to scratch the wound; more flies continue to arrive and lay eggs, with the result that many untreated cases eventually result in the death of the animal.

In the U.S., officials estimated that in the first half of the 20th century screwworms caused yearly losses—primarily to the cattle industry—of more than $100 million.

As a result, the United States Department of Agriculture declared war on the screwworm. Billions of flies were mass produced in factories; the females were killed; and workers zapped the males with radiation to sterilize them. From airplanes, hoards of sterile males were dumped so that they would mate with females, who would therefore fail to reproduce.

By 1966 the screwworm had been driven south of the Rio Grande, and by 1991 Mexico was free of the fly. By 2006 Central America was declared fly free, and currently the Panamanian government and USDA maintain a permanent sterile-male barrier along its border with Colombia in order to prevent the return of the screwworm into Central and North America. However, the fly remains a serious pest in South America.

The so-called "sterile male technique" has been used to control a few other insects, but the potential applications are probably very limited due to the costs, and difficulties, of mass-producing sterile but otherwise hardy males.

FLESH FLIES

Adult flesh flies (Sarcophagidae) often have 3 black stripes on top of their thorax, though some tachinid flies have a similar pattern (exact identification depends on more subtle characteristics).

The larvae hatch from eggs kept within the female, so that the larvae she deposits begin life with a head start. The larvae of many species feed on carcasses, especially those of smaller animals such as rodents, fish, crabs, snails, and insects (the scientific name of the family, Sarcophagidae, means *flesh-eating*). One species (*Eumacronychia sternalis*) seems to have specialized in eating the damaged eggs of sea turtles.

The larvae of other species in the family attack living flesh. The larvae of *Notochaeta bufonivora*, for example, parasitize and kill various types of frogs. Still others are parasitoids (or predators) of earthworms, snails, spider eggs, millipedes, grasshoppers, mantises, cicadas, and beetles; in another group of species, the larvae are kleptoparasites in the nests of solitary bees and wasps. Among the parasitoids are species in the genus *Emblemasoma*; females home in on a singing male cicada, land nearby, squeeze underneath his wings, and inject a larvae into his sound producing organ. The larvae of *Chrysagria alticophaga* are one of the few natural enemies of the flea beetle (p. 143) that so devastates the huge leaves of the poor man's umbrella plant (*Gunnera*).

Species that do not eat flesh often feed on mammal excrement, primarily that of carnivores or omnivores. One species (*Sarcofahrtiopsis thyropterontos*) appears to be a specialist on the droppings of Spix's disc-winged bat, an insectivore that perches inside rolled up leaves of heliconia or banana.

Worldwide, there are about 3,000 named species of flesh fly, and they are especially diverse in tropical America.

Oxysarcodexia. The larvae live in dung.

Tricharaea. Some species are restricted to coastal areas where the larvae feed on dead fish and other carrion.

Oxysarcodexia feeding on bird dropping.

TACHINID FLIES

The flies have more parasitoids among their ranks than any other order except the wasps. Several fly families include parasitoids, though the family Tachinidae has more parasitoids than any other family of flies. In more precise terms, the larvae of all members of this family—at least all members that have been studied—are endoparasitoids, feeding internally on—and eventually killing—primarily other insects, but also on centipedes, scorpions, and spiders. Adult tachinid flies vary greatly in appearance; some species are plain looking, small creatures (2 mm/0.1 in) while others are reddish brown (or black) and large (20 mm/0.8 in). The adult flies feed on honeydew and nectar; mating is often preceded by aggregations of flies on hilltops or tree trunks, although this aspect of their biology is poorly known.

In many species of tachinid fly, the larvae parasitize the caterpillars of moths and butterflies, or those of sawflies. People who feed caterpillars in order to obtain butterflies are often disappointed when a tachinid larva pops out and the caterpillar dies (the whitish tachinid larva soon changes into a shiny reddish-brown puparium).

Other species attack the juvenile and/or adult stages of true bugs, cockroaches, mantises, stick insects, grasshoppers, katydids, crickets, leaf beetles, scarab beetles, and a few other insects. Species in the tropical American genus *Calodexia* fly from leaf to leaf just above army ant swarms, waiting for fleeing cockroaches (many members of the family Conopidae have a similar behavior, but they generally hover above the army ants and attack crickets as well as cockroaches).

Females employ a variety of strategies for transmitting their larvae to potential hosts. Depending on the species, the female sticks an egg directly on (or in) the host; lays thousands of minute eggs on foliage in the hope that at least some of the eggs will be ingested by hosts; or she lays eggs near the host, where the hatched larvae either wait for a passing host or actively search for one. The larva does not kill the host immediately, but rather allows it to feed and grow for a period of time before doing so.

In the mountains of South America and southern Central America, there are orchids in the genus *Telipogon* that have a bristly lip resembling the rear end of certain tachinid flies; when male tachinid flies attempt to copulate with this flower they instead end up transporting orchid pollen. Worldwide there are about 10,000 species of tachinid fly.

Epalpus lapping up sweat on a backpack.

Many tachinid flies, such as this *Xanthoepalpus*, have long bristles on the abdomen.

Unidentified tachinid fly.

Cordyligaster species parasitize caterpillars of small moths.

Eggs of *Winthemia* on caterpillar of owl butterfly (*Caligo*).

Orchid flower (*Telipogon*) resembling the rear end of certain tachinid flies.

BOT FLIES

Adult bot flies (Oestridae) generally lack functional mouthparts and do not feed. The larvae are parasites of mammals. This family is composed of four subfamilies with nearly 200 named species worldwide.

Cuterebrinae is the main subfamily in tropical America, which is home to about 50 species in the genus *Cuterebra*, whose members parasitize mostly rodents (but also opossums and howler monkeys), and *Dermatobia hominis*, which parasitizes a wide range of hosts. Females of *Cuterebra* usually lay eggs on foliage or twigs near mammal burrows and her larvae hatch in response to the heat produced by the bodies of the nearby host animals. The larva, also called a *bot*, enters the host's body through an orifice or wound and then moves to the nasopharyngeal, tracheal, or esophageal regions, where it spends a few days. From there it migrates for a week or so through the body cavities (or under the muscles) until it finally settles somewhere on the body just beneath the skin and forms a swelling. The larva feeds on lymphatic fluid that exudes into the swelling and breaths through a small hole that reaches the surface of the skin. It reaches maturity after 4 to 10 weeks, at which points it leaves the body and pupates in soil. In most cases, the larva has minimal effect on its host, except in cases where large numbers of bots invade the host (one study showed that bot infection reduced rates of reproduction in the host but increased longevity!).

While most bot flies are host specific, *Dermatobia hominis*, the so-called human bot fly (in Spanish, *tórsalo*) infests a wide variety of mammals, not just humans. It occurs from southern Mexico to northern Argentina. The female human bot fly lays her eggs in a most unusual manner; she grabs a mosquito (or some other fly), laying her eggs on its abdomen. When the egg-laden mosquito lands on a mammal, the heat and carbon dioxide that the mammal emits stimulate hatching, and the tiny larva (or larvae) burrows into the skin. After a week or two, a lump in the skin forms, and then a small hole appears, through which the growing larva obtains oxygen. At this point, a human host might be attempted to apply an antibiotic ointment, though that would simply serve to block the larva's breathing hole—and cause it to

wiggle. If the host were to do nothing, the maggot would eventually pop out, leaving a slight scar but no permanent damage. Depending on where it is located on the host's body, it is generally not particularly painful (except psychologically). Nonetheless, most people's first reaction is a frantic attempt get rid of it. The larva has to be gently coaxed out with a little grease over its breathing hole and pressure from below, a simple process, but best performed by someone who has done it before (e.g., a local veterinarian).

Human bot fly adult.

Symptom of the human bot fly. The larva maintains an opening in the skin in order to breathe.

Squeezing the skin from below exposes the larva of the human bot fly.

8

OTHER ARTHROPODS

Insects belong to the phylum Arthropoda, a group of animals characterized by an exoskeleton (a skeleton on the outside of the body) and by legs that have moveable joints (arthropod means "jointed leg"). The exoskeleton is made principally of chitin, a resilient compound that also occurs in some other organisms (e.g., in the cell wall of fungi). In addition to insects, this phylum includes three other groups of arthropod, which are the subject of this chapter: chelicerates (spiders, scorpions, mites, and their kin), myriapods (millipedes and centipedes), and crustaceans (crabs, shrimp, and their kin). Insects have three pairs of legs and one pair of antennae, whereas other arthropods generally have more than three pairs of legs—and only the myriapods have a single pair of antennae (chelicerates lack antennae and crustaceans have two pairs of antennae).

In addition to the insects, and the three groups that are the focus of this chapter, there is yet another group of arthropods that no longer exists. The trilobites were a strictly marine group that dominated the oceans during much of the Paleozoic Era (542 to 251 million years ago) and that went extinct *before* the rise of the dinosaurs. Eurypterids were another group that inhabited the Paleozoic oceans; these were scorpion-like chelicerates, some of which reached a length of 2.5 meters (about 8 feet), making them the largest arthropods that have ever lived. Taking into account the diversity and abundance of these extinct groups, it is apparent that arthropods have always been the dominant group of animals on our planet.

The purpose of this chapter is to provide information on the non-insect arthropods, with a brief glimpse of who's related to whom. Because these "other arthropods" comprise hundreds of thousands (perhaps millions) of species, this chapter necessarily offers an overview only. Moreover, the discussion focuses on terrestrial and freshwater arthropods, with crustaceans, most of which are marine, given just a glancing mention.

ARACHNIDS
(subphylum Chelicerata)

Members of the subphylum Chelicerata have 4 pairs of legs (except for mite larvae, which have 3 pairs, and a group of plant-feeding mites that have just two pairs) and lack antennae. All possess chelicerae, mouthparts that they use to eat, in much the same way that other arthropods use their mandibles. There are three classes of chelicerates. The first two—both containing few species—are composed of strictly marine species. These are the sea spiders (which, despite their name, are not spiders) and the horseshoe crabs (which are not true crabs, as true crabs are crustaceans). Sea spiders are mostly unknown to non-biologists, but horseshoe crabs are much more conspicuous due to their large size and habit of breeding in shallow coastal waters; they are an ancient group with only four surviving species, three in Asia and one on the Atlantic coast of North America (that ranges south to the Yucatan Peninsula.

The third class, the Arachnida, is home to the vast majority (99.99%) of chelicerates. This is primarily a terrestrial group, although several mite species are aquatic. Arachnids are currently divided into 16 orders, but this chapter restricts itself to describing the nine that are most commonly encountered: scorpions, pseudoscorpions, harvestmen, whip spiders, spiders, and four (of the six) orders of mites. While most mites are tiny and difficult to see, they are by far the most species-rich and abundant group of arachnids; though seldom seen, their presence is often felt.

Most groups of arachnids consist primarily of predators, although harvestmen are scavengers and mites vary widely in their diet, depending on the species. Predatory arachnids feed by secreting digestive enzymes into the prey and sucking up the liquefied contents. They can often consume a large amount of food at one sitting and then go for long periods of time without eating. Another biological characteristic of the majority of arachnids, again with the exception of most harvestmen and some mites, is that they lack a penis. As a result, sperm transfer must occur by some other means. Male spiders, for example, use their pedipalps (small appendages on the head) to transfer sperm from their genital orifice to that of the female, while scorpions and some pseudoscorpions deposit the sperm on the ground, where it is retrieved by the female.

In tropical America horseshoe crabs occur only on the Yucatan coast.

SCORPIONS

Scorpions (in Spanish *escorpiones, alacranes*) belong to the order Scorpiones. These nocturnal predators of other arthropods (and sometimes small vertebrates) grab prey with their pincers (pedipalps) and subdue them by injecting venom with the stinger at the end of their tail. They also sting in self defense, as you would know if you have ever put on clothes or shoes that contained a hidden scorpion.

Contrary to popular belief, very few species pose a lethal threat to humans—fewer than 5% of the approximately 2,000 species in the world. Curiously, the most dangerous species are not the largest ones. In the Americas, for example, among the most dangerous are 6 species (found in Mexico) of *Centruroides* and about 4 species (in Brazil) of *Tityus*, none generally more than 5 cm (2 in) in length. Scorpion venom contains a wide array of compounds (especially neurotoxic polypeptides), with correspondingly varying effects (depending on the species and the person), from localized reddening of the skin, to paralysis of the tongue, and to more serious respiratory and cardiovascular problems.

Though many scorpions are associated with desert habitats, they occur in a wide range of places, including tropical rainforests.

Those that hide under stones or bark during the day carry their tails to one side, whereas those that hide in burrows hold their tails up. Scorpions do not see very well and rely more on sensory setae, such as those on their pincers, to perceive the world around them; for this reason, they generally run with their pincers outstretched.

By shining a black light, you can readily locate scorpions at night. Their fluorescent exoskeleton emits a greenish glow when illuminated with ultraviolet light. It is still unclear whether this fluorescence serves some function (it could be just a side effect of light absorption; even Froot Loops fluoresce).

Male scorpions have broader pincers and longer tails than do females. During courtship, the male and female dance and grasp one another's pincers; at some point, the male deposits on the ground a sperm packet that is then picked up by his partner. Most arachnids lay eggs, but scorpions are unusual in that females give live birth to miniature scorpions. The young crawl onto their mother's back, and she carries them around until they have undergone at least one molt.

PSEUDOSCORPIONS

Pseudoscorpions (order Pseudoscorpiones) comprise 26 families and more than 3,500 named species worldwide.

These arachnids have pincer-like appendages (pedipalps), similar to those of scorpions, but lack the stinging tail. Because they are tiny (2-8 mm/0.1-0.3 in), pseudoscorpions are seldom noticed, except when they find their way into houses. Most pseudoscorpions live in leaf litter, rotting wood, fissures in bark or rocks, or in the nests of insects, birds, and mammals. They feed on small invertebrates. Some species subdue prey by injecting venom from a mobile "finger" on their pincers, but they pose no danger to humans.

The female protects the developing eggs in a membranous sac on the underside of her abdomen. When the young hatch, they remain in this sac and feed on a milk-like secretion from their mother's ovaries. After leaving the egg sac, the young ride on the mother for a short time.

Pseudoscorpions secrete silk from their fangs (chelicerae) and use it to make disk-shaped shelters that serve various functions (depending on the species), including providing a place for the female to brood her eggs and a safe haven in which the juvenile stages can molt.

Adults of many species of pseudoscorpion hitch a ride on other animals (mostly insects) in order to colonize new habitats; common carriers include bess beetles, click beetles, longhorn beetles, timber flies, stilt-legged flies, metalmark butterflies, and even mice. A well

Centruroides edwardsii (Buthidae).

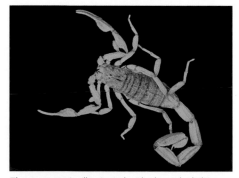

The same species illuminated with ultraviolet light.

Pseudoscorpion carrying eggs (right side of abdomen).

HARVESTMEN

Also called daddy long-legs, harvestmen (order Opiliones) are often confused with a family of spiders (Pholcidae) that are also called daddy long-legs. But harvestmen have undivided bodies—the bodies of spiders are divided into two parts—and do not produce silk. Although the order Opiliones is commonly characterized by species with extremely long legs, it is actually very diverse, and many species in tropical America are rather stout and short-legged.

Most species have dark coloration, a trait most probably not unrelated to the fact that they generally inhabit shady areas and are active mainly at night. Harvestmen constitute the third largest group of arachnids (after mites and spiders). Worldwide, they consist of four suborders (three in tropical America), nearly 50 families, and more than 6,500 named species.

One persistent myth claims that harvestmen are highly venomous even though they lack venom glands. Not having this means of defense, they rely on chemical secretions (mainly phenols and quinones) that give them a pungent smell. Most groups of arachnids are

known example of such transport is that of certain species in the genus *Cordylochernes*; these utilize the harlequin beetle (a species of longhorn beetle, p. 130) as their carrier. Both the beetle larvae and the pseudoscorpions inhabit decaying trunks of fig trees. The pseudoscorpions feed and reproduce in these tree trunks, but when adult harlequin beetles begin emerging from the pupal stage, the pseudoscorpions crawl underneath the wing covers of a beetle. The male pseudoscorpions use their pincers to fight off other males, thereby gaining a monopoly on all the females boarding a particular beetle. Unfortunately for the victorious male, his reproductive monopoly is often incomplete since females may already be carrying sperm from other males by the time they climb onto a beetle.

Relatively few arachnids exhibit social behavior but among the handful of known examples are a few spiders (p. 317) and a species of pseudoscorpion (*Paratemnoides nidificator*) that inhabits the Brazilian tropical savanna. This species lives in colonies on the bark of trees; groups of these pseudoscorpions sometimes cooperatively capture prey that are four times larger than a solitary individual is capable of subduing.

Prionostemma (suborder Eupnoi: Sclerosomatidae).

Eucynorta (suborder Laniatores: Cosmetidae).

Male *Poassa limbata* (Manaosbiidae) defending his nest.

predators that liquefy their prey with enzymes and then ingest the resulting fluid; in contrast, harvestmen ingest solid particles of food; they feed on a variety of small insects, dead organisms, excrement, fallen fruits, and fungi. Some species aggregate in crevices of trees during the day, bobbing up and down if disturbed (in long-legged species), and descend to the ground at night in order to forage. At least some of these species mark their aggregation sites with chemicals, and the same site is used by succeeding generations.

Some harvestmen (most species in the suborders Cyphophthalmi and Eupnoi) have a long ovipositor that they use to hide their eggs in crevices in soil, in tree trunks, or under rocks. Other species (suborder Laniatores) have a short ovipositor and lay their eggs on exposed substrates such as leaves or wood; in these species, the female generally covers each egg with debris. Most harvestmen lay their eggs singly, but females of some species (many families of Laniatores) lay their eggs in a single large clutch and then remain with the eggs (and newly hatched nymphs) in order to

protect them. Such egg guarding by the mother is common in arachnids, but some harvestmen (in at least 5 families of Laniatores) are unique among arachnids in that the father guards the eggs. In some species the male takes over a nest prepared by the female, while in other cases he builds the nest himself. In the latter case, the male builds an open mud nest, then waits for a female to drop by so that he can copulate with her—and then receive her eggs in his nest.

WHIP SPIDERS

Despite their name, whip spiders (order Amblypygi) should not be confused with the true spiders (order Araneae). Also known as tailless whipscorpions, these flat-bodied arachnids have very long front legs that span up to 20 cm (nearly 8 in) long and are splayed out to the side. Although they look somewhat threatening, whip spiders lack a venom gland and are harmless.

Whip spiders generally hide during the day, coming out at night to hunt and mate. Favorite hiding spots are underground burrows (including nests of leafcutter ants or bullet ants), spaces below logs, caves, and crevices within tree

Phrynus pseudoparvulus (Phrynidae), found at base of large tree at night.

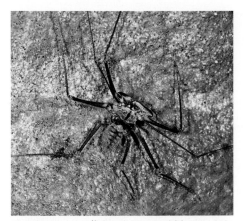

Male *Paraphrynus laevifrons* (Phrynidae), found in a cave.

Female *Phrynus whitei* with her young.

trunks. They feed on crickets, moths, small millipedes, and other arthropods, using their very long front legs as feelers and grabbing the prey with their spiny pedipalps. A species in Tobago preys on freshwater shrimp.

After a prolonged dance near the female, the male deposits a spermatophore on the substrate, then faces the female again and beckons her forward or pulls her over the spermatophore so that she can retrieve it and use the sperm to fertilize her eggs. The female carries her eggs underneath her abdomen until they hatch (after a couple of months), whereupon the young crawl onto their mother's back and reside there for about a week, at the end of which time they molt and disperse.

Worldwide this group consists of 5 families and about 170 species; in tropical America there are 3 families and about 80 species.

SPIDERS

Spiders (in Spanish, *arañas*) belong to the order Araneae, the second largest group of arachnids (after mites) and one of the most diverse groups of animals, with more than 100 families and 45,000 named species worldwide.

The body of a spider is divided into two parts (unlike insects which have three parts): the cephalothorax and the abdomen. The cephalothorax bears eyes (usually eight, and they are not compound eyes like those of insects), a pair of chelicerae, a pair of pedipalps (short, leg-like appendages), and eight legs.

Spiders are predators. They use the fangs at the tips of their chelicerae to inject venom into their prey. Once the prey is immobilized, the spider regurgitates enzymes from its gut onto or into the prey in order to pre-digest it, and then consumes the resulting fluids.

While all spiders produce silk, not all species use the silk to make webs to capture prey; silk is also employed to make sacs that protect the eggs, to line a burrow or other retreat, to wrap up prey, and to produce a dragline that sustains the spider as it moves about—à la Spider-Man. Many other arthropods produce silk (p. 200), but no other group uses it in so many ways. Because of these varying uses, spiders have many kinds of silk glands for producing different types of silk; and the number of silk glands that a spider has varies from species to species. Tarantulas, for example, do not make webs for capturing prey and have just one or two types of silk glands, whereas most orb-weaver spiders have seven. The silk glands produce a mix of proteins that are transported to moveable appendages (spinnerets) on the underside of the abdomen; each spinneret bears numerous tiny spigots from which the silk emerges, and as it does so, the proteins change from liquid to solid, becoming silk fibers.

Among those spiders that build webs for capturing prey, adult females and the juvenile stages of both sexes build webs; but adult males generally do not—after their final molt they leave their web and search for a female in her web. Webs for trapping prey show incredible variation from species to species, both in architecture and silk type (p. 310).

Argiope argentata, an orb-weaver spider (Araneidae). The dorsal surface reflects ultraviolet light, which might serve to attract prey (bees).

The kind of insect that a web captures is a function of web architecture and location. Web-building spiders detect a captured insect by means of vibrations produced in the web (most web-building spiders have poor vision). They subdue prey either with a quick venomous bite or by throwing sticky silk around the prey before biting it. Although some spider species use one or the other strategy, many use wrapping as an optional strategy when confronted with dangerous prey. Occasionally, freshly captured prey are wrapped in silk and left in the web to be consumed later (usually within 24 hours).

Upon reaching adulthood, male spiders build a small platform of silk onto which they ejaculate; they suck up the sperm with syringe-like structures on the tips of their enlarged pedipalps (located between the chelicerae and front legs, the pedipalps look like boxing gloves). In web-building species, the male often courts the female by vibrating her web; males of other spiders often employ visual and acoustic displays. During mating the male inserts the tip of his pedipalp into the genital opening on the underside of the female's abdomen. Because spiders are predatory—and have cannibalistic tendencies—the male must approach the female with caution, especially in web-building species (in which the males are often much smaller than the females). In most species, nonetheless, the female does not eat the male, although the female of a few species of cobweb spider does display this behavior.

Female spiders weave silken sacs in which they lay their eggs. Depending on the species, the egg sacs vary in shape, size, and color. They protect the eggs from adverse environmental conditions, predators, and parasitoids. Egg sacs may be built under a rock, under bark, hidden among foliage, or in the web itself. The female either stands guard over the egg sac until the spiderlings emerge or she carries the sac with her, using either her mouthparts (in the case of daddy long-legs spiders and nursery web spiders, for example) or on her spinnerets (in the case of wolf spiders).

Upon emerging from the egg sac, the spiderlings cluster together for a few days then disperse, either by walking or by ballooning, a process in which the spiderling climbs to a high point on vegetation, lets out a fine silk line that catches the breeze, and then wafts up and away. Sometimes, in the early morning, thousands of ballooning spiderlings can be seen as a delicate carpet of gossamer. Many ballooning spiderlings do not travel far, but some travel very long distances—and spiders are often the first animals to colonize distant oceanic islands.

Although spiders have played the role of villain in more than one movie, fewer than 1% of species pose a danger to humans. There are three main groups of potentially dangerous spiders in tropical America: recluse spiders, also called violin spiders (Sicariidae: *Loxosceles*); black widow spiders (Theridiidae: *Latrodectus*); and Brazilian wandering spiders (Ctenidae: *Phoneutria*). The first two groups occur in widespread locations, while Brazilian wandering spiders have a relatively restricted distribution, occurring from Costa Rica to northern Argentina. Recluse spiders have a necrotic venom that often results in lesions that take a long time to heal. Black widows and Brazilian wandering spiders have neurotoxic venoms that can cause, respectively, painful muscle cramps (especially in the abdomen) and the loss of muscle control (sometimes resulting in breathing problems). Although these three groups of spiders occasionally cause harm to humans, it is also true that medical conditions are frequently misdiagnosed as being caused by spider bites. All in all, arachnophobia is unwarranted.

The spiders of tropical America are placed into two infraorders, Mygalomorphae and Araneomorphae. The latter, which comprises more than 90% of the species that occur in the region, consists of three principal groups of spiders (see table, p. 311): the Haplogyne group, the RTA group, and the Orbiculariae group. In the following section, web-building spiders receive more attention than species that do not build webs, as one tends to encounter them more frequently in the field.

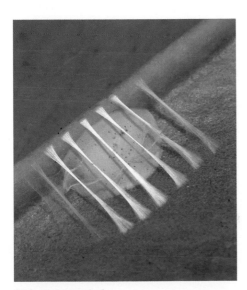

Spider egg sac on leaf.

Mature male *Eriophora* (Araneidae) in cryptic daytime resting position. At night he will search for a female.

SPIDER WEBS

In a cloud forest, an orb web with water droplets.

Many spiders construct silken webs to trap insects and other prey. The webs of some species—including funnel-web tarantulas and many species of daddy long-legs spiders—lack specialized adhesive silk; their webs capture prey merely through means of physical entanglement. Most species that builds webs, however, produce adhesive silk, of two kinds. One consists of extremely thin dry fibers, the other of fibers coated with wet sticky droplets.

Dry adhesive silk is extruded from the cribellum, a perforated plate on the underside of the abdomen. This structure, which is derived from two of the spinnerets, evolved in the ancestors of the infraorder Araneomorphae and is present in a few families of each of the major groups. Most species, however, lack a cribellum, either because through evolution they switched to using silk that is coated with sticky droplets (extruded by normal spinnerets) or because they have become free-ranging hunters and no longer make webs.

The structure of spider webs varies considerably between species. The simplest type of web is built by spiders that hide in burrows in the ground and spin a simple mesh of silk around the opening of their burrow. Other spiders build more extensive sheets of silk over the surface of the ground. Webs that are built above the ground—among the branches of plants, for example—come in two basic forms. Orb webs are the circular webs that most people associate with the term *spider web*. Other spiders, however, build more irregular, messy webs (often called cob webs), which are three-dimensional (orb webs are two-dimensional).

Orb webs, which are only built by certain members of the Orbiculariae group, are among the most elegant (and best studied)

Sheet webs constructed by Linyphiidae on the ground.

of all spider webs. Ranging in diameter from less than 2 cm to almost 2 m, orb webs generally consist of scaffolding—frame lines and radii (like spokes on a wheel)—that is overlaid by a sticky spiral that traps prey. After building the scaffolding, most species first make a temporary non-sticky spiral, beginning at the hub, before laying down the sticky spiral; the non-sticky silk functions as a bridge that protects the spider from getting caught in its own web.

Several species of orb-weavers add web decorations (stabilimenta) to their web, which, depending on the species, can take the form of an X, a simple line, or a circle. There are at least three contending hypotheses about the possible function(s) of this extra component—that it provides the spider with camouflage from predators, makes the web more visible in order to warn off large animals that might damage the web, and that it serves to attract prey to the web.

Table 8-1. Major groups of spiders

Infraorders	Groups	Examples of families
Mygalomorphae		Tarantulas, trapdoor spiders, funnel-web tarantulas*
Araneomorphae	Haplogynae	Recluse spiders*, daddy long-legs spiders*
	RTA	Wandering spiders, wolf spiders, nursery web spiders, long-legged water spiders, jumping spiders, sac spiders, crab spiders, giant crab spiders, tengellid spiders*
	Orbiculariae*	Cribellate orb-weavers, orb-weaver spiders, golden silk orb-weavers, long-jawed orb-weavers, cobweb spiders, sheetweb spiders

* = Groups in which most species build webs for capturing prey.

INFRAORDER MYGALOMORPHAE

Most spiders have fangs that move sideways, but mygalomorph spiders have fangs that move upward and downward. In tropical America, this infraorder comprises ten families, the most species rich being the tarantula family. It also includes three families of trapdoor spider whose members live underground in silk-lined tunnels with an opening covered by a hinged lid; from this retreat the spider darts outs to capture its prey. The funnel-web tarantulas (family Dipluridae) are the only mygalomorphs to build extensive webs for capturing prey.

TARANTULAS

Tarantulas (family Theraphosidae) are hairy, fearsome-looking spiders that are often very large. Indeed, the world's largest spider is a tarantula of the Amazon, the Goliath bird-eating spider (*Theraphosa blondi*), with a leg span of up to 30 centimeters (12 inches); although able to eat small birds and mice, it generally feeds on insects. If threatened, tarantulas are capable of inflicting a painful bite, but their venom is generally not dangerous to humans. They can also defend themselves by using their hind legs to flick urticating (irritating) hairs from the top of the abdomen.

Tarantulas have eight small eyes grouped together at the front of the head, and they hunt not so much by sight as by using sensitive hairs on their body to detect the vibrations and odor of their prey. Most species live on the ground—and many of these hide in silk-lined burrows—but a few inhabit trees.

Most tarantulas take at least 2 to 5 years to reach adulthood. Once they mature, males live for a year or more and rarely molt again; whereas females live much longer (up to 30 years) and usually molt about once a year.

Of all the spiders, female tarantulas are among the very few that molt during the adult stage; during the molt, the female also sheds the lining of the sperm storage organ (spermatheca), which means she has to mate again after each molt. Females lay 50 to 1,000 eggs (depending on the species) in a silken egg sac and guard it for a month or two, during which time they tend to be more aggressive. After hatching, the young spiderlings remain in the nest for some time, living off the remains of their yolk sac before dispersing.

Worldwide, there are approximately 940 named species of tarantula, about half of which occur in tropical America.

Female *Megaphobema mesomelas*.

When inside its burrow *Aphonopelma seemanni* spins silk over the entrance.

Female *Aphonopelma seemanni*.

INFRAORDER ARANEOMORPHAE: HAPLOGYNE GROUP

Members of the Haplogyne group are sometimes called the six-eyed spiders because most species have six eyes instead of the eight that characterize the majority of spider species. In tropical America this group consists of about a dozen families, the most common and species-rich being the daddy long-legs spiders. The Haplogyne group also includes the venomous recluse spiders (family Sicariidae). Most daddy long-legs spiders and recluse spiders build webs to capture prey, but the majority of the species in the other families capture prey by free range hunting or by darting out from silken tunnels.

DADDY LONG-LEGS SPIDERS

Rather confusingly, the name *daddy long-legs* is also used to refer to long-legged species in two groups of non-spider arthropods, the harvestmen (p. 305) and the crane flies (p. 270). Daddy long-legs spiders (family Pholcidae),

however, have a body composed of two parts and are thus readily distinguished from harvestmen, whose bodies consist of a single part; also, daddy long-leg spiders live in webs, a trait shared by neither harvestmen nor crane flies. Note that not all pholcids have long legs (also true of harvestmen).

The most noticeable daddy long-legs spiders, in tropical America and elsewhere, are those that make messy irregular webs under fallen logs, between buttresses of trees, and in other sheltered locations, including buildings and cellars. They generally hang upside down in their web; when disturbed, many species oscillate their bodies. The web itself is not sticky, so when a potential prey makes contact the spider has to quickly cover it in silk in order to immobilize it. A few species invade the webs of other spiders (including those of other daddy long-legs), eating the host, its eggs, or its prey. Several species of daddy long-legs spider live on the ground or on the undersides of leaves, and some of these do not construct a web.

Worldwide, there are more than 1,100 named species of daddy long-legs spider, at least 450 of which occur in tropical America.

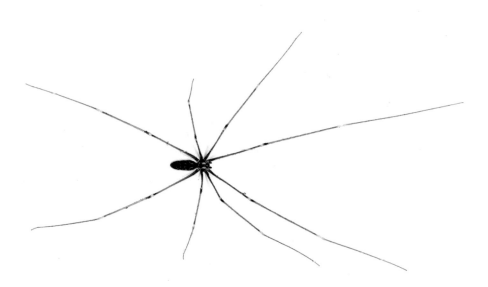

Smeringopus pallidus (Pholcidae), a daddy long-legs spider originally from Africa, but now common in houses throughout the tropics.

INFRAORDER ARANEOMORPHAE: RTA GROUP

Worldwide, the RTA group contains about half of all spider species; it is represented by nearly thirty families in tropical America. Within this group, the vast majority of species have abandoned (in an evolutionary sense) the construction of webs and instead capture prey by free range hunting.

WANDERING SPIDERS, WOLF SPIDERS, AND RELATED FAMILIES

In tropical America, the superfamily Lycosoidea comprises six families, four of which are described in this section: wandering spiders, wolf spiders, nursery web spiders, and long-legged water spiders.

Worldwide, there are nearly 500 named species of wandering spider (family Ctenidae), more than 200 of which occur in tropical America. Many species, including those in the genera *Ctenus* and *Phoneutria*, live primarily on the ground; species in the genus *Phoneutria* are large (leg span up to 13 cm / 5 in) venomous spiders; when threatened they lift the first two pairs of legs high in the air and sway from side to side. Species in the genus *Cupiennius* are associated with certain plants such as aroids, bromeliads, heliconias, and gingers, on which they construct silken retreats, sallying forth after dark to hunt, using substrate vibrations to detect potential prey.

Wolf spiders (family Lycosidae) include about 2,400 named species worldwide, with nearly 300 in tropical America. They have excellent eyesight, generally hunt on the ground, and capture their prey by ambush or pursuit.

Females are easily recognized by their habit of carrying the egg sac attached to the spinnerets at the tip of their abdomen, and after hatching the spiderlings crawl up onto their mother's abdomen and cling to specialized knobbed setae (they remain there until they are old enough to hunt for themselves).

In tropical America, wolf spiders are more frequently seen in open grassy areas, where they can be extremely abundant.

Nursery web spiders (family Pisauridae) include more than 330 named species worldwide, about 60 of which occur in tropical America. They carry the egg sac beneath their head (though it is also attached to the spinnerets by silk threads); just before the spiderlings hatch, the female ties leaves together with silk and suspends the egg sac inside this nursery, where the young spiderlings remain for a short time.

Long-legged water spiders (family Trechaleidae) occur primarily in tropical America, where there are about 100 named species. Most species are apparently nocturnal.

As their name suggests, these spiders have long legs (held out to the sides of the body) and often live along the margins of streams (some species are found away from water, on vegetation).

Aquatic species are very adept at walking on the surface of water and at crawling underwater, where they capture aquatic prey such as small shrimp. Like wolf spiders, the females in this family carry the egg sac on their spinnerets, but the egg sac is hemispherical (rather than spherical) and the young spiderlings ride on the empty sac (rather than the female's abdomen).

Cupiennius getazi (Ctenidae) eating a frog. Members of this family are fast-moving, nocturnal predators.

Cupiennius getazi. Recent research suggests that this genus is unrelated to the other wandering spiders (Ctenidae).

Some species of *Thaumasia* (Pisauridae) hunt for prey on the surface of water.

A few *Thaumasia* species build a web on the upper surface of leaves.

Lynx spider (Oxyopidae: *Peucetia*), another family of Lycosoidea.

Wolf spider (Lycosidae).

JUMPING SPIDERS

Salticidae is the largest family of spiders, both worldwide, with more than 5,400 named species, and in tropical America, where there are nearly 1,500 species. As the name suggests, these spiders are proficient jumpers; to leap, they rely not on large leg muscles but on a sudden increase of blood pressure in the legs.

The most salient characteristic of salticids, nonetheless, is the large pair of eyes in front of their head (they also have 3 smaller eyes on each side of the head) that gives them visual acuity with no known parallel in animals of comparable size. These unique eyes can swivel and possess a telephoto lens system with a four-layered retina, allowing the spider to locate, stalk, chase down, and precisely leap onto active prey. Their predatory behavior is so cat-like that a better name for jumping spiders would be "eight-legged cats" (though they are diurnal hunters, unlike most cats).

Just before jumping, the spider tethers a filament of silk (a dragline) to the surface on which it is poised, thus allowing itself to easily return to its starting point. Some species can subdue prey that much larger than themselves by grabbing the prey with their chelicerae and then dropping off the substrate while remaining attached by the dragline, thereby preventing the prey from making contact with the substrate and resisting. Many jumping spiders occasionally supplement their protein diet with nectar; indeed, one Central American species (*Bagheera kiplingi*) is largely vegetarian (at least in some regions), deftly consuming food bodies produced by swollen-thorn acacias (*Vachellia*) for the benefit of protective ants (p. 193).

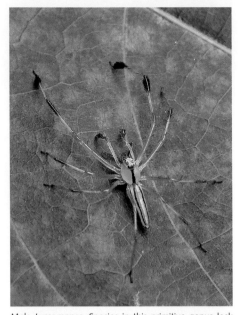

Male *Lyssomanes*. Species in this primitive genus lack the visual acuity of other jumping spiders and they ambush rather than stalk their prey.

Paraphidippus. Females of this large subfamily (Dendryphantinae) generally have paired spots on the abdomen.

Phiale formosa with captured wasp (Vespidae).

INFRAORDER ARANEOMORPHAE: ORBICULARIAE GROUP

It is possible that the ability to make orb webs evolved just once, in the ancestors of Orbiculariae, although there is still some question as to whether this is in fact a single, unified group. Cribellate orb-weavers use dry adhesive silk in their sticky spiral, but the vast majority employ wet sticky droplets. In orb webs containing sticky droplets, the spiral quickly loses its adhesiveness, and therefore many orb-weavers dismantle their webs daily, generally by eating the web (or at least sucking the fluids from it), and then, after resting for a while, building a new one.

Despite the orb web's effectiveness in capturing flying insects, a majority of species in the Orbiculariae group no longer make orb webs. Instead, they make three-dimensional webs; among these species are the cobweb spiders (Theridiidae) and sheetweb spiders (Linyphiidae). (It should be noted that the sheetweb spiders represent the second largest family of spiders in the world, but in tropical America this family is far less diverse.)

Most spiders are solitary and associate with members of their own species only to reproduce. However, a few species of orb-weavers are colonial; in colonial species, individuals link their webs to a larger network of webs, all the while building and defending their own webs within the colony. Some cobweb spiders go a step further by sharing a communal web—cooperating in web construction, prey capture, and sometimes even brood care. Both colonial and communal webs can result in large (several cubic meters in volume), long-lasting, silken tents that are very conspicuous, not only on plants but also on fences, bridges, and power lines.

Worldwide the Orbiculariae group contains about a quarter of all spider species; in tropical America, this group comprises nearly twenty families.

CRIBELLATE ORB-WEAVERS

The family Uloboridae comprises approximately 270 named species worldwide, with about 80 in tropical America. Instead of the sticky droplets used by other orb-weavers, the adhesive spiral of uloborids contains extremely thin, dry fibers that are extruded from a perforated plate (the cribellum) located on the underside of the abdomen. This plate has thousands of tiny spigots. As the thin silken fibers emerge from the spigots, the spider combs them with its hind legs, fluffing them out and thereby giving them a wooly texture. These very fine fibers entangle insect prey without any need for glue. Because of the rather elaborate process involved, cribellate orb-weavers require much more time to construct their web than do other orb-weavers (about 3 hours versus half an hour), which means that it is more cumbersome for these spiders to renew or relocate their webs.

Most species of cribellate orb-weaver build horizontal webs, while the majority of other kinds of orb-weaver build vertical webs. Several species make modifications to the general horizontal structure; some *Uloborus* species, for example, pull the hub of the horizontal web downward to form a cone and then spin a second orb web over the cone's aperture. The

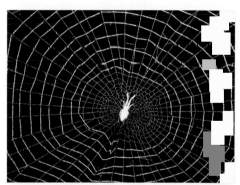

Uloborus at the hub of its web.

Uloborus resting aligned with linear web decoration, apparently providing camouflage.

most extreme variation is shown by species of *Miagrammopes* that make simple webs formed of only a few capture lines.

Cribellate orb-weavers are the only spiders that lack venom glands. Once a prey is tangled in its web, the spider wraps it with an incredible amount of silk (about 100 meters/33 feet in length). After wrapping the prey, which can take up to an hour or more, the spider covers the entire surface of the wrapped prey with digestive fluid (in other spiders digestive fluid is only applied to the feeding spot) and then feeds by sucking up liquefied host remains and a portion of the silk wrapping.

ORB-WEAVER SPIDERS

Many members of this family (Araneidae) have large abdomens that are often brightly colored. Among the most conspicuous orb-weavers in tropical America are species of *Gasteracantha* and *Micrathena*, hard-bodied, spiny spiders that make strong orb webs capable of withstanding high-impact blows. *Cyrtophora citricola*, known as the tropical tentweb orb-weaver, is an Old World species that has been introduced into tropical America, where it is increasingly common.

Some species of *Metepeira*, at least one species of *Cyclosa*, and the orb-weaver *Cyrtophora citricola* are colonial spiders; numerous individuals (sometimes hundreds or even thousands) join their webs together within a common framework, resulting in a large three dimensional structure. However, each spider generally builds and defends its own web within the colonial web.

Parawixia bistriata, which occurs in Brazil and Argentina, displays colonial behavior of a different kind. During the day, the spiders construct a communal web in which to hide; at night, each individual constructs its own orb web, though these are interconnected and the spiders sometimes cooperate in capturing large prey.

Orb webs do not function very well in capturing moths because detachable scales of the moth stick to the web, usually allowing it to escape. Some orb-weaver spiders have overcome this problem by building modified webs, or by attracting and then snaring the moths. The first strategy is exemplified by species in the genus *Scoloderus*, which build ladder-like webs; a moth that has been intercepted slides down the web, leaving a trail of scales until its body becomes denuded of scales and sticks to the web. The second strategy is exemplified by species in the genera *Kaira* and *Mastophora*, which attract male moths by emitting odors that mimic female sex pheromones. Species of *Kaira* hang from a small web and catch moths in a basket formed by its legs, whereas species of *Mastophora* (known as bolas spiders) capture moths by lassoing them with a silken thread that has a sticky ball at its tip. (Spiderlings and adult males of bolas spiders go after smaller prey such as moth flies and simply grab them with their front legs.)

Worldwide there are more than 3,000 named species of orb-weaver spider, about 1,200 of which occur in tropical America (where they form the second largest family, after the jumping spiders).

Argiope argentata occurs throughout tropical America, usually in slightly drier areas.

Cyclosa camouflaged in center of web, with a thick line of prey remains and egg sacs on either side.

Gasteracantha cancriformis varies in color from light yellow to red.

Micrathena sexspinosa female, ventral view. Like other members of the subfamily Micratheninae, it has a hardened ring around the spinnerets.

GOLDEN SILK ORB-WEAVERS (ALSO CALLED GOLDEN ORB-WEB SPIDERS)

The family Nephilidae comprises approximately 60 named species, almost all of which inhabit the Old World tropics.

Facing down, golden silk orb-weavers sit at the hub of the web, monitoring it for vibrations caused by a struggling insect. Instead of first wrapping prey in silk, their initial means of attack is through biting, which means that they are unable to deal effectively with large or aggressive prey. After the bite, they pull the prey out of the web and carry it back to the hub to feed; relatively large or bulky prey items are wrapped in silk before being transported to the hub.

In tropical America there are only 4 species of golden silk orb-weaver, the best known being *Nephila clavipes*, which occurs from the Gulf Coast of the U.S. to Argentina. This species is a large, brightly colored spider with tufts of black hairs below the joints of the legs; males are about one-tenth the weight of adult females.

Species in the genus *Nephila* build only the bottom portion of the web, with the hub therefore being situated near the top. The web of an adult female, which can reach a meter or more in diameter, is constructed of silk that has a golden color (hence the name of these spiders); the spider is able to adjust pigment

Nephila clavipes, ventral view. The species name means "club foot," referring to the clumps of hair on the legs.

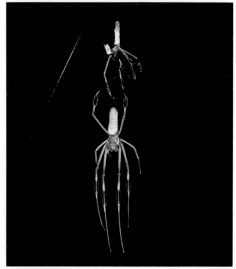

N. clavipes molting.

intensity in the silk relative to the background, suggesting that the color serves some function (e.g., allowing the silk to blend in with the background). The orb web is renewed regularly (not necessarily daily), but if rain has not damaged the web, the spider often rebuilds only a portion of it.

Golden silk orb-weavers (*Nephila*) are frequently victimized by tiny kleptoparasitic spiders (cobweb spiders in the genus *Argyrodes*) that feed on the captured prey.

It has been suggested that *Nephila* attempts to minimize losses due to kleptoparasites both by always bringing prey back to the hub rather than leaving prey wrapped *in situ* (as do some other orb-weavers) and by periodically moving their web.

LONG-JAWED ORB-WEAVERS

The family Tetragnathidae comprises nearly 1,000 named species worldwide, with at least 300 in tropical America. Males of many species have conspicuously enlarged chelicerae (hence the name *long-jawed*), which are used during courtship and copulation. Males in the subfamily Leucauginae, however, have chelicerae that are similar in size to those of females; members of this subfamily generally

Leucauge mariana. Like other long-jawed orb weavers, the female and male usually clasp one another with their jaws (chelicerae) while mating.

have vivid colors and reflective silvery abdominal patterns (these silvery patches, which are also present in various other orb-weavers, are due to an accumulation of guanine, the main nitrogenous excretory product in spiders).

Most long-jawed orb-weavers build orb webs with relatively few radii and spirals; these webs can be horizontal, inclined, or vertical, depending on the species. Several species build webs on vegetation near water.

COBWEB SPIDERS

In contrast to the beautiful symmetry of orb webs, most spiders in the family Theridiidae build nonplanar tangles of silk called, simply, cobwebs. Aesthetics aside, there must be advantages to building cobwebs, since these spiders have evolved from ancestors that made orb webs. One possible advantage is that the three dimensional webs offer greater protection from predatory and parasitic wasps (a spider on a planar orb web is more readily plucked off than one located inside a tangle of silk). A second potential advantage is that the irregular meshwork of silk allows the spiders to utilize fewer sticky droplets; because they are less reliant on these ephemeral droplets cobwebs last much longer than orb webs, and the spiders simply expand and repair them rather than replace them on a regular basis (as occurs with orb webs).

Cobweb spiders are very diverse biologically. Although most cobwebs look pretty much the same to the casual observer, careful observation reveals considerable diversity in architecture between species. Indeed, some species in the subfamily Hadrotarsinae do not build a web at all and specialize in capturing ants.

There are more kleptoparasitic species— and more communal (social) species—in this family of spiders than in any other. Kleptoparasites, exemplified by species in the genus *Argyrodes*, do not build a web of their own, but instead live in the webs of larger spiders; they pilfer small prey caught by their host's web, eat prey killed by the host spider, sometimes consume silk from the host web, and attack and eat the host itself. Communal species, exemplified by eight species in the

Anelosimus female with her egg sac. In most species the spiderlings remain with their mother until adulthood.

Thwaitesia species construct just a few non-sticky lines that serve to attract rather than snare prey.

The brown widow (*Latrodectus geometricus*) has become very widespread, having been accidentally transported by human activity. Its origin is uncertain.

genus *Anelosimus* and by *Theridion nigroannulatum*, are very unusual among spiders in that they cooperate in web construction and prey capture, and their females help each other in brood care; nestmates remain together throughout their lives and mate and reproduce within the natal nest from generation to generation (which results in inbreeding).

Cobweb spiders also show diversity in their sexual habits. The females of some widow spiders (genus *Latrodectus*) eat the males during or shortly after mating. However, sexual cannibalism is actually more common in *Tidarren* and *Echinotheridion*. In all three of these genera the males are much smaller than the females, but in *T. sisyphoides* the male is only 1% the size of the female, and he dies during copulation, uneaten by his mate.

Worldwide there are more than 2,300 named species of cobweb spider. With about 750 species in tropical America alone, this is the third largest family in the region.

MITES

There are probably more species of mite (in Spanish, *ácaros*) than all other arachnids combined; based on current knowledge, mites comprise 540 families and about 60,000 named species, though the true number of species is probably more than a million. Mites are among the most ancient of all terrestrial animals, with fossils dating back to the early Devonian, nearly 400 million years ago. Like other arachnids, mites have chelicerae instead of antennae, and they possess four pairs of legs (except for during the very youngest stages and in gall mites).

Mites are so small that, with the exception of ticks (the largest mites), they are difficult to see with the naked eye. But they are everywhere, from subterranean waters to the human body. Their feeding habits are very diverse; as a group, mites include predators, animal parasites, fungivores, and herbivores. Most feed by piercing and sucking up fluids, but some (order Sarcoptiformes) bite off bits of food particles.

The life cycle of mites generally consists of an egg, prelarva and larva (both stages with 3 pairs of legs), 1 to 3 nymphal stages (with 4

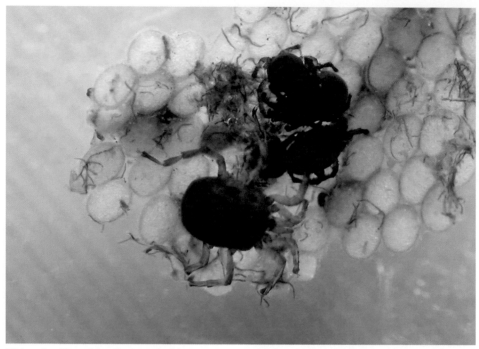

Water mites (Limnesiidae: *Neotorrenticola*).

pairs of legs), and adult. The prelarva (which is also present in other arachnids) does not feed; in some species there are one or two other non-feeding stages. The methods by which males transfer sperm to females varies between groups; in some species (superorder Parasitiformes), the males use their chelicerae to transfer sperm; in other species (suborder Oribatida), the males deposit stalked spermatophores that are later picked up by females; and yet other species (Astigmata) utilize standard copulation.

The mites comprise two superorders, the Parasitiformes and the Acariformes. Because there are simply too many groups of mites to discuss individually, the following discussion is organized by ecological categories rather than taxonomic groups. (It goes without saying that not all ecological categories are included here.)

Table 8-2. Major groups of mites

Superorders	Orders	Examples
Parasitiformes	Ixodida	Ticks
	Mesostigmata	Flower mites, Phytoseiidae, Dermanyssidae, Macronyssidae
Acariformes	Trombidiformes	Most plant-feeding mites, chiggers, water mites, eyelash mites
	Sarcoptiformes: suborder Oribatida	Beetle mites
	Sarcoptiformes: suborder Oribatida: hyporder Astigmata	Scabies, house dust mites, feather mites

MITES IN SOIL

Ancestral mites most likely lived in soil with decaying organic matter, their preferred habitat today. One square meter of forest leaf litter can have tens of thousands of mites. Although the greatest diversity occurs in the upper organic layer, mites penetrate the soil at least as deeply as do plant roots. They also inhabit accumulations of organic material (in epiphytes and tree holes, for example) high up in the canopy. Common inhabitants of leaf litter include predatory mites (Mesostigmata and Trombidiformes) and beetle mites (suborder Oribatida, excluding Astigmata), so named for their beetle-like appearance. The latter

Velvet mite (Trombidiidae). Nymphs and adults are predators of small invertebrates living in soil; the larvae are parasites of other arthropods.

Beetle mite (Oribatida). This is the most diverse group of arthropods living in soil.

comprise 172 families and more than 9,000 named species. Beetle mites feed on detritus, fungi, algae, or small invertebrates (e.g., springtails and nematodes) and play an important ecological role by burrowing inside decaying vegetation, thereby increasing the surface area available for microbial growth. Beetle mites have a pair of glands that secrete a wide range of compounds that function as chemical defenses and alarm signals. Poison dart frogs (Dendrobatidae) probably obtain many of their toxic skin alkaloids by ingesting these mites (as well as certain ants and millipedes).

MITES ON PLANTS

There are at least ten groups in the order Trombidiformes that feed on plants. Among these are the spider mites and gall mites, both of which generally pierce individual plant cells and suck out the contents. Spider mites (superfamily Tetranychoidea) get their name from the silken webbing produced by some species; a mother and her offspring are able to feed in relative security beneath the protection of this webbing. A few spider mites are significant crop pests, but it is worth noting that they were rarely pestiferous before the widespread use of synthetic pesticides. Invisible to the naked eye, gall mites (superfamily Eriophyoidea) are among the smallest of all mites; they have an elongate body with just four legs. Many form galls (p. 280) that generally manifest themselves as either tiny finger-like projections emanating from the leaf surface or as a small hollow area filled with a dense growth of trichomes. Others members of the gall mite group do not cause galls but rather a discoloration of the plant surface; for example, the citrus rust mite (*Phyllocoptruta oleivora*) leaves a rust-colored residue on oranges.

Among members of the order Mesostigmata that occur in tropical America, species in at least a half dozen genera (especially Melicharidae) feed on nectar and pollen, traveling between flowers by hitching a ride on hummingbirds (they run onto their beaks and ride in the nasal cavities), bats, beetles, stingless bees, or butterflies.

In addition to harboring plant-feeding mites, plant foliage also harbors a diversity of fungivorous and predatory mites. Some of the latter (Mesostigmata: Phytoseiidae) have been used in the biological control of spider mites. About half of all trees and shrubs provide sanctuaries (domatia) for predatory and fungivorous mites on the undersides of their leaves, usually in tiny cavities or tufts of trichomes located in the crotches of leaf veins. This is often a mutualistic association whereby the plant potentially gains some protection from pests (such as plant-feeding mites) while the predatory mites gain a safe place in which to lay eggs and molt.

Many Erythraeidae live on plants and are predators (their larvae are generally parasites of other arthropods, p.325).

Gall mites (small white things) in a gall on *Saurauia* (Actinidiaceae).

Galls on *Acalypha* (Euphorbiaceae) leaf produced by gall mites.

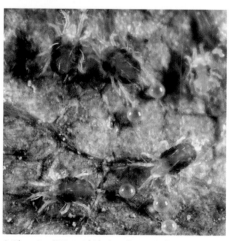

Spider mites (Tetranychidae) on *Bocconia* (Papaveraceae).

Mites riding on the beak of a Green-crowned Brilliant (*Heliodoxa jacula*).

MITES ON ARTHROPODS

Many mites live on insects and other arthropods. Depending on the species of mite, its relationship with its host can be parasitic, beneficial, or neutral (some mites, for example, simply hitch a ride on insects).

Many species in the order Mesostigmata are associated with millipedes, beetles, bees, ants, moths, and other arthropods. Army ants are hosts to an incredible diversity of mite species. Depending on the species of mite, it attaches itself to a specific site on the army ant—eyes, inner curve of jaws, claws of hind legs, upper surface of hind tibia, and other places; a few such mites are parasitic, but the biology of many species is unknown.

Species in the genus *Dichrocheles* (family Laelapidae) feed and breed inside the ears of owlet moths (p. 217), but do so without deafening their host (which would make the moth vulnerable to bats); when host conditions become too crowded, young mated females disembark onto flowers that the moth visits and wait for a new host.

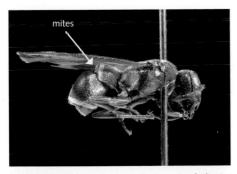

Winterschmidtiidae on *Parancistrocerus declivatus* (Vespidae: Eumeninae).

Larva of Erythraeidae on treehopper (*Entylia*).

Members of the Astigmata generally have a special, non-feeding nymphal stage (hypopus) that is ideally suited for being transported by beetles, wasps, and bees. In turn, some halictid bees (p. 178) and potter wasps (Vespidae: Eumeninae) have a special cavity (acarinarium) on their abdomen for carrying mites (in the family Winterschmidtiidae). As the potter wasp prepares its nest, the mites disembark, suck blood from the wasp's paralyzed prey, reproduce, and feed on the wasp's pupa—all without causing much harm. It is possible that these mites occasionally play a beneficial role by attacking parasitoids or feeding on fungi.

Composed of 50 families, the order Trombidiformes is the most diverse order of parasitic mites (only the larvae are parasites, while the nymphs and adults are free-living predators). A few larvae parasitize vertebrates (e.g., chiggers), but the vast majority parasitize arthropods. Among those that parasitize arthropods are Erythraeidae, which can be seen, for example, on the legs of harvestmen. The water mites (six superfamilies) parasitize the adults of freshwater insects, attaching to them before they emerge from the water; when the adult insect lays eggs in water the fully-fed mite larvae drop off and spend the rest of their lives as predators of insect eggs or larvae, or of microscopic crustaceans.

MITES ON VERTEBRATES

Hardly any terrestrial mammals are free of parasitic mites. This is especially true of bats, which, in all, host 18 families of mite. Dogs and other domestic animals are often infested with scabies mites (family Sarcoptidae), which burrow through the skin and cause mange.

Among the most commonly known mites are the ticks (order Ixodida), which suck blood, primarily from mammals and birds but also from reptiles and toads. This order includes the soft ticks (family Argasidae) and the more commonly encountered hard ticks (family Ixodidae). In the case of most hard ticks, the larva, nymph, and adult feed respectively on three separate hosts. The larva usually feeds on a small mammal; after dropping from its host, that larva molts into a nymph and attaches to a larger mammal; the nymph, in turn, later drops from its host, molts into an adult, and attaches to a yet larger mammal. Adult hard ticks mate on the host; males feed relatively little but females engorge themselves and then drop to the ground to lay thousands of eggs.

The most annoying human parasites are chiggers (family Trombiculidae), which produce intense itching around the ankles, beltline, and wherever else clothing presses tightly against the body. The minute larvae form a feeding tube in the skin, inject digestive enzymes, suck up the dissolved cells and lymph, drop from the body and transform into red mites that feed as predators in the leaf litter; contrary to popular belief, chiggers do not burrow into the skin. Two species in the family Demodicidae are common inhabitants of human hair follicles (especially eyelashes) and sebaceous glands, but they usually cause little harm; indeed many people have them without knowing it.

Birds host a wide diversity of mites, including a few such as *Dermanyssus* (family Dermanyssidae) and *Ornithonyssus* (family Macronyssidae) that suck blood; when large numbers of birds nest on buildings, *Ornithonyssus* mites that they host often leave

them, enter the buildings, and bite the resident humans. Some bird mites burrow into the birds' skin, others inhabit their respiratory tract, and entire food webs of mite fungivores and mite predators occupy the nests. Some of these nest inhabitants have taken up residence in houses. For example, house dust mites (Pyroglyphidae: *Dermatophagoides*) feed on the fungi that grow on flakes of dead human skin and on lint; an allergic reaction to house dust is often the result of inhaling the feces of these mites.

Feather mites are one of the most diverse groups of bird mites, comprising 33 families. Most of them probably cause very little harm to their hosts, although they do seem to cause ill effects in captive birds. Species feed variously on skin flakes, the medulla inside the quill, fluids from the base of the feather, and oil that birds apply to their feathers (and possibly the bacteria and yeasts growing in rancid feather oil). They move off feathers that are about to be molted and go to a more stable location. Many feather mites are quite host-specific, and the number of species inhabiting a particular bird species varies from one or two in small song birds to more than twenty in some parrots.

Tick (Ixodidae: *Amblyomma*) larva. Mite larvae have only three pairs of legs.

Chigger (larva of Trombiculidae) on human skin.

Chiggers (small red dots) on backpack.

MYRIAPODS
(subphylum Myriapoda)

Myriapoda is the smallest subphylum of arthropods. It consists of centipedes, millipedes, and two lesser known groups. This is an ancient group—indeed they are among the first fossils (from the Silurian) of air-breathing animals. Like insects they have a pair of antennae, but unlike insects they have more than 6 legs, usually many more. Most adult insects have compound eyes whereas myriapods usually have simple eyes (the most primitive group of centipedes, Scutigeromorpha, has something similar to compound eyes), and many myriapods lack eyes altogether.

CENTIPEDES

Centipedes belong to the class Chilopoda, which is represented in tropical America by 4 orders and 18 families; worldwide, there are about 3,200 named species. The name *centipede* means "hundred legs," but in fact the number ranges from 30 to nearly 400, a range that overlaps almost exactly with the numbers of legs found in millipedes. A better characteristic for distinguishing centipedes from millipedes is that they have just one pair of legs per body segment whereas millipedes have two pairs per segment.

Centipedes in the orders Scutigeromorpha and Lithobiomorpha have 30 legs; they are born with only 8 to 16 legs but add more as they molt and grow. All other centipedes are born with a full set of legs. Most members of the order Scolopendromorpha have 42 legs; in contrast, members of the order Geophilomorpha (the most diverse order, with 11 families in tropical America) have 54 to 382 legs, depending on the species.

Unlike millipedes, centipedes are fast (especially those with the fewest number of legs), as befits a group of predators. They are generally nocturnal hunters that feed on other invertebrates, although large centipedes occasionally prey on vertebrates; for example, the Amazonian giant centipede (*Scolopendra gigantea*), which occurs in northern South America and is the largest centipede in the world (30 cm or 12 inches in length), has been observed feeding on bats in caves. Centipedes subdue the prey by piercing it with their fangs (modified front legs) and injecting a neurotoxic venom. Some of the larger species produce a painful bite, but they generally do not pose a serious threat to humans.

During mating, the male (in most species) deposits his sperm on a silken web, from which the female retrieves it. In Scolopendromorpha and Geophilomorpha, the female protects her eggs and newborn young by curling around them.

Scutigera coleoptrata (Scutigeromorpha), originally from the Mediterranean region, now common in houses around the world.

Scolopocryptopidae (Scolopendromorpha).

MILLIPEDES

Millipedes belong to the class Diplopoda, the largest group of myriapods. In tropical America, there are 12 orders and about 40 families; worldwide, there are more than 12,000 named species. The word *millipede* means "thousand feet." but none of them have that many legs, the usual number being between 36 and 400, with a maximum of 750.

Unlike centipedes, which have one pair of legs per body segment, millipedes have two pairs of legs on most segments. Moreover, their antennae are generally shorter than those of centipedes.

Millipedes live in leaf litter, soil, rotten wood, and in debris that accumulates in the forest canopy. Species of millipede vary in size and shape, reflecting their slightly differing habits. Many of the elongate cylindrical species push themselves through the soil, like bulldozers. Flat-backed species (e.g., those Polydesmida that have each segment expanded into a lateral plate) work their way through the substrate by wedging themselves into tight spaces, lifting up, pushing, and then lifting again.

Although millipedes eat decomposing vegetation, most of their nutrition comes from eating bacteria and fungi (and the plant material that these microbes break down); many

Platyrhacidae (Polydesmida).

species also eat their own feces. In tropical regions, millipedes are often more important than earthworms in fragmenting leaf litter, thereby increasing rates of decomposition and nutrient release.

The genital openings of both sexes are located toward the front of the body (on the third segment). During courtship (in many species), the male walks along the female's back, and when the female raises her front segments, the male entwines his body around hers. In the vast majority of millipede families, the male has modified legs on the seventh segment with which he transfers sperm to the female. Males of some millipedes ride on the backs of females for several days following copulation. Females of some species make an underground nest in which they lay their eggs (a few species curl around their eggs until they hatch). Young millipedes molt many times, adding legs with each molt; they can require several years to reach adulthood.

Nearly all millipedes have a hardened, calcium-impregnated exoskeleton that serves to protect them from predators. The majority of families also possess a pair defensive glands on most segments and these secrete a diversity of compounds, including benzoquinones (orders Spirobolida and Spirostrepsida), hydrogen cyanide (Polydesmida, the largest order), or alkaloids (order Polyzoniida). If threatened millipedes often coil up into a tight spiral, with the head located in the center, and if further threatened they secrete these noxious chemicals. When coatis (raccoon-like animals) encounter benzoquinone-secreting millipedes, they roll them on the ground to wipe off the secretions; capuchin monkeys and some birds (woodcreepers) rub their bodies with these millipedes to repel mosquitoes and ectoparasites.

Bristly millipedes (order Polyxenida) represent a separate branch (subclass) of the millipede evolutionary tree and look nothing like other millipedes; they are only a few millimeters long, soft-bodied, and covered with tufts of barbed hairs that can entangle an ant. Their biology also differs from that of other millipedes: they feed on algae (that grows on bark or in the soil) and utilize indirect sperm transfer (the male makes a small web on which he deposits sperm and the latter is subsequently picked up by the female).

Order Polydesmida rolled up in defensive position.

Polydesmida.

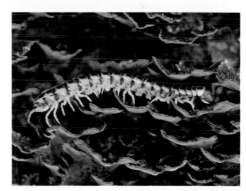

Polydesmida.

Platydesmidae (order Platydesmida) feeding on fungi.

Bristly millipede (Polyxenida).

CRUSTACEANS
(subphylum Crustacea)

The subphylum Crustacea is the only group of arthropods whose members have two pairs of antennae, though in some groups one pair is very reduced or even vestigial (e.g., in terrestrial amphipods and isopods). The body is generally divided into a head, thorax, and abdomen; the number of legs on the thorax varies from group to group but is nearly always more than six, and there are often additional appendages on the abdomen. The majority of species inhabit the oceans, but many occur in freshwater and a few are terrestrial.

The current classification divides crustaceans into 6 classes, the 4 main ones being Branchiopoda (mostly microscopic inhabitants of freshwater), Maxillopoda (marine barnacles and mostly marine copepods, the latter being a major component of zooplankton), Ostracoda (which resemble microscopic clams; though mostly marine, 30% of the species live in freshwater), and Malacostraca (the largest group, consisting of 17 orders).

In this chapter, the focus is on macroscopic crustaceans occurring in freshwater and on land, which belong to the three largest orders of the class Malacostraca: Amphipoda, Isopoda, and Decapoda.

SCUDS, BEACHHOPPERS, SANDHOPPERS, AND LANDHOPPERS
(order Amphipoda)

Amphipods generally have laterally compressed bodies and are tiny, measuring from 5 to 15 mm (0.2-0.6 in). Although they superficially resemble shrimp, they lack a carapace and have 7 pairs of thoracic legs (plus an additional 6 pairs of appendages on the abdomen) rather than the 5 pairs of thoracic legs found on shrimp (and other Decapoda). Females incubate their eggs in the ventral groove between their legs; the eggs hatch directly into juveniles (in shrimp, the eggs first hatch into larvae).

Most amphipods live in the ocean, but some species live in freshwater and a few on land. Freshwater amphipods are sometimes known as scuds, or sideswimmers (for the way they swim). In the tropical America, there are 8 families of scud, with nearly 150 named species. Seven of the families (and slightly more than half of the species) are restricted to subterranean waters and caves. The more readily observable scuds are those associated with the bottom sediment and aquatic vegetation of small streams and springs; in tropical America, these species all belong to the genus *Hyalella* (family Dogielinotidae), a group that originated in South America and includes numerous endemic species in Lake Titicaca. Scuds are omnivorous, feeding primarily on algae and organic detritus; in turn, they are an important food source for fish and waterfowl.

Whereas various groups of amphipods have colonized freshwater, only one family, Talitridae, has colonized terrestrial habitats. Members of this family (250 species worldwide) can be divided into four ecological groups: 1) semi-aquatic inhabitants of estuaries; 2) semi-terrestrial, non-burrowing, coastal species; 3) semi-terrestrial, burrowing, coastal species; and 4) terrestrial, mostly non-burrowing, species found in leaf litter. The second and third groups are called sandhoppers or beachhoppers, the fourth, landhoppers or land shrimp. The name hopper refers to their ability to jump, in some cases nearly a meter; most are nocturnal.

Beachhoppers and sandhoppers feed primarily on algae cast up by the tide. Landhoppers feed on decomposing plant material; they are native to the southern hemisphere, though

Landhopper (Talitridae).

Beachhopper (Talitridae).

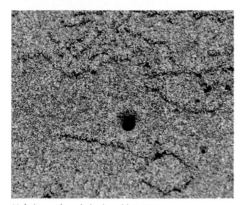

Hole in sand made by beachhopper.

several species occur throughout tropical America—and a few have been widely introduced into various parts of the world. In some landhopper species, large numbers migrate to high ground after heavy rains, and sometimes invade homes, where the inhospitably dry environment kills them off by the thousands, much to the alarm of human occupants.

WOODLICE AND THEIR KIN
(order Isopoda)

Most members of the order Isopoda are 5-30 mm (0.2-1.2 in) in length. They have 7 pairs of legs; in most species, the first leg is modified for grasping. Many have dorsoventrally flattened bodies. As in amphipods, the embryos undergo direct development in the female´s brood pouch, from which they emerge as adult-like juveniles.

Worldwide there are more than 10,000 named species of isopod; about 55% of which are marine species, 10% freshwater, and 35% are terrestrial.

While most marine isopods inhabit the bottom sediments of continental shelves, a few live on the coast. Gribbles (family Limnoriidae), for example, damage jetties and piers by boring into and consuming wood submerged in the sea; they produce their own enzymes for digesting wood, without the aid of gut microbes—unlike shipworms (worm-like clams), which also damage submerged wooden structures, but have microbes that provide some of the necessary enzymes.

Other marine isopods, including *Sphaeroma peruvianum* and *S. terebrans* (family Sphaeromatidae), bore tunnels into the unrooted, aerial prop roots of red mangroves (*Rhizophora*); they do not actually feed on the roots but instead filter out microscopic algae and detritus brought in during high tides. In both gribbles and mangrove root-borers, the offspring often remain in their mother's burrow for several weeks.

A few isopods (family Cirolanidae) that live in the shallow waters of sandy ocean beaches occasionally bite humans when they wade into the water. They feed on a wide diversity of dead and living animals.

Freshwater isopods have evolved numerous times from different marine ancestors (a few are derived from terrestrial ancestors). In tropical America, about 100 species of isopod inhabit freshwater, although they are not commonly encountered, in large part because many of them inhabit subterranean waters and others are parasites of fish or of freshwater shrimps.

There are more species of terrestrial isopod than all other terrestrial crustaceans combined. Terrestrial isopods all belong to the suborder Oniscidea and are called woodlice or sowbugs (in Spanish, *chanchitos*), though species that are capable of rolling up into a ball (e.g., members of the family Armadillidae) are called pillbugs.

Most woodlice feed on decaying leaf litter, both in the soil and in the canopy, and to aid in the digestion of cellulose they harbor endosymbiotic bacteria in their midgut. In many places, woodlice constitute an important part of the soil fauna and make an important contribution

Ischioscia woodlouse (Philosciidae).

to nutrient cycling. Moreover, various other arthropods feed on woodlice; for example, a few species of ant in the genus *Leptogenys* prey specifically on these crustaceans and an entire family of flies (Rhinophoridae) parasitizes them.

Terrestrial isopods appear to have evolved just once, from a marine ancestor. In tropical America, the suborder Oniscidea consists of four main groups: the family Ligiidae (6 species in tropical America), the family Tylidae (4 species), the section Synocheta (2 families and about 20 species), and the section Crinocheta (about 20 families and 500 species). Ligiidae and Tylidae, known as sea slaters and sea roaches, are the most ancient members of the suborder, and most species still inhabit the seashore; species of Synocheta are confined to habitats with high humidity, whereas many species of Crinocheta are capable of living in drier habitats. Increasing terrestriality has involved a number of anatomical and physiological changes; for example, the gills (located on plate-like abdominal appendages) have become more concealed and have been supplemented by internal (lung-like) tracheae in order to minimize water loss during respiration.

SHRIMPS AND CRABS
(order Decapoda)

Decapoda is the largest order of crustaceans, accounting for about 20% of all species, and it includes those that are most familiar to the general public—shrimps, lobsters, and crabs. The head and thorax of these animals are covered by a carapace, which also encloses the gills. The name *Decapoda* means "ten legs" and indeed most members of this order have 10 legs on the thorax, with the front pair often bearing large claws or pincers (chelae). On species in several groups, one claw is larger than the other, the large one used as a crusher and the small as a cutter; the structure of the claws generally reflects the diet of a particular species.

The order Decapoda is divided into two suborders: Dendrobranchiata (prawns), in which females release their eggs into the water after fertilization, and Pleocyemata (all other decapods), in which females incubate the fertilized eggs (stored on the underside of their abdomen) until they are ready to hatch. The terms *prawn* and *shrimp* are frequently used interchangeably, but prawns (Dendrobranchiata) usually have claws on the first three pairs of legs, whereas shrimps (Caridea) have claws on just the first two pairs of legs (sometimes just the second pair). In tropical America, all prawns are marine species, with the sole exception of a freshwater species, *Acetes paraguayensis* (family Sergestidae), which occurs in South America.

The suborder Pleocyemata is divided into 11 infraorders, 7 of which are strictly marine and 4 of which, while principally marine, also include freshwater and terrestrial representatives: Caridea (shrimps), Astacidea (crayfish), Anomura (hermit crabs and one family of

Freshwater shrimp (Palaemonidae: *Macrobrachium heterochirus*).

freshwater crabs), and Brachyura (the vast majority of crabs). Crayfish are not included here because they occur primarily in temperate regions, with two families (Astacidae and Cambaridae) in the northern hemisphere (Cambaridae extends south to Guatemala) and one (Parastacidae) in the southern hemisphere (in Chile and southern Brazil); however, one North American species (*Procambarus clarkii*) has been introduced into many parts of tropical America.

Two of the infraorders mentioned above contain crabs: Anomura, in which the fifth pair of legs are reduced, and Brachyura (true crabs), in which the fifth pair of legs are usually not reduced. True crabs are also characterized by an abdomen that is very reduced and folded under the thorax.

Terrestrial crabs have evolved several times, from distinct marine ancestors. Gills are poorly suited for terrestrial respiration because they generally collapse when exposed to air and lose water quickly through evaporation; terrestrial crabs therefore have reduced gills, and instead respire by means of the surrounding gill (branchial) chamber, which acts as a lung. Both gills and gill chamber function in respiration, but they must be kept moist in order to function properly. To avoid desiccation many terrestrial crabs dig burrows, where access to the water table allows them to replenish their water supply, and where they can also cool off and escape from predators; those that do not dig burrows often hide beneath rocks or, in the case of hermit crabs, inside empty snail shells.

Freshwater crabs spend part of their time out of the water and their respiratory system therefore shows adaptations similar to those found in terrestrial crabs. Freshwater organisms in general require physiological adaptations for living in water having salt concentrations much lower than that of their bodies; these adaptations allow them to prevent the loss of salts and to rid themselves of the excess water that tends to enter their bodies.

Most terrestrial crabs still depend on the sea for reproduction (for this reason they are most accurately referred to as semi-terrestrial). When the eggs are ready to hatch, female crabs release their larvae into the ocean, usually at night, close to the time of high tide on the days of the month with tides of larger amplitude. As a result, newly hatched larvae (zoeae) move rapidly on ebb currents from shallow water into the deeper coastal water where they live and feed as free-swimming plankton for anywhere from a couple of weeks to a couple of months, depending on the species. The larvae then transform into postlarvae (megalopae), settle on the bottom near the shore, transform into tiny juvenile crabs, and come ashore (in some cases it is the postlarvae that come ashore). Given this life cycle, most terrestrial crabs generally do not venture very far inland. Only some freshwater shrimp, the freshwater crabs, and a few sesarmid crabs have severed their ties with the sea.

SHRIMPS

Most true shrimp (infraorder Caridea) are marine creatures, but about a quarter of the species inhabit freshwater. In tropical America the freshwater species belong to five families:

The two largest families are Atyidae and Palaemonidae. Members of the family Atyidae are known as basket shrimp because they use brushes on the first and second pairs of legs to filter out suspended organic matter or sweep up microbial films. The numerous species of Palaemonidae, known as glass shrimp for their transparent bodies, are omnivorous feeders. Some glass shrimp, especially *Macrobrachium rosenbergii* (a native of southeast Asia and Australia), are farmed commercially in many parts of the world, including tropical America. Xiphocarid shrimp shred submerged leaf litter and thus play an important ecological role in streams of the Caribbean islands. In many tropical streams, freshwater shrimp are the dominant animals (in terms of biomass); this is especially true in places where predatory fish are absent, often the case in river pools upstream from waterfalls that serve as a barrier to fish.

Species of freshwater shrimp have three types of life cycle, depending on the given species. In some glass shrimp, females go down river to release their larvae close to saltwater estuaries and then swim back upstream; after completing their larval development in

Table 8-3. Distribution of freshwater species of shrimp

Family	Habitat	Freshwater species in world	Freshwater species in tropical America
Alpheidae	Almost exclusively marine (marine species are called snapping shrimp)	4	1 (caves in Oaxaca, Mexico)
Atyidae	Almost exclusively freshwater	360	20 (throughout the region)
Euryrhynchidae	Exclusively freshwater	7	5 (South America)
Palaemonidae	Marine and freshwater	280	>80 (throughout the region)
Xiphocaridae	Exclusively freshwater	2	2 (Caribbean islands)

saltwater, the juveniles also go upstream. In many species of freshwater shrimp, the females release their larvae in the river, and the larvae are carried downstream to the saltwater estuary, where they complete the larval part of their cycle before returning upstream. In both cases, juveniles generally migrate upstream at night to minimize predation; it should also be noted that dams generally have an adverse affect on these migrations. Finally, some species (including all Euryrhynchidae) have become completely independent of saltwater and pass their entire life cycle in freshwater; this strategy involves laying fewer, but much larger, eggs that hatch either directly into juvenile shrimp (i.e., the entire larval stage occurs inside the egg) or into non-feeding larvae that transform quite quickly into juveniles.

In addition to freshwater shrimps, there is one semiterrestrial shrimp in tropical America, *Merguia rhizophorae* (family Hippolytidae), which inhabits the Atlantic coast from Panama to Brazil. It feeds (mostly at night) on algae and other organic material found on mangrove roots, damp driftwood, and dead coral terraces. Adults of this species begin their lives as males and then, after a series of molts, convert into females; this phenomenon (known as sequential hermaphroditism, or protandry) is more common in crustaceans than in other arthropods (although it occurs in just a small minority of crustaceans).

Macrobrachium hancocki (Palaemonidae).

HERMIT CRABS

Worldwide there are five families of hermit crab, with more than 800 species. The vast majority are marine species. Terrestrial species belong to the family Coenobitidae, which is represented by just two species in the Americas: the Caribbean hermit crab (*Coenobita clypeatus*), which occurs on beaches from southern Florida to Venezuela; and the Ecuadorian hermit crab (*C. compressus*), which occurs on the Pacific coast, from Mexico to Chile.

In most hermit crabs (both marine and terrestrial species) the abdomen is not protected by a calcified exoskeleton; the crabs occupy hollow objects such as empty snail shells, which they generally carry around. Although terrestrial hermit crabs inherited shell-carrying from their marine ancestors, this habit provides several advantages for living on land. For one, the shell can store a small reservoir of water, which means the crabs can forage farther from water sources. The shell also protects the body surface from desiccation, provides a buffer against extreme temperatures, and protects them against predators (when threatened they can withdraw into the shell and cover the opening with their claw). Finally, since hermit crabs utilize the dorsal surface of the abdomen as an additional site for aerial respiration, the cuticle of their abdomen is therefore very thin and vulnerable; but thanks to the shell, the abdomen is well protected.

Nonetheless, dependence on empty snail shells imposes certain constraints. For example, the shell (and the water it contains) is heavy, which means that the crab uses more energy to move about. The availability of empty shells is often a limiting resource; and, since the crabs must seek out larger shells as they grow—and are quite selective in choosing a shell—the search for shells is a recurring and sometimes time-consuming activity. The scarcity of shells can lead to extreme measures; hermit crabs will kill a snail for its shell or attempt to steal a shell from another hermit crab. Some hermit crabs even stockpile empty shells.

The name *hermit* belies the social behavior of these crabs, while the genus name (*Coenobita*) is perhaps more appropriate, a cenobite being a monk or nun who lives in a community (as opposed to a solitary eremite). Indeed, hermit crabs often forage in groups, feeding on excrement, carrion, fallen fruits, and other detritus. A more complicated type of social behavior sometimes occurs when it comes time to change shells: when a large empty shell becomes available, crabs gather around it and queue up in a line from largest to smallest; once the largest crab moves into the vacant shell, each crab in the queue swiftly switches into the newly vacated shell right in front of them.

In order to mate hermit crabs emerge part way from their shells. The female holds the fertilized eggs under her abdomen for about a month and then releases them into the ocean, where the larvae live for about two months. The postlarvae (called megalopae) move to the shore, acquire shells, and eventually metamorphose into juvenile crabs.

A hermit crab (*Coenobita*) inside a snail shell

A pile of hermit crabs (*Coenobita*) feeding on carrion.

FRESHWATER CRABS

Though unrelated, three families of crabs are all called freshwater crabs. The family Aeglidae belongs to the infraorder Anomura (the group to which hermit crabs belong), whereas the families Trichodactylidae and Pseudothelphusidae belong to the infraorder Brachyura (though the latter two families are not closely related). Thus, freshwater crabs have evolved from marine ancestors on at least three separate occasions.

The females of these three families lay large eggs that hatch directly into juvenile crabs (rather than larvae), which means that the females have no need to return to the sea to complete the reproductive cycle. As a result, freshwater crabs are capable of living much farther inland than most terrestrial crabs (which generally have to return to the sea to reproduce). (Some crab species, including those in the family Glyptograpsidae and a few in the family Sesarmidae, also spend much of their lives in freshwater, but they must return to the sea to reproduce, as their eggs hatch into larvae rather than juvenile crabs.)

Aeglidae has more than 60 species, all of which are restricted to southern South America, mostly in temperate regions but with a few species that occur in subtropical regions of southern Bolivia and Brazil. In contrast, the families Trichodactylidae and Pseudothelphusidae are widely distributed throughout tropical America. Trichodactylidae has about 50 species; Pseudothelphusidae consists of about 280 species.

Most members of these three families are omnivorous scavengers, mainly feeding on plant matter, although some feed at least partially on live or dead animals. Adult crabs are capable of leaving the water to feed on terrestrial vegetation and invertebrates.

Pseudothelphusidae, the largest family of freshwater crabs.

Pseudothelphusidae in shallow roadside stream.

FIDDLER CRABS AND GHOST CRABS

Fiddler crabs and ghost crabs belong to the family Ocypodidae. Their eyes are situated at the tips of long stalks that are usually set close together. One claw (cheliped) is larger than the other, in just the males (fiddler crabs), or in both sexes (ghost crabs). (The genus *Ucides* was previously placed in this family, but now occupies a separate family, Ucididae; there are two species in tropical America, one on each coast, and they feed on dead leaves in mangroves.)

Fiddler crabs (genus *Uca*) are mostly diurnal creatures that live in mixed-sex colonies that inhabit either the intertidal zone (the area between low and high tide marks) or the area just above the high tide line. Each crab digs a burrow and defends the area surrounding it. During high tide they seal their burrows from inside, but as the tide recedes they come out to feed on organic detritus deposited on the sand or mud flats. With their small claw (both claws in the case of females), they scoop sediment into their mouth and then flood the oral cavity with water from the gill chamber; lighter edible particles float to the top and are washed into the gut, whereas the indigestible material sinks to the bottom and is spit out as small pellets. Depending on the type of setae on its mouthparts, a given species is specialized for feeding either on sandy sediment with low

Male fiddler crab (*Uca stylifera*). This species occurs on the Pacific coast from El Salvador to northern Peru.

organic content or on muddy sediment with high organic content. Some species find sufficient food near their burrow while others go on excursions to reach areas with more food, generally moving in a group (to minimize the risk to any particular individual).

The large claw of a male fiddler crab is enormous, sometimes weighing half his body weight; as he eats, the rapid movement of the small claw from ground to mouth gives the impression that he is playing the large claw like a fiddle (hence the name). In point of fact, the large claw is used to attract females and in ritualized or actual combat with other males. Within a given species, about half of the males have the right claw enlarged while the rest have the left claw enlarged. If a male loses his large claw (due to an encounter with a predator or combat with another male), he can regenerate it at the next molt; although the new claw is just as large as the original, it is not as strong (which doesn't stop such a male from courting females and bluffing other males).

Fiddler crabs exhibit one of two mating strategies, though several species are capable of employing either one or the other depending on the conditions. In some species,

especially when the population density is low, the male searches for a female and mating takes place on the ground surface near the female's burrow. The male returns to his burrow, and the female continues with her normal life while incubating her eggs. This system can lead to stable neighborhoods, where dominant males attempt to guard neighboring females.

In the other mating strategy, females move through the colony and visit many males before choosing one. Males wave their huge colorful claw vigorously to attract a

Ghost crab (*Ocypode occidentalis*) near its burrow.

Ocypode occidentalis. Ghost crabs can reach speeds up to 4 meters per second.

O. gaudichaudii. Ghost crabs sometimes scrape up sand, put it in their mouth, extract the nutrients, and then spit out small pellets.

coloration, nocturnal habits, and ability to rapidly disappear down their burrows; ghost crabs are members of the genus *Ocypode*, which means, appropriately enough, swift footed.

They are common inhabitants of sandy beaches, where they feed primarily on animal protein. Prey includes invertebrates that live buried in the intertidal zone, especially mole crabs (3 families of Anomura) and small clams; insects; and dead animals washed up on shore. Their widely separated burrows are found from near the high tide line to a distance as great as 400 m (0.25 miles) inland, with those of juveniles occurring closer to the water line. Ghost crab populations often suffer from the damage done to beaches by human recreational activities, and they have disappeared from many places.

There are just 3 species of ghost crab in tropical America.

GECARCINID CRABS

Most gecarcinid crabs (family Gecarcinidae) have one enlarged claw, though not as dramatically enlarged as in fiddler crabs. Some of these crabs are quite large, measuring up to 13 cm (5 in) in width. Young gecarcinid crabs are often very colorful (*Cardiosoma* is popular in the pet-trade), but as they grow older the colors tend to fade. The black land crab (*Gecarcinus ruricola*), a common species in the Caribbean islands, is avidly consumed by the human inhabitants of this region.

Gecarcinid crabs are sometimes called land crabs since they spend most of their lives inland, up to several kilometers (miles) from the coast in some species. They tend to be more active at night and often spend much of the day in burrows that they have excavated. In some coastal forests, these burrows can reach densities of 6 per square meter. Since the crabs feed primarily on leaf litter, fruits, seeds and seedlings, such dense crab populations can have a significant impact on the forest, by reducing, for example, the density and diversity of tree seedlings.

For almost the entire year, the female genital opening is blocked by a rigid calcified covering, and this chastity belt decalcifies for just

female (the choreography of these displays is species-specific). If a male succeeds in attracting a female, she enters his burrow to inspect it. If she is not satisfied, she leaves after a few seconds and continues wandering. When a female accepts a male, she remains in his burrow and the male plugs the entrance from the inside, thereby enclosing the couple underground. Mating occurs in the burrow and the male emerges a day or two later; he then reseals the burrow entrance from the outside, with the female still inside, and leaves in search of another burrow; he may either occupy an empty one, evict a male or female from its burrow, or dig a new hole. The enclosed female remains underground, without feeding, for the next two weeks while she incubates her eggs.

In the Americas, there are about 50 species of fiddler crab, with about 40 species occurring on the Pacific coast and 10 on the Atlantic.

Ghost crabs, which are larger than fiddler crabs, get their name from their pale a few days at the appropriate time of year. Mating in gecarcinid crabs can thus occur only during a short window of time.

After mating, the females incubate their eggs for a couple weeks on the underside of their abdomen; toward the end of this period they migrate to the coast in order to deposit their mature eggs (which are ready to hatch into larvae), then return inland again. These mass migrations, sometimes involving spectacular numbers of individuals, generally occur during a particular time of the year (usually during the rainy season) and are often synchronized with certain phases of the moon. After spending about three weeks in the ocean, the larvae transform into postlarvae (megalope), which come ashore and begin migrating inland (sometimes in huge numbers); at some point these postlarvae transform into tiny juvenile crabs that continue to migrate inland. During both the seaward and inland migration many crabs succumb to predators, and with the rampant spread of coastal resorts, further mortality is caused by motorized vehicles and physical obstructions to migration.

There are 9 species of gecarcinid crab in tropical America: 2 *Cardiosoma*, 3 *Gecarcinus*, and 4 *Johngarthia* (the latter are restricted to small, remote islands in both the Pacific and Atlantic).

Cardiosoma guanhumi, the blue land crab, occurs on the Atlantic coast from Florida to Brazil.

Gecarcinus lateralis in defensive position with raised claws.

Gecarcinus lateralis, the blackback land crab, occurs on the Atlantic coast from southern Texas to Venezuela.

GRAPSID CRABS

Grapsid crabs (family Grapsidae) have a square-shaped body (as seen from above).

Also known as Sally lightfoot, the red rock crab (*Grapsus grapsus*) is a common inhabitant of rocky shores on both the Atlantic and Pacific coasts (including the Galapagos Islands); it feeds mostly on algae, though it is omnivorous. The red mangrove crab (*Goniopsis cruentata*) hides under mangrove roots along the Atlantic coast and feeds on fiddler crabs, mangrove seedlings, and detritus. The mottled shore crab (*Pachygrapsus transversus*) is a small species (less than 2 cm / 1 in) found on both coasts of tropical America; during low tide these crabs emerge from rock crevices to feed on algae, other crabs, and small fish.

There are about 10 species of grapsid crab in tropical America, most of which occur in the intertidal zone or slightly above the high tide line.

Female *Goniopsis pulchra* (Grapsidae) ventilating her eggs (attached beneath the abdomen)

SESARMID CRABS

Like the grapsid crabs, sesarmid crabs have a square-shaped body. Many species dig burrows and, like other terrestrial crabs, most sesarmid crabs have to pass the larval part of their life cycle in the ocean and therefore most species are restricted to coastal areas.

Although generally omnivorous, some sesarmid crabs are at least partially herbivorous. For example, *Armases angustipes* (Atlantic coast of Brazil) and *A. rubripes* (Atlantic coast, from Nicaragua to Argentina) occasionally climb trees near the shore and eat bromeliad flowers. The mangrove tree crab (*Aratus pisonii*) is common on both coasts of tropical America. It is one of the few crabs that is almost exclusively arboreal (it does jump into the water when disturbed); the mangrove tree crab is an agile climber, ascending to the tops of red mangrove trees (*Rhizophora*) to feed on leaves, although it also eats animal matter (e.g., dead fish) when available.

At least ten species of sesarmid crab endemic to Jamaica have a shortened, non-feeding larval stage that occurs in freshwater pools rather than in the ocean. They are thus among the few terrestrial crabs to have entirely severed their ties with the sea. All of these Jamaican species appear to have evolved from a single marine ancestor within the last 4.5 million years. Some inhabit river banks, others the rainforest floor; *Sesarma verleyi* is restricted to caves; and *Metopaulias depressus* spends nearly its entire life in bromeliads that contain water. The latter species, known as the Jamaican bromeliad crab, is unusual in that the female actively cares for her young; prior to releasing her larvae, the female removes detritus from the bromeliad water and then adds empty snail shells to it (the shells provide calcium and make the water less acidic). One to two weeks after being released, the larvae transform into juvenile crabs, and for the next 2-3 months the mother defends these juveniles against predators and provides them with food (snails, millipedes, etc.). Even after they become capable of feeding themselves, young crabs often remain in the bromeliad for another year, assisting their mother in the feeding and protection of the next generation. After this, most of the young crabs disperse (to search for other bromeliads), though one adult female sometimes remains behind, continuing to help, and perhaps someday inheriting the bromeliad when her mother dies.

In tropical America, the family Sesarmidae comprises more than 30 species.

Sesarma rubinofforum. Sesarmid crabs have legs with pointy tips that allow them to scramble over slippery surfaces.

Aratus pisonii, the mangrove tree crab, occurs on the Atlantic coast from Florida to Brazil, and on the Pacific coast from Nicaragua to Peru.

GLOSSARY

Please consult page 4 for illustrations depicting anatomical features,
and page 12 for an explanation of scientific names.

abdomen. The posterior region of the insect body, located behind the thorax.

alkaloids. Naturally occurring chemical compounds that contain at least one nitrogen atom. Alkaloids are produced by many plants and some animals as a chemical defense. Well known examples include caffeine and nicotine.

biological control. The use of a living organism (e.g., a predator or parasite) to reduce the population of another organism (usually a pest species).

camouflage. Resemblance of an organism to its surroundings. The color and pattern of an organism may make it difficult to detect; some organisms are mimetic of an inanimate object such as a leaf or twig. In a few cases an animal covers itself with detritus (known as self-decoration).

cardiac glycosides. Naturally occurring chemical compounds (steroids) produced by some plants (milkweeds, for example) as a chemical defense against herbivores; named for their ability to cause cardiac arrest in humans.

cellulose. The main component of plant cell walls and wood; probably the most abundant biological material on the planet. Most animals cannot digest cellulose (we call it fiber).

chelicerae. The main mouthparts (jaws) of spiders, scorpions, and related arthropods. Hence the subphylum name Chelicerata.

chitin. The main component of the outer skeleton of insects and other arthropods. Chitin contains chains of a glucose derivative.

cocoon. A protective silk covering for the pupa, produced by moth larvae and other insects that undergo complete metamorphosis.

crepuscular. Describes animals that are active primarily during dawn and dusk.

cyanogenic glycosides. Naturally occurring chemical compounds that contain a sugar molecule bound to a cyanide group. When a plant that contains these compounds is attacked by an herbivore, the sugar is removed and toxic hydrogen cyanide is released.

dicots (dicotyledons). Flowering plants whose seeds have two embryonic leaves (cotyledons), as opposed to **monocots**, which have just one. Most plants are dicots.

eusocial. Describes animals (usually a termite, ant, wasp or bee) that live in colonies in which the majority of individuals (the worker caste) do not reproduce; in eusocial insects, generally only the queen lays eggs.

extrafloral nectaries. Plant nectaries (nectar-producing tissue) located outside the flower (e.g., on the leaves), generally serving to attract beneficial arthropods such as predatory ants that fend off plant predators. Only a minority of plants have them.

epiphyte. A plant growing on the branches or trunk of trees and shrubs; most bromeliads and orchids are epiphytes.

femur (plural femora). The first elongate segment of the insect leg, preceding the tibia.

fungivore (adj. fungivorous). An animal that feeds on fungi.

gall. An abnormal plant growth induced by another organism (usually an insect or mite). Unlike a tumor, a gall has a definite structure, and this structure is characteristic of the gall-inducing species.

honeydew. Liquid, sugary excrement of insects that habitually feed on plant (phloem) sap that contains sugar; typically produced by treehoppers, aphids, mealybugs, and related insects. The presence of honeydew usually attracts ants.

inquiline. With respect to galls, an insect or mite that feeds on the gall tissue produced by another species; its presence may or may not result in the death of the gall-inducer. Inquilines are kleptoparasites of the gall-inducer.

iridescence (adj. iridescent). A property of certain surfaces that appear to change color when viewed from different angles (e.g., the surface of a soap bubble).

iridoid glycosides. Naturally occurring chemical compounds (a monoterpene bound to glucose), produced by many plants as a defense against herbivores.

kleptoparasite. An animal that parasitizes the food of another species. It steals the host's food, often resulting in the death of the host. Nest-building wasps and bees are frequent victims of kleptoparasites.

larva. The juvenile stage of insects that undergo complete metamorphosis; the larval stage precedes the pupa. Unlike most adults, the larva usually molts several times. In mites, the first juvenile stage is called a prelarva, the second stage a larva, and the following juvenile stage(s) are called nymphs.

mandibles. Nearly all arthropods have mandibles except for scorpions, spiders, and their kin (Chelicerata). The mandibles are the front pair of mouthparts, typically short and stout, but sometimes highly modified into elongate stylets, as in true bugs, mosquitos, and several other insect groups. In some groups (butterflies and nearly all moths, for example) the mandibles are entirely absent, although the ancestors of these groups once possessed them.

mimicry. Describes a situation in which one organism resembles another organism, often to its benefit. A species may mimic the color or body form of another species, or even its odor.

metamorphosis. A change in body form from one life stage to the next, as when, for example tadpoles change into frogs. Insects with complete metamorphosis undergo a dramatic change; the classic case is that of a caterpillar

(the larva) transforming into a butterfly during the pupal stage. In insects with incomplete metamorphosis (grasshoppers, for example), the change is less drastic, mostly involving the acquisition of functional wings in the adult stage, and without an intervening pupal stage.

monocots (monocotyledons). Flowering plants whose seeds have only one embryonic leaf (cotyledon), as opposed to **dicots**, which have two. Includes aroids, lilies, orchids, grasses, palms, and several others.

nematodes. Nematodes are roundworms; unlike earthworms they lack segments. Although they are a very diverse group of animals, nematodes are seldom appreciated because the vast majority are microscopic in size.

nuptial gift. A food item or other gift given by the male arthropod to the female during courtship or copulation.

nymph. The juvenile stage of those insects that undergo incomplete metamorphosis (i.e., that lack a pupal stage). Unlike most adults, the nymph usually molts several times. See **larva**.

ootheca. A group of eggs surrounded by a protective coating; an egg case. The best known examples are those of cockroaches and mantises.

ovipositor. The egg-laying organ, present in many insects. In some cases it is also used to pierce the substrate into which eggs are laid. In parasitoid wasps it is also used to inject venom into the host. In ants, bees and predatory wasps the ovipositor is used only for injecting venom, and is called a stinger.

parasitoid. An organism that lives in a close parasitic association with another organism. Unlike a parasite, however, a parasitoid always kills its host (parasites sometimes kill their host if there are large numbers present). A predator consumes several individual prey, whereas a parasitoid feeds on just one host.

pedipalps. In scorpions, spiders, and related arthropods (Chelicerata), the pair of appendages in front of the first pair of legs (on either side of the chelicerae).

pheromone. A chemical substance released by an animal that serves to influence the behavior or physiology of other members of the same species. Pheromones play a very important role in the lives of most insects, and can be used to attract sexual partners or send out an alarm signal.

phloem. The tissue in plants that transports sugars (produced by photosynthesis) and other nutrients to various parts of the plant. See also **xylem**.

phytophagous. Describes insects that feed on plants.

polyphagous. Feeding on a diversity of species, for example an herbivore that feeds on various (often unrelated) plant species.

proboscis. Elongated, often tubular mouth-parts used for sucking up food.

pupa. In those insects undergoing complete metamorphosis, the stage during which the larval body is broken down and the adult body forms. In butterflies it goes by the name of chrysalis.

recruitment. In eusocial insects, communication between members of the colony as to the location of a resource (usually food). Examples include ants laying down a chemical trail to the sugar bowl or the waggle dance of the honey bee.

solitary wasps and bees. Those species in which each female builds and provisions her own nest, without help from other females. The majority of nest-building species are solitary, the notable exceptions being those that are eusocial.

spermatophore. A mass containing spermatozoa, produced by males of some animals. In some insects such as katydids the spermatophore is surrounded by an edible treat that the male provides for the female.

spinneret. Silk-spinning organ of a spider or insect; it is usually the larva that possess a spinneret.

spiracles. Small openings in the outer skeleton through which air enters the respiratory system. On most of the body segments there is generally a spiracle on each side.

tarsus (plural tarsi). The last segment of the leg, following the tibia; generally consists of up to five small subdivisions.

terpenoids. A very diverse class of naturally occurring chemical compounds that are produced by many plants to deter herbivores. Well-known examples include the scents given off by eucalyptus and mint.

thorax. In insects, the second region of the body, between the head and the abdomen.

tracheae (singular trachea). The main component of the insect respiratory system. The tracheae are a dense network of small internal tubes that deliver oxygen directly to the tissues (in vertebrates oxygen is delivered via the blood).

tibia. The second elongate segment of the leg; it occurs after the femur and before the tarsus.

tymbal. Specialized part of the outer skeleton capable of producing sound (or ultrasound), either through vibration (as in cicadas) or by buckling (as in tiger moths). Crickets and several other insects lack a tymbal but produce sound by rubbing one part of the body against another part.

tympanum (tympanal organ). Membrane capable of detecting sound; the hearing organ.

ultrasound. High frequency sounds beyond the range of human hearing; bats and some insects are capable of producing such sounds.

urticating hairs or bristles. Barbed hairs that cause physical irritation or bristles that release a stinging venom (e.g., as in certain caterpillars).

warning (aposematic) coloration. Bright colors warning a potential predator that the insect may contain nasty chemical compounds.

xylem. The tissue in plants that transports water and some nutrients to different parts of the plant. Wood is formed from dead xylem tissue. See also **phloem**.

BIBLIOGRAPHY

The following is a representative but not exhaustive list of references;
my apologies to those authors whose publications are not included.

1. Introduction

Battisti, A., G. Holm, B. Fagrell, and S. Larsson. 2011. Urticating hairs in arthropods: their nature and medical significance. Annual Review of Entomology 56:203–220.

Berenbaum, M.R. 1995. Bugs in the system. Insects and their impact on human affairs. Reading: Addison-Wesley. 377 pp.

Capinera, J.L. (ed.) 2008. Encyclopedia of entomology. Second Edition. New York: Springer. 4346 pp.

Chapman, R.F. (Simpson, S.J., and A.E. Douglas, eds.) 2013. The insects. Structure and function. Fifth edition. Cambridge: Cambridge University Press. 929 pp.

Choe, J.C., and B.J. Crespi (eds.) 1997. Mating systems in insects and arachnids. Cambridge: Cambridge University Press. 387 pp.

Choe, J.C., and B.J. Crespi (eds.) 1997. The evolution of social behavior in insects and arachnids. Cambridge: Cambridge University Press. 541 pp.

Costa, J.T. 2006. The other insect societies. Cambridge: Harvard University Press. 767 pp.

Douglas, A.E. 2015. Multiorganismal insects: diversity and function of resident microorganisms. Annual Review of Entomology 60:17–34.

Eisner, T. 2003. For love of insects. Cambridge: Harvard University Press. 448 pp.

Evans, H.E. 1968. Life on a Little-Known Planet. New York: E.P. Dutton. 318 pp.

Grimaldi, D., and M.S. Engel, 2005. Evolution of the insects. New York: Cambridge University Press. 755 pp.

Gullan, P.J., and P.S. Cranston, 2014. The Insects. An outline of entomology. Fifth Edition. Chichester: Wiley-Blackwell. 624 pp.

Hogue, C.L. 1993. Latin American insects and entomology. Berkeley: University of California Press. 536 pp.

Hoorn et al. (17 other authors). 2010. Amazonia through time: Andean uplift, climate change, landscape evolution, and biodiversity. Science 330:927–931.

Kricher, J. 2011. Tropical ecology. Princeton: Princeton University Press. 632 pp.

Laurent, P., J.-C. Braekman, and D. Daloze. 2005. Insect chemical defense. Topics in Current Chemistry 240:167–229.

Lehane, M. 2005. The biology of blood-sucking in insects. Second edition. Cambridge: Cambridge University Press. 321 pp.

Leigh, E.G., A. O'Dea, and G.J. Vermeij. 2014. Historical biogeography of the Isthmus of Panama. Biological Reviews 89:148–172.

Misof, B. et al. (100 other authors). 2014 Phylogenomics resolves the timing and pattern of insect evolution. Science 346:763–767.

Montes, C., A. Cardona, C. Jaramillo, A. Pardo, J.C. Silva, V. Valencia, C. Ayala, L.C. Pérez-Angel, L.A. Rodriguez-Parra, V. Ramirez, and H. Niño. 2015. Middle Miocene closure of the Central American seaway. Science 348:226-229.

Morrone, J.J. 2006. Biogeographic areas and transition zones of Latin America and the Caribbean islands based on panbiogeographic and cladistic analyses of the entomofauna. Annual Review of Entomology 51:467–494.

Mullen, G.R., and L.A. Durden (eds.) 2009. Medical and veterinary entomology. Second edition. Amsterdam: Elsevier. 637 pp.

Quintero, D., and A. Aiello. (eds.) 1992. Insects of Panama and Mesoamerica. Selected studies. Oxford: Oxford University Press. 692 pp.

Rafael, J.A., G.A.R. Melo, C.J.B. Carvalho, S.A. Casari, and R. Constantino (eds.) 2012. Insetos do Brasil. Diversidade e Taxonomia. Riberão Preto: Holos, Editora. 796 pp.

Resh, V.H., and R.T. Cardé (eds.) 2009. Encyclopedia of insects. Second edition. Amsterdam: Elsevier. 1132 pp.

Schoonhoven, L.M., J.J.A. van Loon, and M. Dicke. 2005. Insect-plant biology. Second Edition. Oxford: University Press. 421 pp.

Shaw, S.R. 2014. Planet of the bugs: evolution and the rise of insects. Chicago: University of Chicago Press. 256 pp.

Woodburne, M.O. 2010. The Great American Biotic Interchange: dispersals, tectonics, climate, sea level and holding pens. Journal of Mammal Evolution 17:245–264.

2. Small Orders

Bedford, G.O. 1978. Biology and ecology of the Phasmatodea. Annual Review of Entomology 23:125–149.

Bell, W.J., L.M. Roth, and C.A. Nalepa. 2007. Cockroaches. Ecology, behavior, and natural history. Baltimore: John Hopkins University. 230 pp.

Bignell, D.E., Y. Roisin, and N. Lo (eds.) 2011. Biology of termites: a modern synthesis. Dordrecht: Springer. 576 pp.

Bonada, N., N. Prat, V.H. Resh, and B. Statzner. 2006. Developments in aquatic insect biomonitoring: comparative analysis of recent approaches. Annual Review of Entomology 51:495–523.

Brittain, J.E. 1982. Biology of mayflies. Annual Review of Entomology 27:119–147.

Byers, G.W., and R. Thornhill. 1983. Biology of Mecoptera. Annual Review of Entomology 28:203–228.

Clayton, D.H., and B.A. Walther. 2001. Influence of host ecology and morphology on the diversity of Neotropical bird lice. Oikos 94:455–467.

Clayton, D.H., J.A.H. Koop, C.W. Harbison, B.R. Moyer, and S.E. Bush. 2010. How birds combat ectoparasites. Open Ornithology Journal 3:41–71.

Corbet, P.S. 1999. Dragonflies. Behavior and ecology of Odonata. Ithaca: Cornell University Press. 829 pp.

Córdoba-Aguilar, A. (ed.) 2008. Dragonflies and damselflies. Model organisms for ecological and evolutionary research. Oxford: Oxford University Press. 290 pp.

Drosopoulos, S., and M.F. Claridge (eds.) 2006. Insect sounds and communication. Physiology, behaviour, ecology and evolution. Boca Raton: CRC Press. 532 pp.

Dudley, R. 2000. The biomechanics of insect flight. Form, function, evolution. Princeton: Princeton University Press. 476 pp.

Esquivel, C. 2006. Dragonflies and damselflies of Middle America and the Caribbean. Santo Domingo de Heredia: Instituto Nacional de Biodiversidad. 319 pp.

Gwynne, D.T. 2001. Katydids and bush-crickets. Reproductive behavior and evolution of the Tettigoniidae. Ithaca: Cornell University Press. 317 pp.

Hedwig, B. (ed.) 2014. Insect hearing and acoustic communication. Berlin: Springer-Verlag. 222 pp.

Hopkin, S.P. 1997. Biology of the Springtails (Insecta: Collembola). Oxford: Oxford University Press. 330 pp.

Hynes, H.B. 1976. Biology of Plecoptera. Annual Review of Entomology 21:135–153.

Kamimura, Y. 2006. Right-handed penises of the earwig *Labidura riparia* (Insecta, Dermaptera, Labiduridae): Evolutionary relationships between structural and behavioral asymmetries. Journal of Morphology 267:1381–1389.

Kathirithamby, J. 2009. Host-parasitoid associations in Strepsiptera. Annual Review of Entomology 54:227–249.

Lockwood, J.A. 2004. Locust. The devastating rise and mysterious disappearance of the insect that shaped the American frontier. New York: Perseus Books. 294 pp.

Macadam, C.R., and J.A. Stockan. 2015. More than just fish food: ecosystem services provided by freshwater insects. Ecological Entomology 40 (Suppl. 1):113–123.

Mound, L.A. 2005. Thysanoptera: diversity and interactions. Annual Review of Entomology 50:247–269.

Mound, L.A., and R. Marullo. 1996. The thrips of Central and South America: an introduction (Insecta: Thysanoptera). Memoirs on Entomology, International 6:1–487.

Mugleston, J.D., H. Song, and M.F. Whiting. 2013. A century of paraphyly: A molecular phylogeny of katydids (Orthoptera: Tettigoniidae) supports multiple origins of leaf-like wings. Molecular Phylogenetics and Evolution 69: 1120–1134.

Naskrecki, P. 2000. Katydids of Costa Rica. Vol. 1. Systematics and bioacoustics of the cone-head katydids. Philadelphia: Academy of Natural Sciences. 164 pp. + cd

Penny, N.D. (ed.) 2002. A guide to the lacewings (Neuroptera) of Costa Rica. Proceedings of the California Academy of Sciences 53 (12):161–457.

Redborg, K.E. 1998. Biology of the Mantispidae. Annual Review of Entomology 43:175–194.

Robinson, D.J., and M.J. Hall. 2002. Sound signalling in Orthoptera. Advances in Insect Physiology 29:151–278.

Ross, E.S. 2000. Embia. Contributions to the biosystematics of the insect order Embiidina. Part 1: Origin, relationships and integumental anatomy of the insect order Embiidina. Part 2. A review of the biology of the Embiidina. Occasional Papers of the California Academy of Sciences 149:1–53, 1–36.

Rowell, C.H.F. 2013. The Grasshoppers (Caelifera) of Costa Rica and Panama. The Orthopterists' Society. 611 pp.

Rust, M.K., and M.W. Dryden. 1997. The biology, ecology, and management of the cat flea. Annual Review of Entomology 42:451–473.

Scharf, I., and O. Ovadia. 2006. Factors influencing site abandonment and site selection in a sit-and-wait predator: A review of pit-building antlion larvae. Journal of Insect Behavior 19:197–218.

Simonsen, T.J., J.J. Dombroskie, and D.D. Lawrie. 2008. Behavioral observations on the dobsonfly, *Corydalus cornutus* (Megaloptera: Corydalidae) with photographic evidence of the use of the elongate mandibles in the male. American Entomologist 54: 167–169.

Springer, M., A. Ramírez, and P. Hanson (eds.) Macroinvertebrados de agua dulce de Costa Rica I. Revista de Biología Tropical 58 (Supl.4):1–238.

Tauber, C.A., M.J. Tauber, and G.S. Albuquerque. 2014. Debris-carrying in larval Chrysopidae: unraveling its evolutionary history. Annals of the Entomological Society of America 107:295–314.

Thornton, I.W.B. 1985. The geographical and ecological distribution of arboreal Psocoptera. Annual Review of Entomology 30:175–196.

Ulyshen, M.D. 2015. Insect-mediated nitrogen dynamics in decomposing wood Ecological Entomology 40 (Supp. 1):97–112.

Veracx, A., and D. Raoult. 2012. Biology and genetics of human head and body lice. Trends in Parasitology 28: 563–571.

Watanabe, H., and G. Tokuda. 2010. Cellulolytic systems in insects Annual Review of Entomology 55: 609–632.

Wiggins, G. 2004. Caddisflies: The underwater architects. Toronto: University of Toronto Press. 292 pp.

Windsor, D.M., D.W. Trapnell, and G. Amat. 1996. The egg capitulum of a neotropical walking-stick, *Calynda bicuspis*, induces above ground egg dispersal by the ponerine ant, *Ectatomma ruidum*. Journal of Insect Behavior 9:353–367.

Yanoviak SP, M Kaspari, and R Dudley. 2010 Gliding hexapods and the origins of insect aerial behavior. Biology Letters 5:510–512.

3. True bugs and Their Kin

Ben-Dov, Y., and C.J. Hodgson (eds.) 1997. Soft scale insects: their biology, natural enemies and control. Dordrecht: Elsevier Press.

Burdfield-Steel, E.R., and D.M. Shuker. 2014. The evolutionary ecology of the Lygaeidae. Ecology and Evolution 4:2278–2301.

Byrne, D.N., and T.S. Bellows. 1991. Whitefly biology. Annual Review of Entomology 36:431–457.

Carvalho, G.S., and M.D. Webb. 2005. Cercopid spittle bugs of the New World (Hemiptera, Auchenorrhyncha, Cercopidae). Sofia: Pensoft. 271 pp.

Cassis, G., and R.T. Schuh. 2012. Systematics, biodiversity, biogeography, and host associations of the Miridae (Insecta: Hemiptera: Heteroptera: Cimicomorpha). Annual Review of Entomology 57:377–404.

Denno, R.T., and T.J. Perfect (eds.) 1994. Planthoppers. Their ecology and management. New York: Chapman & Hall. 799 pp.

Dolling, W.R. 1991. The Hemiptera. New York: Oxford University Press. 274 pp.

Douglas, A.E. 2006. Phloem-sap feeding by animals: problems and solutions. Journal of Experimental Botany 57:747–754.

Godoy, C., X. Miranda, and K. Nishida. 2006. Treehoppers of Tropical America. Santo Domingo de Heredia: Instituto Nacional de Biodiversidad. 352 pp.

Greenfield, A.B. 2006. A perfect red: empire, espionage, and the quest for the color of desire. New York: Harper Perennial. 338 pp.

Gullan, P.J., and M. Kosztarab. 1997. Adaptations in scale insects. Annual Review of Entomology 42:23–50.

Hodkinson, I.D. 1989. The biogeography of the Neotropical jumping plant-lice (Insecta: Homoptera: Psylloidea). Journal of Biogeography 16:203–217.

Hodkinson, I.D. 2009. Life cycle variation and adaptation in jumping plant lice (Insecta: Homoptera: Psylloidea) Journal of Natural History 43:65–179.

Jurberg, J., and C. Galvão. 2006. Biology, ecology, and systematics of Triatominae (Heteroptera, Reduviidae), vectors of Chagas disease, and implications for human health. Denisia 19, zugleich Katalogue der OÖ. Landesmuseen Neue Serie 50:1095–1116.

Naskrecki, P., and K. Nishida. 2007. Novel trophobiotic interactions in lantern bugs (Insecta: Auchenorrhyncha: Fulgoridae). Journal of Natural History 41:2397–2402.

Nault, L.R., and J.G. Rodriguez (eds.) 1985. The leafhoppers and planthoppers. New York: Wiley and Sons. 500 pp.

Panizzi, A.R., and J. Grazia (eds.). 2015. True bugs (Heteroptera) of the Neotropics. Dordrecht, Springer. 924 pp.

Redak, R.A., A.H. Purcell, J.R. Lopes, M.J. Blua, R.F. Mizell, and P.C. Andersen. 2004. The biology of xylem fluid-feeding insect vectors of *Xylella fastidiosa* and their relation to disease epidemiology. Annual Review of Entomology 49:243–270.

Reinhardt, K., and M.T. Siva-Jothy. 2007. Biology of the bed bugs (Cimicidae). Annual Review of Entomology 52:351–374.

Ross, L., I. Pen, I., and M.D. Shuker. 2010. Genomic conflict in scale insects: the causes and consequences of bizarre genetic systems. Biological Reviews 85:807–828.

Schaefer, C.W., and A.R. Panizzi. 2000. Heteroptera of economic importance. Boca Raton: CRC Press. 828 pp.

Schuh, R.T., and J.A. Slater. 1995. True bugs of the world (Hemiptera: Heteroptera). Classification and natural history. Ithaca: Cornell University Press. 336 pp.

Spence, J.R., and N.M. Anderson. 1994. Biology of water striders: interactions between systematics and ecology. Annual Review of Entomology 39:101–128.

Stadler, B., and A.F.G. Dixon. 2005. Ecology and evolution of aphid-ant interactions. Annual Review of Ecology and Systematics 36:345–372.

Styrsky, J.D., and M.D. Eubanks. 2007. Ecological consequences of interactions between ants and honeydew-producing insects. Proceedings of the Royal Society B 274:151–164.

Thompson, V. 1997. Spittlebug nymphs (Homoptera: Cercopidae) in *Heliconia* flowers (Zingiberales: Heliconiaceae): preadaptation and evolution of the first aquatic Homoptera. Revista de Biología Tropical 45:905–912.

Wheeler, A.G. 2001. Biology of the plant bugs (Hemiptera: Miridae). Pests, predators, opportunists. Ithaca: Cornell University Press. 507 pp.

Wheeler, A.G., and B.A. Krimmel. 2015. Mirid (Hemiptera: Heteroptera) specialists of sticky plants: adaptations, interactions, and ecological implications. Annual Review of Entomology 60:393–414.

Williams, D.J., and C. Granara de Willink. 1992. Mealybugs of Central and South America. Wallingford: CAB International.

4. Beetles

Chaboo, C.S. 2007. Biology and phylogeny of the Cassidinae Gyllenhal *sensu lato* (tortoise and leaf-mining beetles) (Coleoptera: Chrysomelidae). Bulletin of the American Museum of Natural History 305:1–250.

Choo, J., E.L. Zent, and B.B. Simpson. 2009. The importance of traditional ecological knowledge for palm-weevil cultivation in the Venezuelan Amazon. Journal of Ethnobiology 29:113–128.

Crowson, R.A. 1981. The Biology of the Coleoptera. London: Academic Press. 802 pp.

Eberhard, W., R. Achoy, M.C. Marin, and J. Ugalde. 1993. Natural history and behavior of two species of *Macrohaltica* (Coleoptera: Chrysomelidae). Psyche 100:93–119.

Eberhard, W.G. 1983. Behavior of adult bottle brush weevils (*Rhinostomus barbirostris*) (Coleoptera: Crculionidae). Revista de Biologia Tropical 31:233–244.

Evans, A.M., D.D. McKenna, C.L. Bellamy, and B.D. Farrell, 2015. Large-scale molecular phylogeny of metallic wood-boring beetles (Coleoptera: Buptestoidea) provides new insights into relationships and reveals multiple evolutionary origins of the larval leaf-mining habit. Systematic Entomology 40:385–400.

Farrell, B.D., A. Sequeira, B. O'Meara, B.B. Normark, J. Chung, and B. Jordal. 2001. The evolution of agriculture in beetles (Curculionidae: Scolytinae and Platypodinae). Evolution 55:2011–2027.

Flowers, R.W., and D.H. Janzen. 1997. Feeding records of Costa Rican leaf beetles (Coleoptera: Chrysomelidae). Florida Entomologist 80:334–366.

Franz, N.M., and R.M. Valente. 2005. Evolutionary trends in derelomine flower weevils (Coleoptera: Curculionidae): from associations to homology. Invertebrate Systematics 19:499–530.

Galindo-Cardona, A., T. Giray, A.M. Sabat, and P. Reyes-Castillo. 2007. Bess beetle (Coleoptera: Passalidae): substrate availability, dispersal, and distribution in a subtropical wet forest. Annals of the Entomological Society of America 100:711–720.

García-Robledo, C., G. Kattan, C. Murcia, and P. Quintero-Marín. 2004. Beetle pollination and fruit predation of *Xanthosoma daguense* (Araceae) in an Andean cloud forest in Colombia. Journal of Tropical Ecology 20:459–469.

Hanks, L.M. 1999. Influence of larval host plant on reproductive strategies of cerambycid beetles. Annual Review of Entomology 44:483–505.

Hemp, C., and K. Dettner. 2001. Compilation of canthariphilous insects. Beltträge zur Entomologie 51:231–245.

Herman, L.H. 1986. Revision of *Bledius*. Part IV. Classification of species groups, phylogeny, natural history, and catalogue (Coleoptera, Staphylinidae, Oxytelinae). Bulletin of the American Museum of Natural History 184:1–367.

Jaramillo, J., C. Borgemeister, and P. Baker. 2006. Coffee berry borer *Hypothenemus hampei* (Coleoptera: Curculionidae): searching for sustainable control strategies. Bulletin of Entomological Research 96:223–233.

Jordal, B.H., and L.R. Kirkendall. 1998. Ecological relationships of a guild of tropical beetles breeding in *Cecropia* petioles in Costa Rica. Journal of Tropical Ecology 14:153–176.

Kirkendall, L.R. 2006. A new host-specific ambrosia beetle, *Xyleborus vochysiae* (Curculionidae: Scolytinae), from Central America breeding in live trees. Annals of the Entomological Society of America 99:211–217.

Lewis, S.M., and C.K. Cratsley. 2008. Flash signal evolution, mate choice, and predation in fireflies. Annual Review of Entomology 53:293–321.

Lövei, G.L., and K.D. Sunderland. 1996. Ecology and behavior of ground beetles (Coleoptera:

Carabidae). Annual Review of Entomology 41:231–256.

Navarrete-Heredia, J.L., A.F. Newton, M.K. Thayer, J.S. Ashe, and D.D. Chandler. 2002. Guía ilustrada para los géneros de Staphylinidae (Coleoptera) de México. Guadalajara: La Universidad de Guadalajara. 401 pp.

Pearson, D.L., and A.P. Vogler. 2001. Tiger Beetles. The evolution, ecology, and diversity of the cicindelids. Ithaca: Cornell University Press. 333 pp.

Piel, J., I. Höfer, and D. Hui. 2004. Evidence for a symbiosis island involved in horizontal acquisition of pederin biosynthetic capabilities by the bacterial symbiont of Paederus fuscipes beetles. Journal of Bacteriology 186:1280–1286.

Prischmann, D.A., and C.A. Sheppard. 2002. A world view of insects as aphrodisiacs, with special reference to Spanish fly. American Entomologist 48:208–220.

Ratcliffe, B. 2006. Scarab beetles in human culture. Coleopterists Society Monograph 5:85–101.

Ratcliffe, B.C. 2003. The dynastine scarab beetles of Costa Rica and Panama (Coleoptera: Scarabaeidae: Dynastinae). Bulletin of the University of Nebraska State Museum 16:1–506.

Reichardt, H. 1977. A synopsis of the genera of Neotropical Carabidae (Insecta: Coleoptera). Quaestiones Entomologicae 13:346–493.

Robertson, J.A., J.V. McHugh, and M.F. Whiting. 2004. A molecular phylogenetic analysis of the pleasing fungus beetles (Coleoptera: Erotylidae): evolution of colour patterns, gregariousness and mycophagy. Systematic Entomology 29:173–187.

Roubik, D.W., and P.E. Skelley. 2000. Stenotarsus subtilis Arrow, the aggregating fungus beetle of Barro Colorado Island Nature Monument, Panama (Coleoptera, Endomychidae). Coleopterists Bulletin 55:249–263.

Royale, N.J., P.E. Hopwood, and M.L. Head. 2013. Burying beetles. Current Biology 23:R907–R909.

Saul-Gershenz, L.S., and J.G. Millar. 2006. Phoretic nest parasites use sexual deception to obtain transport to their host's nest. Proceedings of the National Academy of Sciences 103:14039–14044.

Scholtz, C.H., A.L.V. Davis, and U. Kryger. 2009. Evolutionary biology and conservation of dung beetles. Sofia: Pensoft. 544 pp.

Seymour, R.S., and P.G.D. Matthews. 2006. The role of thermogenesis in the pollination biology of the Amazon waterlily Victoria amazonica. Annals of Botany 98:1129–1135.

Short, A.E.Z., and M. Fikáček. 2013. Molecular phylogeny, evolution, and classification of the Hydrophilidae (Coleoptera). Systematic Entomology 38:723–752.

Sousa, W.P., S.P. Quek, and B.J. Mitchell. 2003. Regeneration of Rhizophora mangle in a Caribbean mangrove forest: interacting effects of canopy disturbance and a stem-boring beetle. Oecologia 137:436–445.

Southgate, B.J. 1979. Biology of Bruchidae. Annual Review of Entomology 24:449–473.

Tavakilian, G., A. Berkov, B. Meurer-Grimes, and S. Mori. 1997. Neotropical tree species and their faunas of xylophagous longicorns (Coleoptera: Cerambycidae) in French Guiana. Botanical Review 63:303–355.

Vigneron, J.P., J.M. Pasteels, D.M. Windsor, Z. Vértesy, M. Rassart, T. Seldrum, J. Dumont, O. Deparis, V. Lousse, L.P. Biró, D. Ertz, and V. Welch. 2007. Switchable reflector in the Panamanian tortoise beetle Charidotella egregia (Chrysomelidae: Cassidinae). Physical Review E 76.1=10.

Yee, D.A. (ed.) 2014. Ecology, systematics, and the natural history of predaceous diving beetles (Coleoptera: Dytiscidae). New York: Springer. 468 pp.

Zeh, D.W., J.A. Zeh, and G. Tavakilian. 1992. Sexual selection and sexual dimorphism in the Harlequin beetle Acrocinus longimanus. Biotropica 24:86–96.

5. Wasps, Bees, Ants

Austin, A.D., N.F. Johnson, and M. Dowton. 2005. Systematics, evolution, and biology of scelionid and platygastrid wasps. Annual Review of Entomology 50:553–582.

Barth, F.G., M. Hrncir, and S. Jarau. 2008. Signals and cues in the recruitment behavior of stingless bees (Meliponini). Journal of Comparative Physiology A 194:313–327.

Eltz, T., Y. Zimmermann, J. Haftmann, R. Twele, W. Francke, J.J.G. Quezada-Euan, and K. Lunau. 2007. Enfleurage, lipid recycling and the origin of perfume collection in orchid bees. Proceedings of the Royal Society B 274:2843–2848.

Gotwald, W.H. 1995. Army ants. The biology of social predation. Ithaca: Cornell University Press. 302 pp.

Hanson, P.E., and I.D. Gauld. (eds.) 1995. The Hymenoptera of Costa Rica. Oxford: Oxford University Press. 893 pp.

Hanson, P.E., and I.D. Gauld. (eds.) 2006. Hymenoptera de la Región Neotropical. Memoirs of the American Entomological Institute 77:1–994.

Hölldobler, B., and E.O. Wilson. 1990. The ants. Cambridge: Harvard University Press. 732 pp.

Hölldobler, B., and E.O. Wilson. 2009. The superorganism. The beauty, elegance, and strangeness of insect societies. New York: W.W. Norton & Company. 522 pp.

Hölldobler, B., and E.O. Wilson. 2011. The leafcutter ants. Civilization by instinct. New York: W.W. Norton & Company. 160 pp.

Kapranas, A., and Tena A. 2015. Encyrtid parasitoids of soft scale insects: biology, behavior, and their use in biological control. Annual Review of Entomology 60:195–211.

Keller, L., and E. Gordon. 2009. The lives of ants. Oxford: Oxford University Press. 252 pp.

Kronauer, D.J.C. 2009. Recent advances in army ant (Hymenoptera: Formicidae) biology. Myrmecological News 12:51–65.

Michener, C.D. 2007. The bees of the world. Second Edition. Baltimore: Johns Hopkins University Press. 953 pp.

Moreira, A.A., L.C. Forti, M.A.C. Boaretto, A.P.P. Andrade, J.F.S. Lopes, and V.M. Ramos. 2004. External and internal structure of Atta bisphaerica Forel (Hymenoptera: Formicidae) nests. Journal of Applied Entomology 128:204–211.

O'Neill, K.M. 2001. Solitary wasps. Behavior and natural history. Ithaca: Cornell University Press. 406 pp.

Powell, S., and E. Clark. 2004. Combat between large derived societies: A subterranean army ant established as a predator of mature leaf-cutting ant colonies. Insectes Sociaux 51:342–351.

Quicke, D.L.J. 2015. The braconid and ichneumonid parasitoid wasps. Biology, systematics, evolution and ecology. Chichester: John Wiley & Sons. 681 pp. + 63 plates.

Rico-Gray, V., and P.S. Oliveira. 2007. The ecology and evolution of ant-plant interactions. Chicago: University of Chicago Press. 331 pp.

Ross, K.G., and R.W. Matthews (eds.) 1991. The social biology of wasps. Ithaca: Cornell University Press. 678 pp.

Roubik, D.W. 2006. Stingless bee nesting biology. Apidologie 37:124–143.

Roubik, D.W., and P.E. Hanson. 2004. Orchid bees of tropical America. Biology and field guide. Santo Domingo de Heredia: Instituto Nacional de Biodiversidad. 370 pp.

Schneider, S.S., G. DeGrandi-Hoffman, and D.R. Smith. 2004. The African honey bee: factors contributing to a successful biological invasion. Annual Review of Entomology 49:351–376.

Segoli, M., A.R. Harari, J.A. Rosenheim, A. Bouskila, and T. Keasar. 2010. The evolution of polyembryony in parasitoid wasps. Journal of Evolutionary Biology 23:1807–1819.

Smith, A.R., W.T. Wcislo, and S. O'Donnell. 2008. Body size shapes caste expression, and cleptoparasitism reduces body size in the facultatively eusocial bees Megalopta (Hymenoptera: Halictidae). Journal of Insect Behavior 21:394–406.

Van Driesche, R.G., M.S. Hoddle, and T.D. Center. 2008. Control of pests and weeds by natural enemies: an introduction to biological control. Chichester: Wiley-Blackwell. 484 pp.

Vega, F.E., and H.K. Kaya (eds.) 2012. Insect pathology. Second edition. Amsterdam: Elsevier-Academic Press. 490 pp.

Weiblen, G.D. 2002. How to be a fig wasp. Annual Review of Entomology 47:299–330.

Wirth, W., H. Herz, R.J. Ryel, W. Beyschlag, and B. Hölldobler. 2003. Herbivory of leaf-cutting ants. A case study on Atta colombica in the tropical rainforest of Panama. Berlin: Springer-Verlag. 230 pp.

6. Moths and Butterflies

Aiello, A. 1979. Life history and behavior of the case-bearer Phereoeca allutella (Lepidoptera: Tineidae). Psyche 86:125–136.

Beltrán, M., C.D. Jiggins, A.V.Z. Brower, E. Bermingham, and M. Mallet. 2007. Do pollen feeding, pupal-mating and larval gregariousness have a single origin in Heliconius butterflies? Inferences from multilocus DNA sequence data. Biological Journal of the Linnean Society 92:221–239.

Bernays, E.A., and D.H. Janzen. 1988. Saturniid and sphingid caterpillars: two ways to eat leaves. Ecology 69:1153–1160.

Boggs, C.L, W.B. Watt, and P.R. Ehrlich. (eds.) 2003. Butterflies. Ecology and evolution taking flight. Chicago: University of Chicago Press. 739 pp.

Braby, M.F., and J.W.H. Trueman. 2006. Evolution of larval host plant associations and adaptive radiation in pierid butterflies. Journal of Evolutionary Biology 19:1667–1690.

Burns, J.M., and D.H. Janzen. 2001. Biodiversity of pyrrhopygine skipper butterflies (Hesperiidae) in the Area de Conservación Guanacaste, Costa Rica. Journal of the Lepidopterists' Society 55:15–43.

Castillo-Guevara, C., and V. Rico-Gray. 2002. Is cycasin in Eumaeus minyas (Lepidoptera: Lycaenidae) a predator deterrent? Interciencia (Venezuela) 27:465–470.

Chacón, I., and J. Montero. 2007. Butterflies and moths of Costa Rica. Santo Domingo de Heredia: Instituto Nacional de Biodiversidad. 366 pp.

Connahs, H., G. Rodríguez-Castañeda, T. Walters, T. Walla, and L. Dyer. 2009. Geographic variation in host-specificity and parasitoid pressure of an herbivore (Geometridae) associated with the tropical genus *Piper* (Piperaceae). Journal of Insect Science 9:28

Conner, W.E. (ed.) 2009. Tiger moths and wooly bears: behavior, ecology, and evolution of the Arctiidae. Oxford: Oxford University Press. 303 pp.

DeVries, P.J. 1987. The butterflies of Costa Rica and their natural history. Papilionidae, Pieridae, Nymphalidae. Princeton: Princeton University Press. 327 pp.

DeVries, P.J. 1997. The butterflies of Costa Rica and their natural history. Volume II: Riodinidae. Princeton: Princeton University Press. 288 pp.

Doucet, S.M., and M.G. Meadows. 2009. Iridescence: a functional perspective. Journal of the Royal Society Interface 6:S115–S132.

Dudley, R., R.B. Srygley, E.G. Oliveira, and P.J. DeVries. 2002. Flight speeds, lipid reserves, and predation of the migratory Neotropical moth *Urania fulgens* (Uraniidae). Biotropica 34:452–458.

Gershenzon, J., and N. Dudareva. 2007. The function of terpene natural products in the natural world. Nature Chemical Biology 3:408–414.

Göpfert, M.C., A. Surlykke, and L.T. Wasserthal. 2002. Tympanal and atympanal 'mouth-ears' in hawkmoths (Sphingidae). Proceedings of the Royal Society of London B 269:89–95.

Goyret, J., M. Pfaff, R.A. Raguso, and A. Kelber, A. 2008. Why do *Manduca sexta* feed from white flowers? Innate and learnt colour preferences in a hawkmoth. Naturwissenschaften 95:569–576.

Haber, W.A., and G.W. Frankie. 1989. A tropical hawkmoth community: Costa Rican dry forest Sphingidae. Biotropica 21:155–172.

Hall, J.P.W., D.J. Harvey, and D.H. Janzen. 2004. Life history of *Calydna sturnula* with a review of larval and pupal balloon setae in the Riodinidae (Lepidoptera). Annals of the Entomological Society of America 97:310–321.

Harborne, J.B. 2001. Twenty-five years of chemical ecology. Natural Products Reports 18:361–379.

Hopkins, R.J., N.M. van Dam, and J.J.A. van Loon. 2009. Role of glucosinolates in insect-plant relationships and multitrophic interactions. Annual Review of Entomology 54:57–83.

Janzen, D.H. 2003. How polyphagous are Costa Rican dry forest saturniid caterpillars? *In* Y. Basset, V. Novotny, S.E. Miller, and R.L. Kitching (eds.), Arthropods of tropical forests: spatio-temporal dynamics and resource use in the canopy, pp. 369–379. Cambridge: Cambridge University Press.

Janzen, D.H., W. Hallwachs, and J.M. Burns. 2010. A tropical horde of counterfeit predator eyes. Proceedings of the National Academy of Sciences 107:11659–11665.

Kawahara, A.Y., and D. Adamski. 2006. Taxonomic and behavioral studies of a new dancing *Beltheca* Busck (Lepidoptera: Gelechiidae) from Costa Rica. Proceedings of the Entomological Society of Washington 108:253–260.

Kristensen, N.P. (ed.) 1999. Handbook of zoology, volume IV, part 35. Lepidoptera, moths and butterflies. Volume 1. Evolution, systematics, and biogeography. Berlin: Walter de Gruyter. 490 pp.

Kristensen, N.P. (ed.) 2003. Handbook of zoology, volume IV, part 36. Lepidoptera, moths and butterflies. Volume 2. Morphology, physiology, and development. Berlin. Walter de Gruyter. 564 pp.

Monteiro, A. 2015. Origin, development, and evolution of butterfly eyespots. Annual Review of Entomology 60:253–271.

Nijhout, H.F. 1991. The development and evolution of butterfly wing patterns. Washington, D.C.: Smithsonian Institution Press. 297 pp.

Nishida, R. 2002. Sequestration of defensive substances from plants by Lepidoptera. Annual Review of Entomology 47:57–92.

Peña, C., and N. Wahlberg. 2008. Prehistorical climate change increased diversification of a group of butterflies. Biology Letters 4:274–278.

Rhainds, M., D.R. Davis, and P.W. Price. 2009. Bionomics of bagworms (Lepidoptera: Psychidae). Annual Review of Entomology 54:209–226.

Robbins, R.K., and A. Aiello. 1982. Foodplant and oviposition records for Panamanian Lycaenidae and Riodinidae. Journal of the Lepidopterists' Society 36:65–75.

Rota, J., and D.L. Wagner 2006. Predator mimicry: Metalmark moths mimic their jumping spider predators. PLoS ONE 1(1):e45.

Rota, J., and D.L. Wagner. 2008. Wormholes, sensory nets and hypertrophied tactile setae: the extraordinary defence strategies of *Brenthia* caterpillars. Animal Behaviour 76:1709–1713.

Rydell, J., S. Kaerma, H. Hedelin, and N. Skals. 2003. Evasive response to ultrasound by the crepuscular butterfly *Manataria maculata*. Naturwissenschaften 90:80–83.

Seago, A.E., P. Brady, J.-P. Vigneron, and T.D. Schultz. 2009. Gold bugs and beyond: a review of iridescence and structural colour mechanisms in beetles (Coleoptera). Journal of the Royal Society Interface 6:S165–S184.

Shawkey, M.D., N.I. Morehouse, and P. Vukusic. 2009. A protean palette: colour materials and mixing in birds and butterflies. Journal of the Royal Society Interface 6:S221–S231.

Srygley, R.B., and R. Dudley 2008. Optimal strategies for insects migrating in the flight boundary layer: mechanisms and consequences. Integrative and Comparative Biology 48:119–133.

Stamp, N.E., and T.M. Casey (eds.) 1993. Caterpillars. Ecological and evolutionary constraints on foraging. New York: Chapman & Hall. 586 pp.

Tyler, H.A., K.S. Brown Jr., and K.H. Wilson. 1994. Swallowtail butterflies of the Americas: a study in biological dynamics, ecological diversity, biosystematics, and conservation. Gainesville: Scientific Publishers. 376 pp.

7. Flies and Their Kin

Aluja, M., and A.L. Norrbom (eds.) 1999. Fruit flies (Tephritidae): phylogeny and evolution of behavior. Boca Raton: CRC Press. 984 pp.

Armitage, P.D., P.S. Cranston, and L.V.C. Pinder (eds.) 1995. The Chironomidae: biology and ecology of non-biting midges. Dordrecht: Springer. 572 pp.

Arroyo-Rodríguez, V., J. Puyana-Eraso, A. Bernecker-Lücking, and P. Hanson. 2007. Observations of Geranomyia recondita (Diptera: Tipuloidea: Limoniidae) larvae feeding on epiphyllous liverworts in Costa Rica. Journal of the New York Entomological Society 114:170–175.

Bain, R.S., A. Rashed, V.J. Cowper, F.S. Gilbert, and T.N. Sherratt. 2007. The key mimetic features of hoverflies through avian eyes. Proceedings of the Royal Society of London B: Biological Sciences 274:1949–1954.

Barrett, A.D.T., and S. Higgs. 2007. Yellow fever: a disease that has yet to be conquered. Annual Review of Entomology 52:209–229.

Blanco, M.A., and G. Barboza. 2005. Pseudocopulatory pollination in Lepanthes (Orchidaceae: Pleurothallidinae) by fungus gnats. Annals of Botany 95:763–772.

Brown, B.V., A. Borkent, J.M. Cumming, D.M. Wood, N.E. Woodley, and M.A. Zumbado (eds.) 2009. Manual of Central American Diptera: Volume 1. Ottawa: National Research Council, pp. 1–714.

Brown, B.V., A. Borkent, J.M. Cumming, D.M. Wood, N.E. Woodley, and M.A. Zumbado (eds.) 2010. Manual of Central American Diptera: Volume 2. Ottawa: National Research Council, pp. 715–1442.

Colwell, D., and K. Milton. 1998. Development of Alouattamyia baeri (Diptera: Oestridae) from howler monkeys (Alouatta palliata) on Barro Colorado Island, Panama. Journal of Medical Entomology 35:674–680.

Condon, M.A., S.J. Scheffer, M.L. Lewis, and S.M. Swensen. 2008. Hidden Neotropical diversity: greater than the sum of its parts. Science 320:928–931.

Crosskey, R.W. 1990. The natural history of blackflies. New York: John Wiley & Sons. 711 pp.

Dick, C.W., and B.D. Patterson. 2006. Bat flies: obligate ectoparasites of bats. In S. Morand, B. Krasnov, and R. Poulin (eds.), Micromammals and macroparasites: from evolutionary ecology to management, pp. 179–194. Tokyo: Springer-Verlag.

Disney, R.H.L. 1994. Scuttle flies: the Phoridae. London: Chapman & Hall. 467 pp.

Eberhard, W.G. 2001. The functional morphology of species-specific clasping structures on the front legs of male Archisepsis and Palaeosepsis flies (Diptera, Sepsidae). Zoological Journal of the Linnean Society 133:335–368.

Eizemberg, R., L.T. Sabagh, and R.S. Mello. 2008. First record of myiasis in Aplastodiscus arildae (Anura: Hylidae) by Notochaeta bufonivora (Diptera: Sarcophagidae) in the Neotropical area. Parasitology Research 102:329–331.

Feener, D.H., and B.V. Brown. 1992. Reduced foraging of the tropical fire ant Solenopsis geminata (Hymenoptera: Formicidae), in the presence of parasitic phorid flies, Pseudacteon spp. Annals of the Entomological Society of America 85:80–84.

Feener, D.H., and B.V. Brown. 1993. Oviposition behavior of an ant-decapitating fly, Neodohrniphora curvinervis (Diptera: Phoridae) and defense behavior by its leaf-cutting ant host, Atta cephalotes (Hymenoptera: Formicidae). Journal of Insect Behavior 6:675–688.

Feinstein, J., L. Purzycki, S. Mori, V. Hequet, and A. Berkov. 2008. Neotropical soldier flies (Stratiomyidae) reared from Lecythis poiteaui in French Guiana: Do bat-pollinated flowers attract saprophiles? Journal of the Torrey Botanical Society 135:200–207.

Ferreira, R.L.M., A.L. Henriques, and J.A. Rafael. 2002. Activity of tabanids (Insecta: Diptera: Tabanidae) attacking the reptiles Caiman crocodilus (Linn.) (Alligatoridae) and Eunectes murinus (Linn.) (Boidae), in the Central Amazon, Brazil. Memórias do Instituto Oswaldo Cruz 97:133–136.

Gagné, R.J. 1994. The gall midges of the Neotropical region. Ithaca: Cornell University Press. 352 pp.

Gibson, G., and I. Russell. 2006. Flying in tune: sexual recognition in mosquitoes. Current Biology 16:1311–1316.

Goff, M.L. 2000. A fly for the prosecution. How insect evidence helps solve crimes. Cambridge: Harvard University. 225 pp.

Halstead, S.B. 2008. Dengue virus-mosquito interactions. Annual Review of Entomology 53:273–291.

Headrick, D.H., and Goeden, R.D. 1998. The biology of nonfrugivorous tephritid fruit flies. Annual Review of Entomology 43:217–241.

Honigsbaum, M. 2003 (2001). The fever trail. In search of the cure for malaria. New York: Farrar, Straus and Giroux. 315 pp.

Hull, F.M. 1973. Bee flies of the world. The genera of the family Bombyliidae. Washington, D.C.: Smithsonian Institution Press. 687 pp.

Kyle, J.L., and E. Harris. 2008. Global spread and persistence of dengue. Annual Review of Microbiology 62.71–92.

Lakes-Harlan, R., and G.U.C. Lehmann, 2015. Parasitoid flies exploiting acoustic communication of insects – comparative aspects of independent functional adaptations. Journal of Comparative Physiology A 201: 123–132.

Lavigne, R.J. 2003. Evolution of courtship behaviour among the Asilidae (Diptera), with a review of courtship and mating. Studia Dipterologica 9:703–742.

LeBas, N.R., and L. Hockham. 2005. An invasion of cheats. The evolution of worthless nuptial gifts. Current Biology 15:64–67.

Marshall, S.A. 2012. Flies. The natural history and diversity of Diptera. Buffalo: Firefly Books. 616 pp.

Nishida, K., G. Rotheray, and F.C. Thompson. 2002. First non-predaceous syrphine flower fly (Diptera: Syrphidae): A new leaf-mining *Allograpta* from Costa Rica. Studia Dipterologica 9:421–436.

O'Grady, P., and R. DeSalle. 2008. Out of Hawaii: the origin and biogeography of the genus *Scaptomyza* (Diptera: Drosophilidae). Biology Letters 4:195–199.

Pape, T., D. Dechmann, and M.J. Vonhof. 2002. A new species of *Sarcofahrtiopsis* Hall (Diptera: Sarcophagidae) living in roosts of Spix's disk winged bat *Thyroptera tricolor* Spix (Chiroptera) in Costa Rica. Journal of Natural History 36:991–998.

Parrella, M.P. 1987. Biology of *Liriomyza*. Annual Review of Entomology 32:201–224.

Pates, H., and C. Curtis. 2005. Mosquito behavior and vector control. Annual Review of Entomology 50:53–70.

Patterson, B.D., C.W. Dick, and K. Dittmar. 2007. Roosting habits of bats affect their parasitism by bat flies (Diptera: Streblidae). Journal of Tropical Ecology 23:177–189.

Pipkin, S.B., R.L. Rodríguez, and J. León. 1966. Plant host specificity among flower-feeding Neotropical *Drosophila* (Diptera: Drosophilidae). American Naturalist 100:135–156.

Pritchard, G. 1983. Biology of Tipulidae. Annual Review of Entomology 28:1–22.

Raman, A., C.W. Schaefer, and T.M. Withers, T.M. (eds.) Biology, ecology, and evolution of gall-inducing arthropods. Enfield: Science Publishers. 817 pp.

Ready, P.D. 2013. Biology of phlebotomine sand flies as vectors of disease agents. Annual Review of Entomology 58:227–250.

Rotheray, G.E., M. Zumbado, E.G. Hancock, and F.C. Thompson. 2000. Remarkable aquatic predators in the genus *Ocyptamus* (Diptera, Syrphidae). Studia Dipterologica 7:385–398.

Sancho, E. 1988. *Dermatobia*, the Neotropical warble fly. Parasitiolgy Today 4: 242–246,

Sherman, R.A., M.J.R. Hall, and S. Thomas. 2000. Medicinal maggots: an ancient remedy for some contemporary afflictions. Annual Review of Entomology 45:55–81.

Slansky, F. 2007. Insect/mammal associations: effects of cuterebrid bot fly parasites on their hosts. Annual Review of Entomology 52:17–36.

Stireman, J.O., J.E. O'Hara, and D.M. Wood. 2006. Tachinidae: evolution, behavior, and ecology. Annual Review of Entomology 51:525–555.

Ureña, O., and P. Hanson. 2010. A fly larva (Syrphidae: *Ocyptamus*) that preys on adult flies. Revista de Biología Tropical 58:1157–1163.

Weng, J.-L., and G. Rotheray. 2008 (2009). Another non-predaceous syrphine flower fly (Diptera: Syrphidae): pollen feeding in the larva of *Allograpta micrura*. Studia Dipterologica 15:245–258.

Yeates, D.K., and D. Greathead. 1997. The evolutionary pattern of host use in the Bombyliidae: a diverse family of parasitoid flies. Biological Journal of the Linnean Society 60:149–186.

8. Other Arthropods

Blackledge, T.A., M. Kuntner, and I. Agnarsson. 2011. The form and function of spider orb webs: Evolution from silk to ecosystem. Advances in Insect Physiology 41:175–262.

Duffy, J.E., and M. Thiel (eds.) 2007. Evolutionary ecology of social and sexual systems: crustaceans as model organisms. Oxford: Oxford University Press. 520 pp.

Edgecombe, G.D., and G. Giribet. 2007. Evolutionary biology of centipedes (Myriapoda: Chilopoda). Annual Review of Entomology 52:151–170.

Foelix, R.F. 2010. Biology of spiders. Third edition. Oxford: Oxford University Press. 432 pp.

Friend, J.A., and A.M.M. Richardson. 1986. Biology of terrestrial amphipods. Annual Review of Entomology 31:25–48.

Giribet, G., and P.P. Sharma. 2015. Evolutionary biology of harvestmen (Arachnida, Opiliones). Annual Review of Entomology 60:157–175.

Greenaway, P. 2003. Terrestrial adaptations in the Anomura (Crustacea: Decapoda). Memoirs of Museum Victoria 60:13–26.

Hopkin, S.P., and H.J. Read. 1992. The biology of millipedes. Oxford: Oxford University Press. 223 pp.

Hormiga, G., and C.E. Griswold. 2014. Systematics, phylogeny, and evolution of orb-weaving spiders. Annual Review of Entomology 59:487–512.

Kight, S.L. 2009. Reproductive ecology of terrestrial isopods (Crustacea: Oniscidea). Terrestrial Arthropod Reviews 1:95–110.

Krantz, G.W., and D.E. Walter. (eds.) 2009. A manual of acarology. Third edition. Lubbock:Texas Tech University Press. 704 pp.

Levi, H.W., and L.R. Levi. 1990. Spiders and their kin. New York: Golden Press. 160 pp.

Linton, S.M., and P. Greenaway. 2007. A review of feeding and nutrition of herbivorous land crabs: adaptations to low quality diets. Journal of Comparative Physiology B 177:269–286.

Lucrezi, S., and T.A. Schlacher. 2014. The ecology of ghost crabs. Oceanography and Marine Biology: Annual Review 52:201–256.

Okabe, K., and S. Makino. 2008. Parasitic mites as part-time bodyguards of a host wasp. Proceedings of the Royal Society B 275:2293–2297.

Peretti, A.V. 2002. Courtship and sperm transfer in the whip spider *Phrynus gervaisii* (Amblypygi, Phrynidae): a complement to Weygoldt's 1977 paper. Journal of Arachnology 30:588–600.

Pinto-da-Rocha, R., G. Machado, G., and G. Giribet (eds.) 2007. Harvestmen. The biology of Opiliones. Cambridge: Harvard University Press. 597 pp.

Polis, G.A. (ed.) 1990. The biology of scorpions. Stanford: Stanford University Press. 587 pp.

Proctor, H.C. 2003. Feather mites (Acari: Astigmata): ecology, behavior, and evolution. Annual Review of Entomology 48:185–209.

Saporito, R.A., M.A. Donnelly, R.A. Norton, H.M. Garraffo, T.K. Spande, and J.W. Daly. 2007. Oribatid mites as a major dietary source for alkaloids in poison frogs. Proceedings of the National Academy of Sciences 104:8885–8890.

Sierwald, P., and J.E. Bond. 2007. Current status of the myriapod class Diplopoda (millipedes): taxonomic diversity and phylogeny. Annual Review of Entomology 52: 401–420.

Tizo-Pedroso, E., and K. Del-Claro. 2007. Cooperation in the neotropical pseudoscorpion, *Paratemnoides nidificator* (Balzan, 1888): feeding and dispersal behavior. Insectes Sociaux 54:124–131.

van Berkum, F.H. 1982. Natural history of a tropical, shrimp-eating spider (Pisauridae). Journal of Arachnology 10:117–121.

Vetter, R.S., and G.K. Isbister. 2008. Medical aspects of spider bites. Annual Review of Entomology 53:409–429.

Vollrath, F., and P. Selden. 2007. The role of behavior in the evolution of spiders, silks, and webs. Annual Review of Ecology, Evolution and Systematics 38:819–846.

Walter, D.E., and H.C. Proctor. 2013. Mites: ecology, evolution and behaviour. Life at a microscale. Second Edition. Dordrecht: Springer. 494 pp.

Weldon, P.J., C.F. Cranmore, and J.A. Chatfield 2006. Prey-rolling behavior of coatis (*Nasua* spp.) is elicited by benzoquinones from millipedes. Naturwissenschaften 93:14–16.

Weldon, P.J., J.R. Aldrich, J.A. Klun, J.E. Oliver, and M. Debboun, M. 2003. Benzoquinones from millipedes deter mosquitoes and elicit self-anointing in capuchin monkeys (*Cebus* spp.). Naturwissenschaften 90:301–304.

Weng, J.-L., G. Barrantes, and W.G. Eberhard. 2006. Feeding by *Philoponella vicina* (Araneae, Uloboridae) and how uloborid spiders lost their venom glands. Canadian Journal of Zoology 84:1752–1762.

Weygoldt, P. 1969. The biology of pseudoscorpions. Cambridge: Harvard University Press. 145 pp.

Yip, EC., and L.S. Rayor. 2014. Maternal care and subsocial behaviour in spiders. Biological Reviews 89:427–449.

Zeh, J.A., D.W. Zeh, and M.M. Bonilla. 2003. Phylogeography of the harlequin beetle-riding pseudoscorpion and the rise of the Isthmus of Panamá. Molecular Ecology 12:2759–2769.

Zeil, J., and J.M. Hemmi. 2006. The visual ecology of fiddler crabs. Journal of Comparative Physiology A 192:1–25.

ACKNOWLEDGMENTS

This book would not have been possible without the guidance and assistance of biologists in countries near and far. For help in identifying specimens (whether actual or in photographs) and for providing additional information about some species, we thank the following people:

David Adamski, Ingi Agnarsson, Hugo Aguilar, Art Borkent, Stephan Boucher, Stephanie Boucher, Michael Braby, Harry Brailovsky, John Brown, Daniel Burckhardt, John Burger, Roberto Cambra, Caroline Chaboo, Isidro Chacón, Laura Chavarría, Guillermo Chaverri, Shawn Clark, Oskar Conle, James Coronado, Charles Covell, Mauro Daccordi, Don Davis, Yosuke Degawa, Lewis Deitz, Phil DeVries, Edwin Dominguez, William Eberhard, G. B. Edwards, Tom Eichlin, Marc Epstein, Bernardo Espinoza, Carlos Esquivel, Neal Evenhuis, Eric Fisher, Wills Flowers, David Furth, Ray Gagne, Bolívar Garcete-Barrett, Alfonso García Aldrete, Jon Gelhaus, Carolina Godoy, Geert Goemans, Tom Goldschmidt, Phillipe Grandcolas, Penny Gullan, César Gullén, Bill Haber, Winnie Hallwachs, Christer Hansson, Sam Heads, Frank Hennemann, Thomas Henry, Luis Hernandez, Federico Herrera, Henry Hespenheide, Motoki Hoshi, Bernhard Huber, Ron Huber, Daniel Janzen, Paul Johnson, David Kavanaugh, Akito Kawahara, Alexander Knudson, Jim Lewis, James Lloyd, Jorge Lobo, John Longino, Stuart McKamey, Dave Marshall, Steve Marshall, Juan Mata, Ramon Mello, Ximo Mengual, Peter Mondale, Julian Monge, Carlos Morales, Laurence Mound, Ichiro Nakamura, Piotr Naskrecki, José Luis Navarrete-Heredia, Gino Nearns, Masaru Nishikawa, Allen Norrbom, John Noyes, Charles O'Brien, Thomas Pape, Menno Pepijn, Nicholás Pérez, Jenny Phillips, Marc Pollet, Jens Prena, Alonso Ramírez, Ed Riley, Bob Robbins, Emanuel Rodríguez, Jadranka Rota, David Roubik, Kazuhiko Sakurai, Ana Catalina Sánchez, Christoph Schubart, Scott Shaw, Satoshi Shimano, Kojiro Shiraiwa, Andrew Short, Jay Sohn, Chris Simon, Jay Sohn, Alma Solis, Angel Solís, Francisco Sosa, Monika Springer, Charlie Staines, Ian Swift, Stefano Taiti, Chris Thompson, Vinton Thompson, Vic Townsend, Natalia Vandenberg, Alejandro Vargas, Rita Vargas, Danny Vasquez, Fred Vencl, Carlos Víquez, David Wagner, William Villalobos, David Wahl, Matthew Wallace, Andy Warren, Ingo Wehrtmann, Monty Wood, Norman Woodley, Rodrigo Zeledón, and Ronald Zuñiga.

A special thanks goes to Angel Solís, who came through with an extraordinary effort on the beetle chapter when the deadline was upon us.

We would also like to thank MINAET and the conservation areas (SINAC) for granting us permits and allowing us to conduct field research in Costa Rica.

Paul would like to give special thanks to John McCuen of Zona Tropical Press, who made this book possible, both by initiating the idea and also through his diligent work on editing the text. I would also like to express my gratitude to colleagues, students, and staff in the School of Biology at the University of Costa Rica for providing a stimulating and friendly environment in which to study tropical biology. Particular thanks are due to Bill Eberhard for the scientifically rigorous discussions he promotes among our group of entomologists. In addition to two anonymous reviewers, I thank the following persons for reading early versions of sections pertaining to their specialty: Brian Brown, Guillermo Chaverri, Bill Eberhard, Wills Flowers, Niko Franz, Carolina Godoy, Ray Gagné, Larry Kirkendall, Bert Kohlmann, Juan Mata, Laurence Mound, Piotr Naskrecki, Bernard Pacheco, Adrian Pinto, Alonso Ramírez, Monika Springer, Vinton Thompson, Dave Walter, Ingo Wehrtmann, and Rodrigo Zeledón. Any errors, however, are the responsibility of the author. Finally, I would like to especially thank my wife, Carolina Godoy, and daughter, Lorraine Hanson, for patiently tolerating, even forgiving, my time away from family while working on this book.

Kenji would like to thank the following field stations, institutions, and projects for providing research support: The Organization for Tropical Studies (OTS), the Museum of Zoology at the University of Costa Rica (UCR), Reserva Biológica Manuel Alberto Brenes, Reserva Ecológica Leonelo Ovideo (UCR), Estación Biológica Cerro de la Muerte, Estación Biológica Monteverde, Children's Eternal Rainforest, Zona Protectora El Rodeo (Universidad para la Paz), Centro Agronómico Tropical de Investigación y Enseñanza (CATIE), EARTH University, Finca Café Cristina, Las Brisas Nature Reserve, Instituto Nacional de Biodiversidad (INBio), Smithsonian Institution National Museum of Natural History, Project ALAS, and the Miconia project (UCR-USDA Forest Service). Kenji would also like to thank the following people for help in taking photos of insects in the field or in the lab: David Adamski, Hugo Aguilar, Guillermo Chaverri, Adrián Damaceno, Inez Januszcak, Juan Jiménez, Ricardo Jiménez, Mirjam Knoernschild, Ken Kobayashi, Humberto Lezama, Ximena Miranda, Rossy Morera, Ricardo Murillo, Patricia Ortiz, Federico Paniagua, Natalia Ramírez, Fernanda Retana Alvarado, Josué Castro Rodríguez, Ruth Salas, and Scott Whittaker.

PHOTO CREDITS

Front cover, clockwise from upper left: dobsonfly (*Chloronia* sp.) (Piotr Naskrecki/Minden Pictures); fulgorid planthopper (Fulgoridae) (Piotr Naskrecki/Minden Pictures); large-headed ant (*Daceton armigerum*) (Piotr Naskrecki/Minden Pictures); male harlequin beetle (*Acrocinus longimanus*) (Kenji Nishida); (*Charidotella* sp.) (Kenji Nishida); stink bug (*Edessa* sp.) (Piotr Naskrecki/Minden Pictures); horse fly (*Tabanus fulmineus*) (Piotr Naskrecki/Minden Pictures); *Metopoceris gemmans* (Kenji Nishida); emerald cicada (*Zamara smaragdina*) (Piotr Naskrecki/Minden Pictures).

Spine: leaf beetle (*Alurnus ornatus*) (Piotr Naskrecki/Minden Pictures).

Back cover, clockwise from upper left: *Papilio* (*Pterourus*) *garamas* (Kenji Nishida); *Edessa arabs* (Kenji Nishida); *Morpho cypris* (Kenji Nishida); *Chrysina aurigans* (Kenji Nishida); tortoise beetle (*Echoma clypeata*) (Piotr Naskrecki/Minden Pictures); *Stilodes undecimlineata* (Kenji Nishida); flag-footed bug (*Anisocelis flavolineata*) (Piotr Naskrecki/Minden Pictures); jumping spider (*Lurio solennis*) (Kenji Nishida); turtle ant (*Cephalotes*) (Kenji Nishida).

p. ii: *Euthyrhynchus floridanus* (Pentatomidae)
p. vi: *Rhodochlora* (Geometridae)
p. 1: *Sematura* (Sematuridae)
p. 13: *Paratropes bilunata* (Ectobiidae)
p. 59: *Edessa irrorata* (Pentatomidae)
p. 97: *Stolas lebasi* (Chrysomelidae)
p. 153: *Camponotus sericeiventris* (Formicidae)
p. 196: *Rothschildia triloba* (Saturniidae)
p. 269: Male *Plagiocephalus latifrons* (Ulidiidae)
p. 302: Spirostreptida

We gratefully acknowledge the following people for allowing us to use their photographs:

Ricardo Alvarado: p. 315 (wolf spider); **Ingo Arndt**: p. 6 (complete metamorphosis of morpho butterfly: older larva; adult emerging from pupa; recently emerged adult), p. 68 (bed bug), p. 206 (*Perola*), p. 211 (*Oxytenis larva*), p. 247 (monarch butterflies); **Stephen Boucher**: p. 106 (larvae of *Passalus punctatostriatus*); **Luciano Capelli**: p. 220 (larva of *Thagona tibialis*); **Ernesto Carman**; p. 216 (*Urania fulgens*); **Gill Carter**: p. 239 (*Semomesia croesus*), p. 245 (*Baeotus beotus*); **Bill Eberhard**: p. 317 (both photos of *Uloborus*); **Luiz Carlos Forti**: p. 190 (*Atta bisphaerica*), p. 191 (*Atta bisphaerica*); **Nico Franz**: p. 151 (*Cyclanthura*); **Kim Garwood**: p. 226 (*Jemadia scomber*, *Yaguna spatiosa*), p. 239 (*Rhetus dysoni*), p. 251 (*Prepona praeneste*); **Pablo Gutiérrez**: p. 100 (ventral view of larva of water penny beetle), p. 294 (two photos of Streblidae); **Marvin Hidalgo**: p. 31 (*Harroweria*), p. 123 (aggregation of *Stenotarsus subtilis*); **Larry Kirkendall**: p. 148 (bark beetles, two photos), p. 149 (two photos of bark beetles), p. 314 (*Cupiennius getazi* eating a frog); **Rólier Lara**: p. 332 (*Macrobrachium heterochirus* shrimp), p. 337 (male fiddler crab); **James E. Lloyd**: p. 119 (female Phengodidae emitting light); the late **Keiji Morishima**: p. 95 (mortar and pestle), p. 96 (cacti and dried insects); **Ricardo Murillo**: p. 275 (female *Forcipomyia* on *Battus polydamas*), p. 229 (*Parides iphidamas*), p. 236 (larvae of *Eumaeus godarti*), p. 250 (larvae of *Mechanitis polymnia*), p. 252 (larval head of *Memphis pithyusa*), p. 264 (larva of *Hamadryas amphinome*); **Yumiko Nagano**: p. 140 (*Diabrotica*, top right), p. 229 (*Papilio cresphontes*), p. 268 (larva and pupa of *Chlosyne janais*, two photos), p. 296 (*Chrysomya*); **Piotr Naskrecki**: p. 29 (*Acanthodiphrus conspersus*), p. 31 (*Copiphora rhinoceros*), p. 33 (*Amphiacusta saba*), p. 45 (adult *Capucina*), p. 49 (*Cerastipsocus*), p. 87 (*Enchophora*, two photos), p. 89 (Ciriacreminae), p. 145 (*Brentus* cf. *anchorago*), p. 189 (*Atta cephalotes* with hitchhiker and *Acromyrmex* queen, two photos), p. 266 (*Marpesia petreus* pupa, photo inset), p. 301 (human bot fly adult), p. 303 (horseshoe crab); **Yoshiaki Oikawa**:

p. 164 (larva of Polysphinctini), p. 213 (larva of *Hemeroplanes triptolemus*), p. 288 (*Neodohrniphora curvinervis, Apocehphalus ritualis*); **Bernald Pacheco**: p. 61 (water boatman); **Laura Sánchez**: p. 40 (colony of webspinners); **Monika Springer**: p. 61 (*Rhagovelia*); **Aiko Takahashi**: p. 108 (*Phyllophaga* on *Trichilia*); **Roy Toft**: p. 334 (*Macrobrachium hancocki*); **Rita Vargas**: p. 340 (female *Goniopsis pulchra*); **Danny Vasquez**: p. 22 (Coenagrionidae females laying eggs), p. 27 (larva of *Leptonema*), p. 62 (male Belostomatidae carrying eggs); **Dan Wade**: p. 267 (*Junonia evarete*); **Ingo Wehrtmann**: p. 339 (*Cardiosoma guanhumi*).

We also thank the **Smithsonian Institution** for allowing us to use the following photographs: p. 8 (mimicry photos), p.130 (*Tetraopes, Neoptychodes trilineatus*), p. 132 (*Caryedes brasiliensis*), p. 147 (two photos of *Rhynchophorus palmarum*), p. 166 (sugarcane borer), p. 251 (*Agrias amydon philatelica*), p. 252 (*Agrias aedon narcissus*), p. 290 (*Toxotrypana curvicauda*).

INDEX

Abedus 61
Acalymma 140
Acalyptini 149, 151
Acanthaceae 244, 268
Acanthocephala 73
Acanthodiphrus conspersus 29
Acanthopidae 42
Acanthoscelides obtectus 132
Acetes paraguayensis 332
Acharia horrida 6, 205
Achilidae 74
Aconophora 84
Acontista multicolor 42
Acraeini 11, 258
Acraga 206
Acrididae 37
Acridinae 37
Acrocinus longimanus 130, 305
Acrolepia 197
Acromyrmex 188, 189
Acropyga 195
Acrotelsa collaris 16
Actinidiaceae 324
Adelpha 261
Adelpha fessonia 261
    leucophthalma 261
    serpa celerio 261, 262
    tracta 261
Aedes 272
Aedes aegypti 9, 273, 274
    albopictus 9, 274
Aeglidae 336
Aegithus 122
Aenolamia 79
Aepytus 198
Aeshnidae 20
African honey bee 182, 183
African oil palm 149
Aganacris insectivora 30
Agaonidae 160–163
Agaristinae 217
Agelaia 155
Ageniellini 170
Agrias aedon narcissus 252
    amydon philatelica 251
Agromyzidae 292, 293
Agrosoma bispinella 82
Alderflies 25
Aleyrodidae 87, 90, 91, 124, 159, 160, 289
Algae 14, 16, 19, 23, 26, 35, 49, 62, 98, 100, 104, 208, 270, 275, 277, 287, 323, 328, 330, 331, 334, 340
Alkaloids 124, 192, 212, 216, 224, 259, 323, 328
Alkaloids, pyrrolizidine 120, 220, 224, 246, 249
Allergic reaction 9, 183, 326
Allochroma sexmaculatum 143
Allograpta centropogonis 288
Alpheidae 334
Alticini 141-143
Alurnus ornatus 133
Alydidae 73
Amaranthaceae 201, 276
Amazon region 11, 25, 112, 146, 186, 195, 238, 243, 312
Amazon water lily 112
Amazonian giant centipede 327
Ambates 149
Amblycera 50
Amblyomma 326
Amblyopinus 104
Amblypygi 306, 307
Ambrosia beetles 148, 149
Amphiacusta 32
Amphiacusta saba 33
Amphimoea walker 212
Amphipoda 330, 331
Ampulicidae 171, 172
Anacaena 98
Anacardiaceae 88, 131, 144, 145, 216
Anacroneuria 23
Anartia fatima 245
Anastrepha 291
Anatomy of insects 4, 5
Anaxipha agaea 32
Andes mountains 11, 58, 258, 271
Anelosimus 321
Anetia 246
Anisolabididae 41
Anisoscelis 72
Annonaceae 89, 121, 228, 235
Anochetus 186
Anomala 109
Anopheles 9, 272, 273
Anopleura 50
Anovia punica 125
Antaeotricha 204
Ant-decapitating flies 287, 288
Anteros kupris 240
Anthanassa ardys 268
Antheraea godmani 210
Anthicidae 129
Anthonomini 149, 151
Anthonomus grandis 149
    monostigma 151
Anthoptus epictetus 227
Anthrax 284
Antianthe expansa 87
Antiblemma 219
Antirrhea 256
Antirrhea philoctetes tomasia 257
    pterocopha 256
Antitumor properties 129
Ant-like beetles 129
Antlions 53
Ants 9, 39, 87, 124, 175, 176, 183-195, 325
Anurogryllus 32
Apaturinae 197
Aphelinidae 160
Aphidiinae 165
Aphids 5, 87, 88, 91, 92, 124, 125, 165, 279, 289
Aphis nerii 88, 91, 92, 125
    spiraecola 92
Aphonopelma seemanni* 312
Aphrophoridae 78
Apiaceae 228
Apidae 177-183
Apiomerini 64
Apiomerus pictipes 64
Apis mellifera 182, 183
Apocephalus ritualis 288
Apocynaceae 71, 88, 92, 130, 212, 224, 246, 248, 249
Apterostigma 188, 189
Aquatic insects 7, 18-27, 37, 61, 62, 98–100, 208, 269-277, 283, 286, 289, 294
Araceae 111, 112, 121, 143, 149, 151, 314
Arachnida 3, 200, 303-326
Araliaceae 88, 89, 227
Araneae 307-321
Araneidae 54, 164, 170, 308, 309, 318, 319
Aratus pisonii 340, 341
Arawacus togarna 236
Archaeognatha 2, 3, 16
Archaeoprepona amphimachus 251
    demophoon 252
Archimandrita tesselata 45
Archipsocidae 49
Archipsocus 49
Archisepsis diversiformis 292
Archonias brassolis approximata 8
    brassolis negrina 8
Arctiinae 8, 220-222, 224

Argasidae 325
Argia 22
Argidae 154
Argiope argentata 308, 318
Argynnini 258
Argyrodes 320
Argyroeides notha 220
Arilus gallus 63
Aristolochiaceae 224, 228, 230, 287
Aristolochic acids 224
Armadillidae 331
Armases angustipes 340
    rubripes 340
Army ants 104, 180, 186-188, 190, 299, 325
Arthropoda 2, 5, 302
Ascalapha odorata 218
Ascalaphidae 53, 54
Ascia monuste 233
Asexual reproduction 5, 39, 91-93, 147, 192, 195
Asilidae 285
Asopinae 69, 70
Aspisoma 117
Assassin bugs 63-65
Asteraceae 71, 85, 86, 90, 120, 135, 150, 224, 237, 241, 246, 249, 258, 267, 281, 291-293
Asthma medicine 126
Astigmata 322, 325
Astraptes 228
Astraptes alardus latia 226
Atta 188-191, 288
Atta bisphaerica 190, 191
    cephalotes 189, 190
Attelabidae 144, 145
Atyidae 333, 334
Auchenorrhyncha 59, 74-87
Avocado 67, 80, 149, 230, 279
Azteca 193, 194
Aztecs 95-96, 118, 200, 243
Backswimmers 61, 62
Bacteria 5, 7, 44, 48, 59, 74, 79, 80, 87, 88, 104, 105, 118, 185, 188, 190, 192, 195, 220, 291, 293, 295, 296, 328, 331
Baeotus 267
Baeotus beotus 245
Baetidae18, 19
Baetodes 19
Bagheera kiplingi 316
Bagworm moths 199
Baleja 81
Baleja flavoguttata 81
Balloon setae 240
Ballooning 199, 309
Bamboos 149, 253, 254
Banana 133, 146, 206, 254, 298

Barbinola costaricensis 82
Baridinae 149-151
Bark beetles 148, 149, 152
Bark lice 3, 49
Baronia brevicornis 228
Bartonella bacilliformis 271
Basket shrimp 333, 334
Bat flies 294
Bats 5, 7, 17, 29, 42, 54, 58, 105, 197, 198, 209, 211, 220, 253, 294, 298, 323, 325, 327
Battus 228
Battus polydamas 230, 275
Beach 101, 102, 104, 172, 275, 285, 330, 331, 335-338
Beachhoppers 330, 331
Beans 124, 132, 140
Beauties 267
Bed bug 60, 68
Bee flies 284
Bees 57, 177-183, 325
Beeswax 182
Beetle mites 322, 323
Beetles 2, 3, 17, 97-152
Beirneola 82
Belostoma 61
Belostomatidae 61, 62
Belotus 121
Beltheca oni 203
Bembicina 172
Bemisia tabaci 90, 91
Benzoquinones 328
Bess beetles 105, 106
Biblidinae 245, 262-264
Bicellychonia 116
Bignoniaceae 114
Biological control 7, 67, 124, 152, 165, 166, 324
Bioluminescence 115, 118, 119, 278
Biomass 8, 48, 175, 185
Birds 7, 8, 11, 17, 29, 49, 50, 58, 103, 187, 196, 211, 225, 246, 253, 272, 275, 282, 294, 297, 304, 312, 325, 326, 328
Biting midges 275, 276
Bittacidae 57
Blaberidae 44, 45
Blaberus 44
Blaberus giganteus 45
Black flies 275, 294
Black land crab 338
Black scavenger flies 292
Black widow spiders 309
Black witch 218
Blackback land crab 339
Blastobasidae 203
Blattaria 3, 5, 6, 42, 44, 45, 87, 157, 171, 299

Blattidae 45
Bledius 104
Blepharoneura 291
Blissidae 71
Blister beetles 128, 129, 224
Blisters 104, 129, 180
Blood feeders 9, 50, 58, 65, 68, 269, 271-275, 282, 294, 295, 325, 326
Blow flies 295-297
Blue land crab 339
Blues 235
Bocydium 86
Bolas spiders 318
Bolbonota 86, 160
Bolla cupreiceps 226
Bombardier beetles 101
Bombus 177
Bombycidae 200, 210
Bombyliidae 284
Bombyx mori 200
Boraginaceae 67, 136, 137, 193, 224, 246, 249
Brachiacantha 125
Brachinus 101
Brachygastra 174
Brachys insignis 114
Bracon 165
Braconid wasps 164-166
Braconinae 165
Bradypodicola 209
Bradysia 278
Bradysia floribunda 277
Brassicaceae 224, 231
Brassolini 254, 255
Brassolis 254
Brassolis isthmia 255
Brazilian wandering spiders 309
Brenthia 207, 208
Brentidae 145
Brentus 145
Bristly millipedes 328, 329
Broad-bodied leaf beetles 139, 224
Broad-headed bug 73
Broad-nosed weevils 147, 148
Bromeliads 11, 18, 20, 29, 60, 98, 101, 146, 149, 151, 206, 254, 270, 272, 276, 292, 314, 340
Brown lacewing 54, 55
Brown widow 321
Bruchinae 132
Brush-footed butterflies 244-268
Bryocorinae 66
Bubonic plague 58
Buckeye butterflies 224, 267
Buenoa 62
Bullet ant 185, 186
Bumble bees 17, 175, 176, 177

Buprestidae 113, 114
Burrower bug 70
Burseraceae 74, 88, 131, 141, 216
Buthidae 304
Butterflies 8, 225-268, 276
Butterfly mud-puddling 225,
    227, 229, 232, 233, 235, 250,
    260, 265, 268
Cacao 66, 276
Cactaceae 95, 96, 177, 211
Caddisflies 3, 18, 26, 27
Caelifera 28
Caenia 120
Caerois 256
Calamoceratidae 27
Calcium oxalate 204, 212
Calephelis 241
Calephelis iris 241
Caligo 254, 300
Caligo atreus 255
    telamonius 255
Calleida scintillans 102
Callicore 262
Calligrapha fulvipes 139
Callimantis antillarum 43
Calliphoridae 295-297
Callona rutilans 131
Calophaena ligata 102
Calopterygidae 22
Calpini 218, 219
Calycopis 235, 236
Camouflage 53-55, 78, 102, 104,
    109, 127, 311, 317, 318
    Bird droppings 164, 204,
    237, 261
    Caterpillar excrement 86,
    138
    Leaves 29-31, 37, 39, 43,
    198, 206, 208, 212, 216, 234,
    250, 252, 261, 264, 266
    Moss-covered twig 74, 86,
    261, 262
    Moss/lichen-covered bark
    31, 78, 130, 131, 210, 213,
    214, 262, 264
    Plant spines 83-85
    Self adornment 42, 55
    Twigs 39, 42, 203, 215, 218
    Wood fragment 216, 228
Campanulaceae 288
Campodeoidea 15
Camponotus 87, 183, 195
Camponotus femoratus 195
    senex 195
    sericeiventris 184
Campopleginae 164
Campsomeris 168
Campylocentrus 85
Canavanine 132

Cantharidae 120, 121
Cantharidin 129
Canthon 106
Capparidaceae 231
Capucina 45
Carabidae 101, 102
Cardiac glycosides 71, 224, 246
Cardiosoma 338, 339
Cardiosoma guanhumi 339
Caricaceae 246
Carineta 78
Carrion feeders 103, 104, 107,
    181, 198, 250, 238, 267, 287,
    292, 295, 298, 335, 336, 338
Carrion beetles 103
Carrion's disease 271
Caryedes brasiliensis 132
Cassia 237
Cassidinae 133-137
Castilla elastica 51
Castilleja 289
Castniid moths 8, 206
Cat flea 58
Catachlorops 282
Catasticta cerberus 232
    eurigania straminea 232
    flisa 233
    sisamunus 233
    teutila 232
Caterpillars 154, 196-268
    Attracting ants 235, 238, 242
    Carnivorous 238
Catocyclotis adelina 242
Caves 32, 294, 306, 307, 327,
    330, 334, 340
Cecidomyiidae 161, 279-281
Cecropia 49, 140, 148, 149, 194,
    244, 245, 267
Celastrina gozora 235
Cellulose 16, 48, 105, 106, 331
Centipedes 2, 200, 299, 327
Centruroides 303
Centruroides edwardsii 304
Cephaloleia 133, 134
Cephaloleia nigropicta 134
Cephalotes 191, 192, 193
Cephalotes specularis 192
Cephisus siccifolius 78
Ceraeochrysa montollana 54
Cerambycidae 130, 131
Cerastipsocus 49
Ceratitis capitata 291
Ceratopogonidae 275, 276
Cercopidae 78, 79
Cercopoidea 59, 74, 78, 79, 293
Cerotoma 140
Chagas disease 9, 65
Chalcidoid wasps 156, 159-163
Chalcolepidius bomplandi 115

Chalodeta lypera 241
Charaxinae 250-252
Charidotella egregia 133, 135
Charidotis 137
Charoxus 104
Chauliognathus 120, 121
Checkered beetles 168
Checkerspots 224, 267-268
Chelicerata 2, 3, 302, 303
Chelymorpha 136
Chetone 220
Chetone angulosa 8
Chiggers 9, 322, 325, 326
Chigoe flea 9, 58
Chikungunya 9
Chilopoda 2, 200, 299, 327
Chinch bugs 71
Chironomidae 276, 277
Chitin 5, 302
Chlamisini 138
Chlorocoris isthmus 70
Chloronia 25
Chlosyne janais 268
    theona 267
Choeradodis 42, 43
Cholus 149, 150
Choreutidae 207, 208
Chromacris psittacus 36
Chrysagria alticophaga 298
Chrysanthrax 284
Chrysididae 167
Chrysina 109
Chrysina aurigans 105, 109, 110
    boucardi 109
    optima 2
Chrysobothris 114
Chrysomelidae 132-143
Chrysomelinae 139, 160
Chrysomya 296
Chrysopidae 54, 55
Cicada 5, 33, 34, 59, 74, 77, 78,
    298
Cicadellidae 57, 59, 74, 80-82,
    87, 167
Cicadidae 5, 33, 34, 59, 74, 77,
    78, 298
Cicindela 101, 102
Cicindela macrocnema 102
Cicindelinae 101, 102
Cimex hemipterus 68
    lectularius 68
Cimicidae 68
Ciriacreminae 89
Cirolanidae 331
Cithaerias pireta 252
Citrus 80, 90, 202, 218, 323
Citrus greening disease 88
Citrus rust mite 323
Cixiidae 74

Cladochaeta 293
Cladonota 86
Classification 12
Clastoptera 79, 293
Clastopteridae 78, 79
Cleaning behavior 30, 45, 80, 99, 161, 168, 188
Cleistolophus 148
Cleridae 168
Click beetles 114-116, 118
Clogmia albipunctata 271
Cloning (polyembryony) 159, 160
Cobweb spiders 320, 321
Coccidae 93
Coccinellidae 91, 124, 125
Coccoidea 87, 93-96, 124, 159, 194, 195, 289
Cocconotini 29
Coccotrypes rhizophorae 148
Cochliomyia hominivorax 296, 297
Cockroach wasps 171
Cockroaches 3, 5, 6, 42, 44, 45, 87, 157, 171, 299
Cocoon 9, 52, 53, 56, 58, 164, 165, 167, 168, 200, 204, 210, 217, 222, 287
Coeini 245, 267
Coelomera 140
Coenagrionidae 22
Coenaletidae 14
Coenobita 335
Coenobita clypeatus 335
Coenobita compressus 335
Coffee 24, 80, 152
Coffee berry borer 152
Coffeikokkos copeyensis 158
Colaspis 133
Colaspoides batesi 133
Colax 217
Coleoptera 2, 3, 17, 97-152
Coliadinae 231, 233, 234
Collembola 2, 3, 14, 15, 243, 323
Colobura 267
Colobura dirce 225, 244
Colors of insects
  Iridescence 109, 113, 209, 215, 243, 249, 250, 256, 282
  Pigments 14, 95, 133, 196, 213, 224, 243
  Structural 243, 292
Colpoptera 76
Combretaceae 220, 238
Compound eye 5, 307, 327
Compsus 148
Condylostylus 287
Conocephalinae 31
Conoderini 149-151

Conopidae 299
Conotelus 121
Convolvulaceae 86, 133, 134, 136, 137, 143, 177
Copaxa 211
Copaxa syntheratoides 210
Copepods 2, 330
Copidosoma 159, 160
Copiphora rhinoceros 29, 31
Coptocycla 137
Coptocycla leprosa 136
Copulation 32, 46, 71, 176, 272, 320-322, 328
  Alternative entry 56, 68
  Dangers for male 42, 276, 321
  Double penis 18, 41
  For nine hours 220
  In flight 18, 20, 168, 286
  Indirect sperm transfer 14, 303, 304, 322, 327, 328
  Nuptial gift 29, 44, 129, 220, 246, 249, 286, 290
  With flowers 277, 299, 300
Cordulegastridae 20
Corduliidae 21
Cordyligaster 300
Cordylochernes 305
Coreidae 71-73
Corixidae 61, 62
Corn 74, 80, 140, 146, 217
Cornitermes 46
Cornops aquaticum 37
Coronigonalia 80
Corrachia leucoplaga 240
Corydalidae 25
Corydalus 25
Cosmetidae 306
Cosmopolites sordidus 146
Cotesia flavipes 166
Cotton 149
Cotton boll weevil 149
Crabronidae 171, 172
Crabs 298, 332, 333, 335-341
Crackers 262-264
Crambidae 208, 209
Crane flies 270, 276
Cratosomus 150
Crayfish 332, 333
Creeping water bugs 61, 62
Crematogaster 87, 142, 191
Crematogaster ampla 192
  levior 195
Creonpyge creon 227
Cribellate orb-weavers 317-318
Crickets 5, 28, 32-34, 56, 171, 172, 299, 307
Crustacea 2, 302, 325, 330-341
Cryphocricos 62

Cryptinae 163
Cryptocephalinae 138
Cryptocephalus trizonatus 138
Cryptoses 209
Ctenidae 309, 314, 315
Ctenocephalides canis 58
  felis 58
Ctenosoma 101
Ctenostoma longipalpe 102
Ctenus 314
Cuckoo wasps 167
Cucurbitaceae 85, 140, 177, 208, 224, 258, 291
Cucurbitacins 224
Culex 272
Culex quinquefasciatus 272
Culicidae 272
Culicoides 275
Culicoides furens 275
Cupido 235
Cupiennius 314
Cupiennius getazi 314, 315
Curculionidae 146-152
Curculioninae 149, 151
Curculionoidea 1, 97, 144-152
Cutworms 217, 218
Cyanide 224, 259, 328
Cybister festae 99
Cycads 122, 224, 235, 236
Cycasin 224
Cyclanthaceae 149
Cyclanthura 149, 151
Cyclocephala 112
Cyclocephalini 111
Cycloneda sanguinea 125
Cyclosa 318
Cydia deshaisiana 207
Cydnidae 70
Cylindrical leaf beetles 138
Cynea cynea 227
Cynipoidea 158
Cyphonia 86
Cyrestinae 265-266
Cyrtacanthacridinae 38
Cyrtomenus teter 70
Cyrtophora citricola 318
Dactylopiidae 95-96
Dactylopius coccus 95, 96
Daddy long-legs 270, 305
Daddy long-legs spiders 309, 310, 313
Daggerwings 265, 266
Dalbulus maidis 80
Dalceridae 206
Dalcerides 206
Dalla eryonas 227
Damselflies 17, 20, 22
Danaini 8, 246-248
Danaus 246

Danaus plexippus 246-248
Dance flies 286
Dargida 218
Darkling beetles 126, 127
Darninae 86
Decapoda 332-341
Deer flies 282
Delphacidae 57, 74
Deltochilum mexicanum 107
Deltochilum valgum 107
Demodicidae 326
Dendromyrmex 195
Dengue fever 9, 273, 274
Depressariidae 204
Deraeocorinae 66
Derbidae 74
Dermanyssidae 322, 326
Dermanyssus 326
Dermaptera 3, 41
Dermatophagoides 326
Detritivora hermodora 239
Diabrotica 140
Diactor bilineatus 71
Diaethria 262
Diaethria astala 264
Diaethria eupepla 264
Diaphania 208
Diaphania plumbidorsalis 209
Diapheromeridae 39
Diaphorina citri 88
Diareusa 76
Diaspididae 93, 94
Dichotomius annae 107
Dichrocheles 325
Diclidophlebia lucens 88
Dictyla monotropidia 67
Dictyopharidae 167
Digging behavior 32, 53, 70, 77,
    78, 101, 104, 106, 107, 168,
    177, 184, 285, 331, 333, 336-
    338, 340
Dilobitarsus bidens 115
Dilobopterus 81
Dineutus 99
Dinoponera 186
Dioclea 132
Diogmites 285
Dione juno 258
Dione moneta 259
Dioptinae 216, 217
Diplopoda 2, 63, 107, 119, 224,
    287, 298, 307, 323, 325, 328,
    329, 340
Diplura 2, 3, 15
Dipluridae 312
Diptera 3, 269-301
Dircenna klugii 249
Disersus 100
Dismorphia theucharila 231

theucharila fortunata 8
Dismorphiinae 231
Disonycha reticollis 141
Disonycha trifasciata 142, 143
Diyllus maximus 30
Dobsonflies 3, 25
Dogielinotidae 330
Dolichoderinae 194
Dolichopodidae 287
Domatia 324
Dorylinae 186-188
Doxocopa cyane 197
Dragonflies 4, 17, 20, 21, 276
Drake, Francis 273, 274
Drosophila melanogaster 293
    willistoni 293
Drosophilidae 293
Druciella 198
Dryas iulia moderata 260
Dryinidae 167
Drymophilacris 38
Dryophthorinae 146, 147
Dryopidae 98
Dung beetles 106, 107
Dung feeders 106, 107, 209, 289,
    292, 295, 298
Dyes 95, 96
Dynamine 262
Dynastes hercules 111, 112
Dynastinae 111, 112
Dynastor 254
Dysmicoccus brevipes 93
Dysschema jansonis 8
Dystus puberulus 70
Dytiscidae 98, 99
Earthworms 46, 116, 298, 328
Earwigs 3, 41
Eccritosia zamon 285
Eccritotarsini 66
Echinargus isola 235
Echinotheridion 321
Echinothrips 51
Eciton burchellii 187
    hamatum 187, 188
Ectatomma 86
Ectatomma ruidum 184
Ectatomminae 184
Ectobiidae 44, 45
Ectrichodiinae 63
Edessa arabs 69
Egg case (ootheca) 28, 42-45,
    157
Egg mass 23, 25, 26, 61, 64, 69,
    70, 76, 83, 86, 94, 154, 207,
    258
Elaeidobius kamerunicus 149
Elateridae 114-116, 118
Elmidae 100
Elmoparnus 98

Eloria 219
Embates 149
Embioptera 3, 40
Emblemasoma 298
Emerald ash borer 113
Emerald moths 213, 214
Emesinae 63, 64
Emesis lucinda 239
Empididae 286
Encephalitis 273
Enchophora sanguinea 87
Encyrtidae 159, 160
Encyrtus infelix 160
Endomychidae 124
Ennya chrysura 85
    pacifica 85
Ensifera 28
Ensign wasps 157
Entiminae 147, 148
Entomobryidae 14, 15
Eois 213
Epalpus 299
Ephemeroptera 3, 18, 19
Epia 210
Epicauta 128
Epilachna abrupta 125
Epilachnini 124, 125
Epimecis 214
Epiphytes 185, 195, 323
Epipleminae 215, 216
Epiponini 174
Epipyropidae 75
Episactidae 38
Episactus tristani 38
Erateina 214
Erebid moths 218-222
Erechtia 85
Eresia ithomioides alsina 267
Eriophora 170, 309
Eriophyoidea 323, 324
Erotylidae 122, 123
Erythraeidae 324, 325
Erythrodiplax fervida 21
    funerea 21
Escovopsis 188
Eucharitidae 159, 160
Eucheira socialis 200
Euchroma gigantea 113
Eucnemidae 115
Eucynorta 306
Eueides 258
Euglossa 179
Euglossini 178-180
Euglyphis 209
Eulaema 179
Eulepidotis 219
Eulophidae 161
Eumacronychia sternalis 298
Eumaeus 224, 235

Eumaeus godarti 236
Eumeninae 173, 174, 325
Eumolpinae 133
Eumorpha satellitia 213
Euphalerus 89
Euphorbiaceae 52, 70, 207, 215, 238, 251, 252, 262, 324
Euplectrus 161
Euptychia westwoodi 253
Eurhinus magnificus 151
Eurtytides 228
Eurybia 238
Eurybiini 238
Eurycotis biolleyi 45
Eurypterids 302
Euryrhynchidae 334
Eurytomidae 156
Euschistus heros 69
Euselasia bettina 240
    chrysippe 200, 240
Euselasiinae 238, 240
Eusocial insects 8, 46, 47, 172-178, 180-195
Euthyrhynchus floridanus 70
Evaniella 157
Evaniidae 157
Evolutionary tree 3, 97, 185, 249, 328
Exochrysis 167
Exophthalmus 147
Extinct arthropods 2, 4, 53, 215, 302
Eyelash mites 322, 326
Fabaceae 39, 71, 74, 77, 79, 83, 84, 88, 89, 124, 132, 133, 140, 144, 204, 211, 216-218, 228, 231, 237, 239, 241, 256, 257
Fairyfly 159
False click beetles 115
False eye spots 75, 210-213, 230, 235, 236, 252-257, 267
False head 79, 207, 236
Fast flight 211
Feather mites 322, 326
Females that resemble larvae 119, 199
Ferns 92, 278, 279
Fiddler crabs 336-338
Fidena trapidoi 282
Fig trees 51, 70, 71, 76, 86, 90, 104, 130, 160-163, 165, 208, 246, 266, 305
Fig wasps 160-163
Finlay, Carlos 274
Fire ants 192
Fireflies 116-118, 224
Fish 7, 19, 61, 99, 298, 330, 331, 333, 340

Flannel moths 9, 204
Flatidae 76
Flea beetles 141-143
Fleas 3, 7, 58
Flesh flies 298
Flies 3, 269-301
Flight 17
Flour beetle 127
Flower flies 288, 289
Flower mites 322
Fluorescent exoskeleton 304
Forcipomyia 275, 276
Forensic entomology 295
Forficulidae 41
Formic acid 101, 180, 195, 216
Formicinae 195
Frankliniella diversa 51
    insularis 51
Freshwater crabs 336
Fritillaries 258
Froghoppers 78, 79
Frogs 7, 29, 61, 129, 272, 293, 298, 314, 323
Fruit flies 165, 290, 291
Fulgora 74, 75
Fulgora lampetis 75
    laternaria 75
Fulgoridae 74-76, 87
Fulgoroidea 57, 74-76, 87, 167
Fungal infection, prevention of 41, 88, 107, 325
Fungus cultivation 148, 188-191
Fungus feeders 14, 15, 40, 49, 51, 74, 104, 107,114, 121-124, 126, 144, 148, 188-191, 198, 199, 253, 254, 256, 277-280, 287, 293, 306, 323, 324, 326, 328, 329
Fungus gnats 277, 278
Fungus moths 198, 199
Funnel-web tarantulas 310, 312
Galapagos 192, 340
Galerita 101
Galerucinae 140-143, 224
Gall formers 7, 51, 52, 88, 89, 114, 151, 156, 158, 159, 161-163, 165, 203, 208, 269, 276, 277, 279-281, 291, 321, 323, 324,
Gall midges 279-281
Gall mites 323, 324
Gall wasps 158
Gargaphia 67
Gasteracantha 318
Gasteracantha cancriformis 319
Gecarcinid crabs 338, 339
Gecarcinus 339
Gecarcinus lateralis 339
    ruricola 338

Gelastocoridae 61
Gelechiidae 203, 204
Gelechioidea 203, 204
Geometer moths 213, 214, 224
Geometridae 213-215, 224
Geophilomorpha 327
Geranomyia recondita 270
Gerridae 60, 61
Gesneriaceae 249
Ghilianella 64
Ghost crabs 337, 338
Ghost moths 197, 198
Giant Amazon ants 186
Giant silk moths 9, 209-211
Giant water bugs 61, 62
Gibbomyzus pteridophytorum 92
Gills 18-20, 23, 62, 98, 332, 333, 336
Glass shrimp 333, 334
Glasswings 224, 249, 250
Gloveria 209
Gloveria psidii 200
Glowworm beetles 118, 119
Glucosinolates 224
Glyphipterigidae 197, 201
Glyptapanteles 165
Golden silk orb-weavers 319, 320
Goliath bird-eating spider 312
Gomphidae 20
Gomphocerinae 37
Gonatocerus 160
Gondwana 10, 11
Goniopsis cruentata 340
    pulchra 340
Gonodonta 218, 219
Gracillariidae 201, 202
Grallipeza 290
Graphocephala albomaculata 82
    coccinea 82
    coronella 82
Grapsid crabs 340
Grapsus grapsus 340
Grasses 71, 74, 79, 218, 226, 227, 253, 254
Grasshoppers 4, 28, 33, 37-38, 63, 128, 284, 298, 299
Great American Biotic Interchange 12
Green lacewing 54, 55
Gribbles 331
Ground beetles 101, 102
Gryllidae 5, 28, 32-34, 56, 171, 172, 299, 307
Gryllotalpidae 32
Gryon 159
Guava 138, 144
Guayaquila 86
Gunneraceae 134, 141, 143, 298

Gymnandrosoma aurantianum 207
Gynaikothrips ficorum 51
  uzeli 51
Gypsy moth 219
Gyrinidae 99
Gyrinus 99
Haemagogus 272
Hairstreaks 235-237
Halictid bees 175, 176, 178, 325
Halobates 60
Halysidota 222
Hamadryas 262, 264
Hamadryas amphinome 264
  amphinome mexicana 263
  februa ferentina 263
  guatemalena 264
Handsome fungus beetles 124
Hardingia maximespina 217
Harlequin beetle 130, 305
Harlequin bug 224
Harmonia axyridis 124, 125
Harpactorini 63, 64
Harpalinae 101, 102
Harroweria 31
Harvestmen 305, 306, 325
Hawk moths 211-213, 224
Hedylidae 225
Hegemona 127
Heilipus 149, 150
Heilipus areolatus 150
Heliconia 66, 71, 79, 133, 134, 137, 217, 254, 284, 298, 314
Heliconiini 258-260
Heliconius 258, 259
Heliconius charithonia 259
  erato 258
  erato petiverana 260
  erato venustus 8
  hecale 223
  melpomene aglaope 8
Heliothrips haemorrhoidalis 52
Helleriella 149
Hellgrammites 25
Hemerobiidae 54, 55
Hemeroplanes triptolemus 212, 213
Hemileucinae 210, 211
Hemiptera 3, 59-96
Hemoglobin 62, 277
Hepialidae 197, 198
Heraclides 229, 230
Heraclides torquatus 8
Hermetia illucens 283, 284
Hermeuptychia harmonia 253
Hermit crabs 14, 335
Hesperiidae 225-228
Hesperocharis costaricensis 234

Hetaerina 22
Heterochroma 218
Heteronemia grande 39
Heteroptera 57, 59-73, 247, 299
Heterosternus oberthuri 105
Hexapoda 2
Heza 64
Hilarographa 207
Hipocrita drucei 221
Hippoboscidae 294
Hippolytidae 334
Hirtodrosophila batracida 293
Historis 245, 267
Historis acheronta 245
Hitchhiker ants 189, 190, 287, 288
Hitching a ride 56, 128, 159, 304, 305, 323-325
Homoeocera 221
Honey 182
Honey bee 175, 176, 182, 183
Honeydew 83, 87, 88, 91, 93
  As food 23, 54, 87, 120, 185, 191, 194, 195, 287, 289, 290, 299
  Used to attract prey 289
Hoplocopturus varipes 151
Hoplomutilla xanthocerata 169
Horned scarabs 111, 112
Hornets 173
Hornworms 212
Horse flies 282
Horseshoe crabs 303
House dust mites 322, 326
House fly 294, 295
Hovering 178, 179, 212, 282, 284, 288, 293, 299
Human bot fly 9, 301
Humboldt, Alexander 118
Hummingbirds 11, 17, 87, 178, 323, 324
Hyalella 330
Hyalurga 220
Hybosciara gigantea 278
Hybotidae 286
Hydrophilidae 98
Hydropsychidae 18, 26, 27
Hylaeogena 113, 114
Hymenaea courbaril 74
Hymenoptera 3, 153-195
Hyperinae 151
Hyphinoe 85
Hypna clytemnestra 251
Hypolimnas misippus 267
Hypothenemus hampei 152
Hypselonotus concinnus 72
Iaspis andersoni 236
Ichneumonid wasps 163, 164

Idarnes 163
Impatiens 227
Inbreeding 148, 152, 162, 176, 321
Inchworms 161, 213-215, 224
Indian paintbrush 267
Inga 39, 144, 216, 231
Intertidal zone 14, 104, 270, 336, 338, 340
Introduced arthropods 51, 68, 88, 92, 124, 166, 173, 192, 194, 202, 219, 267, 274, 291, 292, 294, 296, 313, 318, 321, 327, 331, 333
Iphiclus catillifer 123
Iphirhina 79
Iresine 201, 276
Iridoid glycosides 224, 267
Isa diana 205
Ischioscia 332
Ischnocera 50
Ischnocodia annulus 136
Isobutyric acid 228
Isopoda 331, 332
Isoptera 3, 16, 46-48
Isotes marginella 140
  sexpunctata 141
Issidae 76
Ithomia heraldica 249, 250
  patilla 249
Ithomiini 8, 249, 250
Ixodida 322, 325, 326
Jamaican bromeliad crab 340
Japygoidea 15
Jemadia scomber 226
Johngarthia 339
Jumping bristletails 3, 16
Jumping plant lice 88, 89, 279, 280
Jumping spiders 3, 207, 208, 316
June beetles 108
Junonia 267
Junonia evarete 267
Junonini 267
Kaira 318
Kalotermitidae 46, 47
Kapala 159
Katydids 4, 5, 28-31, 33, 34, 56, 159, 172, 276, 299
Keroplatidae 118, 277, 278
Kerria lacca 96
Kerriidae 96
Kissing bugs 7, 9, 65
Kisutam syllis 236
Kleptoparasite 128, 167, 170, 171, 177, 287, 298, 320
Labania minuta 165
Labidus praedator 187
Laboulbeniales 124

Lace bugs 67
Lacquer 93, 96
Ladoffa 82
Lady beetles 91, 124, 125
Laelapidae 325
Lagocheirus kathleenae 131
Lake Titicaca 330
Lampria 285
Lamprosomatinae 138
Lampruna rosea 222
Lampyridae 116-118
Landhoppers 330, 331
Languriinae 122, 123
Lantana 67, 258, 260
Laothus oceia 237
Largest body 111
Largest wingspan 218, 219
Lasiocampidae 200, 209
Latrodectus 309, 321
Latrodectus geometricus 321
Lauraceae 67, 80, 88, 144, 149,
    150, 205, 222, 228, 230,
    252, 279
Leaf beetles 132-143
Leaf butterflies 250-252
Leaf miner flies 292, 293
Leaf miner moths 201, 202
Leaf miners 7, 113, 114, 133,
    134, 151, 159, 201, 202, 203,
    208, 277, 278, 288, 292, 293
Leaf tiers 144, 203, 207, 208, 226,
    227, 252
Leafcutter ants 175, 176, 188-
    191
Leafcutter bees 177
Leaf-footed bugs 71-73
Leafhoppers 57, 59, 74, 80-82,
    87, 167
Leaf-rolling weevils 144, 145
Lebia 101
Lebiini 101, 102
Leishmania 271
Leishmaniasis 9, 271
Leistotrophus versicolor 104
Lek (see swarm) 249, 283
Lemoniadina 239
Leodonta tellane chiriquensis
    231
Lepanthes glicensteinii 277
Lepidoptera 3, 196-268
Lepismatidae 16
Leptobasis vacillans 22
Leptoceridae 27
Leptocircini 228
Leptoconops 275
Leptogenys 332
Leptoglossus zonatus 73
Leptonema 26, 27
Leptopyrgota 291

Leptoscelis tricolor 71
Lestidae 20
Lestrimelitta 181
Lethocerus 61, 62
Letis mycerina 218
Leucanella hosmera 210
Leucauge mariana 320
Leuciris 214
Leucoagaricus gonglyophorus
    188
Leucochrysa 55
Libellulidae 17, 20, 21
Lice 3, 7, 50, 104
Lichenomorphus 31
Lichens 16, 29, 31, 42, 130, 199,
    213, 220, 270
Lieinix 231
Limacodidae 6, 204-206
Limenitidinae 261, 262
Limenitis 261
Limnesiidae 322
Limnocoris 62
Limnoriidae 331
Linyphiidae 311, 317
Liposcelis 49
Liriomyza huidobrensis 292
    sativae 292
    trifolii 292
Lissoderes 149
Listroscelidinae 29
Lithobiomorpha 327
Liturgusa 42
Liturgusidae 42
Live birth (viviparity) 5, 92, 304
Liverworts 42, 270
Lixinae 151
Lixus 151
Llaveia axin 95, 96
Locusts 37, 38
Long tongue 179, 211, 212, 238
Longevity 7, 190, 312
Longhorn beetles 48, 130, 131
Long-jawed orb-weavers 320
Long-legged flies 287
Long-legged water spiders 314
Loranthaceae 201, 231, 237
Louse flies 294
Loxosceles 309
Lubber grasshoppers 28, 36
Lucibufagins 116
Lucilia sericata 296
Lurio solennis 3
Lutzomyia 271
Lutzomyia longipalpis 271
Lycaenidae 165, 235-237
Lycidae 44, 120
Lycorea 246
Lycorea halia 8
Lycosidae 314, 315

Lygaeidae 71, 224
Lymantriinae 219, 220
Lynx spider 315
Lyssomanes 316
Lytta 128, 129
Macadamia nut borer 207
Macraspis 110
Macrobrachium hancocki 334
    heterochirus 332
    rosenbergii 333
Macrodactylus 108
Macrohaltica 141, 143
Macronyssidae 322, 326
Macrostemum 18
Maggot therapy 295, 296
Mahanarva 79
Malaria 9, 273, 274
Males, haploid 51, 90, 148, 153,
    176, 183, 184
Mallophora 285
Malpighiaceae 67
Malpighian tubules 4, 53
Malvaceae 113, 121, 138, 139,
    142
Mammals 11, 12, 49, 50, 58, 65,
    103, 106, 225, 262, 271, 272,
    275, 282, 294, 297, 301, 304,
    325, 326
Manaosbiidae 306
Manataria maculata 252, 253
Manduca florestan 213
Mange 325
Mango 276, 293
Mangrove tree crab 340, 341
Mangroves 60, 148, 275, 331,
    334, 336, 340
Mantidae 43
Mantises 3, 42, 43, 57, 129, 298,
    299
Mantisflies 56
Mantispidae 56
Mantodea 3, 42, 43, 129, 298,
    299
Mantoida 43
Mantoididae 43
Marantaceae 102, 133, 238, 254
Marellia remipes 37
Markia hystrix 31
Marpesia marcella valetta 265
    merops 266
    petreus 266
    zerynthia dentigera 265
Martialis heureka 185
Maruina 270, 271
Masarinae 173
Mastophora 318
Maternal care 32, 41, 83-86, 122,
    133, 153, 304-306, 307, 314,
    321, 340

Mayans 126
Mayflies 3, 18, 19
Mealybugs 93, 94, 124, 159, 194, 195
Mechanitis menapis saturata 250
   polymnia 250
Mecoptera 3, 57
Megachilidae 177
Megaloblatta 44
Megaloblatta blaberoides 45
Megaloprepus caerulatus 22
Megalopta genalis 178
Megaloptera 3, 25
Megalopyge 204
Megalopygidae 204
Megaphobema mesomelas 312
Megapodagrionidae 20
Megascelis 133
Megaselia scalaris 287
Megasoma elephas 111, 112
Melanagromyza rosales 292
Melanchroia 215
Melastomataceae 52, 86, 88, 151, 203, 217, 219, 240
Melete leucanthe 232
Meliaceae 108, 133
Melicharidae 323
Meliponini 180, 181
Melipotis fasciolaris 219
Melitaeini 267-268
Meloe 129
Meloidae 128, 129, 224
Melolonthinae 108
Membracidae 83-87, 160, 238, 325
Membracis dorsata 84
   mexicana 84
Memphis pithyusa 252
   proserpina 251
Merguia rhizophorae 334
Merosargus 283, 284
Mesembrinella 296
Mesembrinellidae 296
Mesomphaliini 134-137
Mesopteron 120
Mesosemia telegone 239
Mesostigmata 322-325
Metaleptea brevicornis 37
Metallic wood-boring beetles 113, 114
Metalmark butterflies 238-242
Metalmark moths 207, 208
Metamorphosis 3, 6, 28, 53, 57, 91, 97, 153, 196, 269
Metepiera 318
Methylbutyric acid 228
Metopaulias depressus 340
Metresura 41
Mexican jumping bean 207

Miagrammopes 318
Miconia 88, 151, 205, 217, 237, 240
Micrathena 318
Micrathena sexspinosa 319
Microgastrinae 164-166
Microparsus olivei 92
Micropeza 290
Micropezidae 290
Micropterigidae 198
Microstigmus 172
Migration 20, 38, 215, 231, 246-249, 267, 334, 339
Milkweed butterflies 224, 246-248
Milkweeds 71, 88, 130, 246, 248
Milky latex 212
Millipedes 63, 107, 119, 224, 287, 298, 307, 323, 325, 328, 329, 340
Mimetica crennulata 30
Mimicry (see camouflage) 8, 207
   Chemical 72, 185, 318
   Models 120, 168
   Of ants 71, 73, 86, 104, 149, 290
   Of bees 64, 285, 288, 289
   Of beetles 44
   Of jumping spider 207, 208
   Of Lepidoptera 207, 216, 220, 225, 228, 231, 238, 249, 258, 267
   Of snakes 211-213, 230
   Of stinging wasps 30, 56, 63, 163, 133, 220, 221, 270, 283, 288, 290
Mimoides 228
Miridae 66
Mischocyttarus 174
Mistletoes 231
Mite leaf miner 202
Mites 9, 23, 124, 182, 279-281, 287, 303, 321-326
Mixothrips 52
Mole cricket 32
Molippa nibasu 211
Molting 5, 6, 14, 15, 18, 43, 46, 48, 51, 78, 80, 133, 269, 304, 307, 312, 319, 324-328, 334, 337
Molytinae 149, 150
Mompha 203
Momphidae 203
Monarch butterfly 246-248
Moncheca 29
Moncheca elegans 31
Monkeys 11, 29, 106, 272, 301, 328
Monophlebidae 93-96, 125
Moraceae (see fig trees) 88, 89, 265, 266

Morphini 256, 257
Morpho 256
Morpho cypris 257
   helenor 6
   helenor narcissus 243, 257
   menelaus 257
Mosquitos 9, 17, 272-274
Mosses 213, 261, 270, 287
Moth attraction to lights 197
Moth flies 270, 271, 294
Moths 197-222
Mottled shore crab 340
Mud daubers 171, 172
Musca domestica 294, 295
Muscidae 294, 295
Mushrooms 104, 118, 121, 123, 278
Mutillidae 168, 169
Mutualism 7, 185, 193-195, 324
Mycetophilidae 277, 278
Mymaridae 84, 159, 160
Myndus crudus 74
Myriapoda 2, 302, 327-329
Myricaceae 92
Myrmecopsis 221
Myrmelachista 242
Myrmeleon 53
Myrmeleontidae 53
Myrmicinae 188-192
Myrtaceae 88
Myscelia cyaniris 245
Myxomycete amoebae 120
Nannotrigona 181
Naprepa houla 217
Nascus 228
Nasutitermes 46-48
Natada 205
Naucoridae 61, 62
Neacoryphus bicrucis 71
Neanuridae 14
Nematodes 7, 40, 146, 273, 275, 280, 323
Nemobiinae 32
Nemorimyza 293
Neocococytius cluentius 212
Neocordulia batesii 21
Neodohrniphora curvinervis 288
Neoponera 186
Neoptera 17
Neoptychodes cretatus 131
   trilineatus 130
Neorileya 156
Neostictoptera 218
Neotorrenticola 322
Nephila clavipes 319, 320
Nephilidae 319, 320
Nephrotoma 270
Neptopsyche 27
Neruda 258

Nest 41, 46, 47, 107, 155, 169-195, 306, 328
Netelia 163, 164
Net-winged beetles 44, 120
Neuroptera 3, 53-56
Nezara viridula 69
Nica flavilla canthara 264
Nicrophorus quadrimaculatus 103
Nitidulidae 121
Noctuidae 165, 217, 218, 224
Nolidae 218
Nomamyrmex esenbeckii 188
Non-biting midges 276, 277, 294
No-see-ums 275
Notochaeta bufonivora 298
Notodontidae 216, 217
Notonectidae 61, 62
Number butterflies 262, 264
Nursery web spiders 314, 315
Nycteribiidae 294
Nymphaeaceae 112
Nymphalidae 244-268
Nymphalinae 267, 268
Nymphalini 244, 267
Nymphidiini 238, 239, 242
Oaks 158, 209, 224
Ocypode 338
Ocypode gaudichaudii 338
    occidentalis 337, 338
Ocypodidae 336-338
Odonata 3, 17, 20-22
Odontomachus 184, 186
Odontoptera carrenoi 75
Oebalus poecilus 69
Oecanthinae 32
Oedemeridae 129
Oedipodinae 37, 38
Oenomaus ortygnus 235
Ogdoecosta catenulata 134
Oiketicinae 199
Oleria paula 8
Omaspides convexicollis 137
Omolabus conicollis 144
    corvinus 144
Omophoita 142
Omphalea 215
Onchocerciasis 275
Oniscidea 331, 332
Oospila 214
Opiliones 305, 306
Opsiphanes 254
Opsiphanes bogotanus 255
Orb-weaver spiders 54, 164, 170, 307, 308, 318, 319
Orchid bees 178-180
Orchids 11, 146, 149, 178, 211, 277, 299, 300

Oressinoma typhia 253
Oribatida 322, 323
Ornidia 289
Ornithonyssus 326
Orophus 31
Orthemis 21
Orthemis discolor 17
Ortheziidae 94
Orthoptera 3, 28-38
Orthotylini 66
Osmeterium 228, 230
Osoriinae 104
Oval leaf beetles 133
Ovipositor 4, 15, 20, 28-30, 32, 83, 153-155, 159, 163, 164, 306
Owl butterflies 254, 255, 300
Owlet moths 217, 218, 224, 325
Owlflies 53, 54
Oxelytrum 103
Oxelytrum discicolle 103
Oxeoschistus tauropolis 253
Oxycheila 101
Oxydia 215
Oxyopidae 315
Oxyporus 104
Oxysarcodexia 298
Oxyteninae 209, 211
Oxytenis modestia 211
Oxytrigona 180
Ozophora 71
Pachycoris torridus 70
Pachygrapsus transversus 340
Paederus 104
Painted ladies 267
Palaemonidae 332-334
Paleoptera 17
Palm weevils 146, 147
Palms 74, 112, 121, 133, 146, 149, 204, 206, 217, 254-256
Palpada 289
Panchlora 44, 45
Panthiades bathildis 236
Pantophthalmidae 283
Pantophthalmus bellardii 283
    planiventris 283
Pantoteles 149
Papaveraceae 51, 92, 218, 324
Papaya fruit fly 290, 291
Paper wasps 57, 155, 173-176, 187, 188, 208, 247
Papilio 228
Papilio anchisiades 229
    cresphontes 229
    thoas 229, 230
    garamas 229
    menatius cleotas 230
Papilionidae 8, 228-230

Papilionini 228-230
Paragrallomyia 290
Paranapiacaba 140
Parancistrocerus declivatus 325
Parandra polita 131
Paraphidippus 316
Paraphrynus laevifrons 307
Paraponera clavata 185, 186
Paraponerinae 185, 186
Parasa macrodonta 205
Parasites 7, 23, 50, 56, 58, 65, 93, 182, 188, 271, 273, 276, 293, 294, 301, 321, 323-326, 328, 331
Parasitoid wasps 155-170
Parasitoids 7, 39, 91, 101, 155-161, 163-170, 202, 224, 269, 292, 298, 299, 332
    Ants 160, 287, 288
    Aphids 88, 165, 279
    Beetles 101, 157, 160, 165, 168, 196, 284, 285, 287, 291, 298
    Caterpillars 160, 161, 163-165, 284, 299, 300
    Cicadas 298
    Cockroaches 171, 299
    Crickets, grasshoppers 298, 299
    Eggs 69, 80, 83, 84, 156, 157, 159, 160, 167
    Gall midges 161
    Mantises 298, 299
    Millipedes 287, 298
    Other parasitoids 156
    Planthoppers 167
    Scale insects 93, 160
    Spiders 164, 169, 170, 309, 320
    True bugs 299
    Wasps and bees 167-169, 284, 287, 288
    Whiteflies 160
Paratemnoides nidificator 305
Paratropes bilunata 44, 45
Parawixia bistriata 318
Parides 228
Parides iphidamas 229
Paromenia isabellina 81
Passalidae 105, 106
Passalus punctatostriatus 106
Passifloraceae 71, 72, 258, 259
Passion-vine butterflies 224, 258-260
Paternal care 61, 62, 306
Paulinia acuminata 37
Peach palm 149
Peanut-headed bug 74, 75

Pediculus humanus 50
Pedipalp 303, 304, 307, 308
Pelecinid wasps 157
Pelecinus polyturator 157
Pentatomidae 60, 69, 70
Pentobesa 216
Pepsis 30, 63, 169, 170
Pergidae 154
Peridinetus 149
Peridroma semidolens 218
Perigonia stulta 213
Perisceptis carnivora 199
Perlidae 23
Perola producta 204
Perreyia tropica 154
Peucetia 315
Phaea 130
Phalangogonia sperata 110
Phalangopsinae 32, 33
Phaneropterinae 28, 30, 31, 34
Phanocles costaricensis 39
Pharaxonothinae 122
Phasmida 3, 39, 276, 299
Pheidole 191
Phelypera distigma 151
Phengodidae 118, 119
Phereoeca uterella 198
Pheromones 5, 26, 57, 69, 91,
116, 185, 197, 199, 220, 238,
246, 249, 250, 279, 290, 318
Phiale formosa 316
Philophyllia 30
Philornis 294
Phlaeothripidae 52
Phobetron 204
Phobetron hipparchia 206
Phoebis sennae 234
Pholcidae 313
Phoneutria 309, 314
Phoridae 190, 287, 288
Photinus 117
Photonic crystals 243
Photuris 116
Photuris crassa 117
Phrenapates 126
Phrictus quinquepartitus 75
Phrygionis polita 214
Phrynidae 306, 307
Phrynus pseudoparvulus 306
whitei 307
Phthiraptera 3, 7, 50
Phylinae 66
Phyllanthaceae 215
Phyllocnistis 201
Phyllocnistis citrella 202
maxberryi 201
Phyllocoptruta oleivora 323
Phylloicus 27

Phyllophaga 108
Phyllovates 43
Phymacysta 67
Physonota alutacea 137
Phytoseiidae 322, 324
Pieridae 8, 200, 231-234
Pierinae 231-233
Piezodorus guildinii 69
Piezogaster chontalensis 71, 72
Pillbugs 331
Pimplinae 164
Pincher wasps 167
Pineapple 93, 235, 295
Piperaceae 79, 149, 213, 228,
230, 280, 281
Pisauridae 314, 315
Plant bugs 66
Plant chemical defenses 223, 224
Planthoppers 57, 74-76, 87, 167
Plant-sap (phloem) feeders 59,
74, 80, 83, 87, 88, 91, 93
Plant-sap (xylem) feeders 59,
74, 77-80
Plasmodium 273
Platycoelia humeralis 110
Platydesmidae 329
Platygastridae 159
Platyphyllini 30
Platyrhacidae 328
Pleasing fungus beetles 122, 123
Plecoptera 3, 23
Plesiochrysa 55
Poassa limbata 306
Poecilopompilus 170
Poecilopsocus 49
Polistes 174
Polistes erythrocephala 173
Polistinae 173-176
Pollen wasps 173
Pollination 7, 51, 111, 112, 121,
122, 149, 161-163, 175, 177,
178, 180, 211, 258, 269, 276,
277, 289
Polybia 174-176
Polycyrtus melanoleucus 163
Polydesmida 328, 329
Polyglypta 85
Polygonaceae 216
Polyspinctini 164
Polyxenida 328, 329
Polyzoniida 328
Pomace flies 293
Pompilidae 169, 170
Ponerinae 184, 186
Potato 139, 204
Potter wasps 173, 174, 325
Praeacedes atomosella 198
Predaceous diving beetles 98, 99

Predators 7, 15, 25, 29, 41, 42,
51, 53, 61-63, 66, 69, 70, 98,
99, 101, 104, 115, 116, 120,
170-174, 185, 276, 282, 285-
287, 303, 307-321, 323-327
Cockroach eggs 157
Eusocial insect larvae 186-188
Grasshopper eggs 128, 284
Millipedes 107, 119
Snails 116, 117
Spider eggs 56, 298
Spiders 20, 169-170
Sternorrhyncha 54, 91, 124,
125, 289
Predators cooperating 186-188,
305, 318, 320, 321
Predatory wasps 170-174
Prepona praeneste isabelae 251
Primulaceae 214
Prionostemma 305
Procambarus clarkii 333
Processionary larvae 154, 210,
220, 222, 238, 240, 278
Prochoerodes 214
Proconiini 80
Proctolabinae 38
Prolegs 154, 196, 204, 213, 216,
217
Prominent moths 216, 217
Pronophila timanthes 254
Prosapia 79
Protesilaus 228
Protographium 228
Protopolybia 174
Protozoans 7, 9, 48, 65, 105, 118,
271, 273
Protura 2, 3, 15
Pselaphacus 122
Psephenidae 100
Pseudacysta perseae 67
Pseudatteria volcanica 207
Pseudococcidae 93, 94
Pseudolechriops 149
Pseudolmedia mollis 89
Pseudomethoca chontalensis 168
Pseudomops 44
Pseudomyrmecinae 193
Pseudomyrmex 193
Pseudomyrmex spinicola 72
Pseudophasmatidae 276
Pseudophera 80
Pseudophilothrips 52
Pseudophyllinae 29, 30
Pseudoplusia includens 165
Pseudoscorpions 304, 305
Pseudostigmatidae 20
Pseudothelphusidae 2, 336
Pseudoxycheila tarsalis 102

Psocidae 49
Psocoptera 3, 49
Psychidae 199
Psychoda 271
Psychodidae 9, 270, 271
Psylloidea 88, 89, 279, 280
Ptecticus 283
Pteronymia simplex 249
Pthirus pubis 50
Ptilosphen viriolatus 290
Pulex irritans 58
Pulicidae 58
Pygmy grasshopper 35
Pygmy mole cricket 35
Pyraloidea 208, 209
Pyrgotidae 291
Pyrgus adepta 226
Pyroglyphidae 326
Pyrophorini 114, 115, 118
Pyrophorus 115, 118
Pyrota 128
Raleigh, Sir Walter 243
Ranzovius 66
Raphidioptera 13
Rapid movement 186, 207, 338
Rat-tailed maggots 289
Recluse spiders 309
Recruitment 181, 185, 192
Red mangrove crab 340
Red rock crab 340
Reduviidae 9, 63-65
Reduviinae 63, 64
Reed, Walter 274
Regurgitation 103, 177, 212, 307
Rekoa marius 237
    palegon 237
Reptiles 7, 29, 225, 228, 272, 282,
    287, 294, 325
Resins 7, 64, 178, 180
Rhagovelia 61
Rhetus arcius 238
    dysoni 239
Rhinophoridae 332
Rhinostomus barbirostris 146
Rhinotermitidae 46
Rhizophoraceae 148, 331
Rhodochlora 214
Rhuda difficilis 217
Rhynchophorus palmarum 146,
    147
Rhyparochromidae 71
Rhyssomatus 150
Rice 69, 74, 127, 146
Rickettsia typhi 50
Rifargia 216
Riffle beetles 100
Riffle bugs 60, 61
Riodinidae 238-242

Ripipterygidae 35
Ripipteryx 35
Ripipteryx limbata 35
Robber flies 285
Rodents 11, 58, 103, 104, 271,
    298, 301, 304, 312
Romalaeidae 28, 36
Root feeders 32, 70, 74, 77, 108,
    109, 114, 115, 130, 133, 140,
    141, 147, 148, 195, 197,
    207, 208
Rosaceae 143
Rothschildia lebeau 211
    triloba 210
Rove beetles 104
Rubiaceae 66, 85, 114, 154, 211,
    212, 227, 258, 259, 261, 281
Rutaceae 88, 228
Rutelinae 105, 109, 110
Salticidae 3, 207, 208, 316
Salyavata variegata 63
Sand flies 9, 270, 271
Sand wasps 171, 172
Sandhoppers 330, 331
Sap beetles 121
Sapindaceae 228, 262
Sarcina purpurascens 220
Sarcofahrtiopsis thyropterontos
    298
Sarcophagidae 298
Sarcoptidae 325
Sarcoptiformes 322
Saturniidae 209-211
Saturniinae 209-211
Satyrinae 252-257
Satyrs 252-254
Sawflies 154
Scabies mites 322, 325
Scalaris 75
Scale insects 87, 93-96, 124, 159,
    195, 289
Scales on body 14, 15, 113, 147,
    196, 206, 207, 243, 250,
    286, 318
Scaptomyza 293
Scaptotrigona 181
Scarab beetles 105-112, 285
Scarabaeinae 106, 107
Scarabaeoidea 105-112
Scatopyrodes 131
Scelionid wasps 159
Sceliphron 171
Sceloenopla 134
Schistocerca 38
Schistocerca cancellata 38
    interrita 38
    piceifrons 38
Sciaridae 277-278

Sclerosomatidae 305
Scoliid wasps 168
Scoloderus 318
Scolopendra gigantea 327
Scolopendromorpha 327
Scolopocryptopidae 327
Scolytinae 148, 149, 152
Scolytodes 148
Scorpionflies 3, 57
Scorpions 9, 303, 304
Screwworms 296, 297
Scuds 330
Scutelleridae 70
Scutigera 327
Scutigeromorpha 327
Scymnus 125
Sea spiders 303
Sebastiania pavoniana 207
Seed beetles 132
Seed bugs 71, 224
Seed feeders 71, 132, 148-151,
    156, 191, 207
Selaginella 253
Sematura 215
Sematuridae 215
Semiotus 114
Semiotus cuspidatus 115
Semomesia croesus 239
Sepsidae 292
Sequential hermaphroditism 334
Sergestidae 332
Serritermitidae 46
Sesarma rubinofforum 341
Sesarma verleyi 340
Sesarmid crabs 340, 341
Setabis lagus 238
Shield-backed bug 70
Shining leaf chafers 105, 109,
    110
Shrimps 333, 334
Shrimps compared with prawns
    332
Sialidae 25
Sicariidae 309
Sigmatomera seguyi 270
Silk 14, 26, 27, 40, 49, 53, 195,
    198-200, 202, 204, 207-209,
    226, 228, 231, 234, 254, 261,
    275, 277, 286, 304, 307-314,
    316-320, 323, 327
Silkmoth 200
Silphidae 103
Silverfish 3, 16
Simopelta 186
Simuliidae 275
Simulium quadrivittatum 275
Siphonaptera 3, 7, 58

Siproeta epaphus 244
  stelenes 243
Sisters 261-262
Sitophilus 146
Sitotroga cerealella 203
Skeletonizing leaf beetles 140,
  141, 224
Skippers 225-228
Sleeping aggregation 258, 260,
  265, 290
Sloths 11, 12, 106, 209, 271
Slug caterpillars 9, 204-206
Smerdalea 86
Smeringopus pallidus 313
Smyrna 267
Snails 29, 87, 107, 116, 298, 333,
  335, 340
Snakes 103, 212, 213
Snout moths 208, 209
Solanaceae 36, 38, 87, 90, 124,
  125, 224, 249
Soldier beetles 120, 121
Soldier flies 283, 284
Solenopsis 192
Solenopsis geminata 192
  invicta 192
Songs 28, 29, 32-34, 54, 77
Sound perception 4, 28, 42, 197,
  211, 220, 253, 325
Sound production 5, 28, 33,
  36, 37, 64, 74, 77, 78, 106,
  114, 131, 218, 220, 262, 272,
  283, 290
Sowbugs 331, 332
Soybeans 69
Spaethiella 137
Spanish empire 95, 96
Spanish fly 129
Spermatophore 14, 29, 30, 129,
  220, 246, 259, 304, 307, 322
Sphaeroma peruvianum 331
  terebrans 331
Sphaeromatidae 331
Sphecidae 171, 172
Sphecomyiella 291
Sphex 172
Sphingidae 211-213, 224
Spider egg sac 56, 309, 312, 314,
  318, 321
Spider mites 323
Spider wasps 169-170
Spider webs 20, 54, 66, 196, 199,
  270, 310, 311, 313, 317-321
  Communal web 317, 318,
  320, 321
  Stabilimentum 311, 317
Spiders 3, 20, 129, 200, 212, 220,
  303, 307-321

Spilomelinae 208
Spinneret 200, 307, 309, 310,
  314, 319
Spiracle 4, 28, 33, 36
Spirobolida 328
Spirostrepsida 328
Spittlebugs 59, 74, 78, 79, 293
Spodoptera frugiperda 217
Spongiphoridae 41
Springtails 3, 14, 15, 243, 323
Stable fly 294, 295
Stagmomantis 43
Staphylinidae 104
Stem borers 123, 141, 149, 150,
  203, 206, 208
Stenotarsus subtilis 124
Sterculiaceae 151
Sterile male technique 297
Sternorrhyncha 59, 88-96
Stick insects 3, 39, 167, 276, 299
Stilodes undecimlineata 139
Stilt-legged flies 290
Stingless bees 175, 176, 180,
  181, 323
Stink bug 60, 69, 70
Stolas costaricensis 134
  labasii 135
Stomoxys calcitrans 294, 295
Stoneflies 3, 23
Straight-snouted weevils 145
Stratiomyidae 283, 284
Stratocles unicolor 276
Streblidae 294
Strepsiptera 3, 56, 57
Strymon 235
Sugarcane 77, 79, 166, 206
Sugarcane borer 166, 208
Sulfurs 231, 233, 234
Swallowtails 8, 228-230
Swarm 18, 26, 174, 176, 181,
  183, 269, 276, 282, 283, 286
Sweat seekers 178, 180, 181, 250,
  269, 299
Swollen-thorn acacias 72, 149,
  193, 316
Sycophaginae 163
Symbolia 287
Synale cynaxa 227
Synergus 158
Syngria 216
Syrphidae 288, 289
Tabanidae 282
Tachinidae 299, 300
Tachygerris 60
Taeniopoda 28
Taeniopoda reticulata 36
Tailless whipscorpions 306, 307
Talitridae 330, 331

Tannins 210, 224
Taphura 77
Tapinoma melanocephalum 194
Tarantula hawk 169, 170
Tarantulas 170, 307, 312
Tegosa anieta 268
Telchin atymnius 206
  diva 206
Teleonemia scrupulosa 67
Temenis laothoe 245
Tenebrionidae 126, 127
Tent caterpillars 209
Tenthredinidae 154
Tephritidae 290, 291
Termites 3, 16, 46-48, 175, 176,
  224
Termitidae 46, 47
Terpenoids 133, 139, 140, 224,
  228
Terpsis quadrivittata 136
Tetragnathidae 320
Tetragonisca 181
Tetranychoidea 323, 324
Tetraopes 130
Tettigoniidae 4, 5, 28-31, 33, 34,
  56, 159, 172, 276, 299
Thagona tibialis 220
Thasus 72
Thaumasia 315
Thecla 237
Theope virgilius 239
Theraphosa blondi 312
Theraphosidae 312
Thereus lausus 237
Theridiidae 309, 320, 321
Theridion nigroannulatum 321
Theritas hemon 236
  mavors 237
Thripidae 51, 52
Thrips 3, 51, 52
Thwaitesia 321
Thysania agrippina 218, 219
Thysanoptera 3, 51, 52
Thysanopyga 215
Ticks 104, 322, 325, 326
Tidarren 321
Tidarren sisyphoides 321
Tiger beetles 101, 102
Tiger moths 220-222, 224
Tigridia 267
Timber flies 283
Timulla 168
Tinacrucis 207
Tineidae 198, 199
Tineoidea 198, 199
Tineola bisselliella 198
Tingidae 67
Tipulidae 270, 276

Titanus giganteus 130
Tithraustes 217
Tithrone roseipennis 42
Tityus 303, 304
Toad bugs 61
Tomato 91
Tomoplagia 291
Tortoise beetles 133-137
Tortricid moths 207
Tortyra 208
Torymidae 161
Torymus 161
Toxomerus marginatus 289
Toxotrypana curvicauda 290, 291
Tracheae 4, 62, 332
Trap-jaw ants 184, 186
Trechaleidae 314
Treehopper 83-87, 160, 238, 325
Triatominae 9, 65
Tribolium castaneum 127
Tricharaea 298
Trichochermes 89
Trichodactylidae 336
Trichoptera 3, 26-27
Tridactylidae 35
Trigona 180
Trigona necrophaga 181
Trigonidiinae 32
Trilobites 2, 302
Trioza 89
Triozidae 89
Trogidae 285
Troidini 224, 228-230
Trombiculidae 9, 326
Trombidiformes 322, 323, 325
Trombidiidae 323
Tropaeolum 278
Tropidacris 36
Trosia nigropunctigera 204
True bugs 57, 59-73, 87, 247, 299
Trypanosoma cruzi 65
Tryphoninae 163
Trypoxylon 171, 172
Tunga penetrans 58
Turtle ants 191, 192
Turtles 225, 277, 285, 298
Turuptiana obliqua 222
Tussuck moths 219, 220
Twirler moths 203
Twisted wing parasites 56, 57
Tylomus fuscomaculatus 150
Tymbal 33, 74, 77, 78, 220

Typhlocybinae 80
Typhus 50
Uca 336-338
Uca stylifera 337
Udamoselis 90
Uloboridae 317, 318
Uloborus 317
Ulomoides dermestoides 126
Umbonia ataliba 83
   crassicornis 84
Urania boisduvalii 215
   brasiliensis 215
   fulgens 215, 216
   leilus 215
   poeyi 215
   sloanus 215
Uraniidae 215, 216
Urbanus dorantes 226
Urticating hairs 9, 204, 210, 256, 312
Utetheisa 222
Vachellia collinsii 72, 193, 316
Vanessa cardui 267
Varnish 96
Veliidae 60, 61
Velvet ants 168, 169
Velvet mite 323
Venadicodia caneti 205
Venom 9, 153, 155, 168, 170, 171, 173, 183, 192, 303, 304, 307-309, 312-314, 327
Vespidae 173, 174
Vestistilus 85
Vettius aurel 227
   coryna conka 227
Veturius sinuaticollis 106
Viburnum 261
Victorini 244, 245, 267
Violin spiders 309
Viruses 7, 9, 80, 91, 156, 273, 274
Vitaceae 151, 154, 212, 213
Vochysiaceae 148
Waldheimia interstitialis 154
Wandering spiders 314, 315
Warning coloration 8, 71-73, 78, 79, 121, 216, 220, 223 249, 258,
Warning display 221
Wasmannia auropunctata 192
Wasps 9, 155-174
Water boatmen 61, 62
Water mites 322, 325

Water penny beetles 100
Water quality 24
Water scavenger beetles 98
Water striders 60, 61
Wax 74-76, 88-91, 93, 124, 125, 150, 151, 180, 182, 209, 226
Wax moth 208, 209
Webspinners 3, 40
Weevils 1, 97, 144-152
Whip spiders 306, 307
Whirligig beetles 99
White witch 218
Whiteflies 87, 90, 91, 124, 159, 160, 289
Whites 231-234
Winteraceae 94
Winterschmidtiidae 325
Winthemia 300
Wire worm 114
Wolf spiders 314-315
Wood feeders 46, 48, 105, 113, 130, 148, 149, 270, 283, 331
Woodlice 331, 332
Wyeomyia 272
Xanthoepalpus 300
Xanthomis grandis 222
Xiphocaridae 333, 334
Xyleborus vochysiae 148
Xylella 80
Xylophanes 211, 212
Xylophanes chiron 212
   letiranti 212
   thyelia 212
Yanguna spatiosa 226
Yeasts that aid digestion 54
Yellow fever 9, 272-274
Yellow jackets 173, 175
Yersinia pestis 58
Zammara 74, 77, 78
Zaretis ellops 250, 252
Zegara columbina 8
Zelurus formosus 63
   spinidorsis 63
Zingiberaceae 149, 238
Zonocopris gibbicollis 107
Zopheridae 126, 127
Zopherus jansoni 127
Zoraptera 3, 40
Zulia 79
Zygenoidea 204, 205
Zygentoma 2, 3, 16
Zygogramma violaceomaculata 139

ABOUT THE AUTHOR

Paul E. Hanson was born in Dawson, Minnesota. He obtained a B.A. in biology from the University of Minnesota (Morris), M.Sc. in entomology from the University of Minnesota (St. Paul), and Ph.D. in entomology from Oregon State University. Since 1987 he has been a professor in the School of Biology at the University of Costa Rica, where he teaches courses in entomology, advises graduate students, and carries out research. His principal areas of research include the taxonomy and biology of parasitic wasps, biological control, and gall-forming insects. In addition to numerous scientific papers, he has coauthored three previous books: *The Hymenoptera of Costa Rica* (1995), *Orchid Bees of Tropical America* (2004), and *Hymenoptera de la Región Neotropical* (2006).

ABOUT THE PHOTOGRAPHER

Kenji Nishida, an entomologist and photographer, was born in Osaka, Japan, in 1972. His encounters with nature, especially with small creatures, began before his first birthday. In 1998, he began studying insects at the University of Costa Rica, where he obtained his M.Sc. Kenji continues to do research on insects in Costa Rica, specializing in butterflies and moths and in gall-forming insects. You can learn more about him by visiting his page on National Geographic Japan: http://natgeo. nikkeibp.co.jp/nng/article/20110512/269653/.